Plant Embryology
Classical and Experimental

Plant Embryology
Classical and Experimental

H.P. Sharma

Alpha Science International Ltd.
Oxford, U.K.

Plant Embryology
Classical and Experimental
302 pgs. | 165 figs. | 8 tbls.

QK
665
.S449
2009

H.P. Sharma
Department of Botany
Ranchi University, Ranchi

Copyright © 2009

ALPHA SCIENCE INTERNATIONAL LTD.
7200 The Quorum, Oxford Business Park North
Garsington Road, Oxford OX4 2JZ, U.K.

www.alphasci.com

All rights reserved. No part of this publication may be reproduced, stored in a retrieval system or transmitted in any form or by any means, electronic, mechanical, photocopying, recording or otherwise, without prior written permission of the publisher.

ISBN 978-1-84265-525-2
Printed in India

―――――― **DEDICATED** ――――――
**To the Memory of
My Beloved Parents
Late Smt. Bachkali Devi
&
Late Sri Devi Prasad**

Preface

Embryology simply deals with the study of the events starting from microsporogenesis, megasporogenesis, pollination and fertilization till the development of embryo. Experimental embryology, on the other hand, covers the role of internal factors, such as nutrients and phytohormones; external factors, such as light, temperature, humidity etc., on the cultures of ovary, ovule, embryo, anther or microspores, which facilitates in the understanding of plant growth and development.

The history of embryology is very old and can be traced back to the third century B.C., when Theophrastus described pollination of date palm in his book. However, in seventeenth century only, embryology got impetus with the invention of microscope. Camerarius (1694), for the first time described anther as male and ovary with its style as female. The discovery of double fertilization by Nawaschin (1898) proved to be a turning point. Embryological studies gained great momentum in twentieth century and scientific events in 1970's opened avenues for genetic engineering which prompted scientists to undertake genetic manipulation of the embryological processes.

It is, therefore, essential to bring all recent information within the folds of a single book which could meet the requirement of students. The experience gathered during my long years of teachings has prompted me to take up this subject as I feel that without the knowledge of the fundamental aspects of reproductive biology we can not move into any area of plant study. However, there are large numbers of books available in the market, discussing the fundamental or classical embryological studies while information relating to the experimental aspect is very short. These books are either very short or too comprehensive and research oriented, which create confusion or disinterest amongst the students. Therefore, it is required that various chapters discussing the fundamental aspects be presented in a simple form describing all the relevant events. Experimental or the applied section also needs a thorough discussion necessitated inclusion of a few new chapters which are relevant in the light of progress made in this area in the past decades.

This book has been written with a view to provide up to date information about the subject for both under-graduate and post-graduate students of the Universities. This book deals with all the three aspects of embryology i.e. classical or descriptive, comparative or phylogenetic and experimental embryology. There are altogether 22 chapters, divided into two parts. Part-A comprises fifteen chapters discussing the classical and comparative embryology, which has been presented in simple and concise form. The Part-B comprises seven chapters discussing the experimental embryology.

A few new chapters have been incorporated into this section, such as 'Somatic Embryogenesis' and 'Synthetic Seeds' which are quite relevant in the present day. The Appendix shows some simple embryological experiments which could be performed easily in the laboratory. Important contributions have been highlighted in the reference section.

The material has been collected from different sources and compiled systematically for explicit presentation. Different chapters have been presented in a simple manner with up to date information; the diagrams are hand sketches and have been redrawn from authentic books or published papers for their clarity. I express my gratitude to these authors and they have been duly acknowledged at proper places. The schematic diagrams have also been added in many chapters with a view to make the topic understandable at a glance. Some important references have been given at the end of the book.

I express my heart felt thanks to my colleagues, Dr. (Mrs) Radha Sahu, Dr. R.K. Pandey, Dr. A.K. Srivastava, Dr. A.K. Choudhury, Dr. H.B. Sahu and Dr. Jyoti Kumar for their co-operation during the preparation of this book.

I am highly indebted to my teacher Dr. K.K. Nag, former Head of the Department of Botany, Ranchi University, Ranchi and former Vice-Chancellor to different Universities as he always encouraged and remained my model teacher. I am also grateful to my teacher Dr. N. Dayal, former Head of the Department of Botany, Ranchi University, Ranchi for his co-operation and valuable suggestions from time to time for the improvement of the text. The author is greatly indebted to Mr. P.P. Das, Artist-cum-Photographer of the Department, who sketched most of the diagrams.

I am indebted to Dr. P.K. Tiwary, Senior Research Officer, Regional Sericulture Research Station, Central Silk Board and Dr. Ramesh Chandra, Lecturer, Department of Pharmacy, Birla Institute of Technology, Mesera for their painstaking work of going through the manuscript. However, all errors and naiveties are mine.

My thanks to Dr. N.K. Dubey, University Professor, Department of Botany, BHU, Varanasi and Dr. P.K. Behera, Berhampur University, Berhampur for their valuable suggestions. My heartfelt thanks to Dr. J.V.V. Dogra, former HOD, T.M. Bhagalpur University, who contributed in a major way in giving final shape to this text.

I am especially grateful to Mr. N.K. Mehra, Director, Narosa Publishing House for his forbearance and all support as without his co-operation the work could not be presented for publication.

The response of the students will be the final verdict. I would very much appreciate receiving suggestions and criticism from different quarters. I would be especially thankful if errors are pointed out with page numbers and research contribution in the form of reprints as this will be most helpful in preparation of an improved revised edition.

Finally, it is a pleasure to thank my wife Dr. Malti Sharma, sons Anurag Sharma and Abhinav Sharma, who remained as my source of energy and always stood by me with great patience.

H.P. Sharma

Contents

Preface ... vii

PART A: CLASSICAL EMBRYOLOGY

1. **Historical Account and Present Status** ... 1.3
2. **The Flower** ... 2.1
   - Parts of Flower ... 2.2
   - Seed ... 2.5
   - Fruit ... 2.6
3. **Microsporangium** ... 3.1
   - Development of Anther ... 3.2
   - Sporogenous Tissues ... 3.7
4. **Male Gametophyte** ... 4.1
   - Microspore ... 4.1
   - Vegetative and Generative Cell ... 4.1
   - Division of the Generative Cell ... 4.4
   - Pollen Wall ... 4.5
   - Unusual Type of Pollen Development ... 4.11
5. **Megasporangium** ... 5.1
   - Types of Ovule ... 5.2
   - Parts of Ovule ... 5.3
   - Megasporogenesis ... 5.9
   - Some Abnormal and Reduced Ovules ... 5.10
6. **Female Gametophyte** ... 6.1
   - Development of Embryo Sac or Megasporogenesis ... 6.1
   - Aberrant and Unclassified Type ... 6.7

Mature Embryo Sac	6.8
Storage Materials	6.12
Embryo Sac Haustoria	6.12
Evolutionary Tendencies	6.12

7. Pollination — 7.1
- Self-Pollination or Autogamy — 7.1
- Cross-Pollination or Allogamy — 7.2
- Contrivances for Cross-Pollination — 7.7
- Advantages and Disadvantages of Self-and Cross-Pollination — 7.11

8. Fertilization — 8.1
- Structure of Stigma — 8.2
- Structure of Style — 8.5
- Post-Pollination Events — 8.8
- Double Fertilization — 8.15
- Unusual Features — 8.16

9. Sexual Incompatibility — 9.1
- Self-Incompatibility — 9.1
- Mechanism of Self-Incompatibility — 9.9
- Pollen-Pistil Interaction — 9.10
- Methods of Overcoming Incompatibility — 9.14
- Biologiocal Significance of Incompatibility — 9.19

10. The Endosperm — 10.1
- Nuclear Endosperm — 10.1
- Cellular Endosperm — 10.3
- Helobial Endosperm — 10.4
- Endosperm Haustorium — 10.5
- Variants of Endosperm — 10.10
- Structure of Endosperm — 10.12
- Cytology of Endosperm — 10.14
- Morphological Nature of Endosperm — 10.14

11. The Embryo — 11.1
- Embryogeny — 11.1
- Ultrastructure and Histochemical Studies of the Organogenetic Part of Embryo — 11.8
- Nutrition of the Embryo — 11.14
- Unclassified and Abnormal Embryos — 11.15
- Unorganized and Reduced Embryos — 11.15

12. Polyembryony — 12.1
Types of Polyembryony — 12.1
False Polyembryony — 12.7
Some Special Cases — 12.7
Twins and Triplets — 12.8
Experimental Induction of Polyembryony — 12.10
Causes of Polyembryony — 12.10
Role of Polyembryony in Plant Breeding and Horticulture — 12.11

13. Apomixis — 13.1
Types of Apomixis — 13.1
Organization of Embryo Sac — 13.7
Embryogenesis — 13.8
Genetics of Apomixis — 13.8
Significance of Apomixis — 13.9

14. The Seed — 14.1
Parts of Seed — 14.3
Labyrinth Seeds — 14.5
Seed Dispersal — 14.5
Seed Dormancy — 14.8
Conditions for Germination — 14.10
Types of Germination — 14.11
Importance of Seed — 14.11

15. Embryology in Relation to Taxonomy — 15.1
Embryology in Solving Taxonomical Problems — 15.3
Conclusion — 15.9

PART B: EXPERIMENTAL EMBRYOLOGY

16. Technique of Tissue Culture — 16.3
Culture Medium — 16.3

17. Somatic Embryogenesis — 17.1
Types of Somatic Embryogenesis — 17.2
Technique — 17.2
Factors Affecting Somatic Embryogenesis — 17.4
Importance of Somatic Embryogenesis — 17.6

18. Haploid Production — 18.1
Anther Culture — 18.1
Pollen Culture — 18.6
Induction of Gynogenesis — 18.10

Production of Haploids by Bulbosum Technique	18.10
Applications of Haploidy	18.11
Limitation of Haploid Culture	18.13
19. *In Vitro* Pollination and Fertilization	**19.1**
In Vitro Pollination	19.2
In Vitro Fertilization	19.5
Application of *In Vitro* Pollination and Fertilization	19.9
20. Embryo Culture	**20.1**
Technique	20.1
Embryo-Endosperm Transplant	20.3
Culture Medium	20.4
Applied Aspects of Embryo Culture	20.7
21. Culture of Different Parts of Pistil and Seed (Organ Microculture)	**21.1**
Ovary Culture	21.1
Importance of Ovary Culture	21.3
Ovule Culture	21.4
Importance of Ovule Culture	21.4
Nucellus Culture	21.6
Endosperm Culture	21.6
22. Synthetic Seeds	**22.1**
Technique	22.2
Appendix: Experiments in Embryology	*A.1–A.6*
Selected References	*SR.1–SR.5*
Subject Index	*SI.1–SI.18*

PART A

Classical Embryology

Chapter 1. Historical Account and Present Status
Chapter 2. The Flower
Chapter 3. Microsporangium
Chapter 4. Male Gametophyte
Chapter 5. Megasporangium
Chapter 6. Female Gametophyte
Chapter 7. Pollination
Chapter 8. Fertilization
Chapter 9. Sexual Incompatibility
Chapter 10. The Endosperm
Chapter 11. The Embryo
Chapter 12. Polyembryony
Chapter 13. Apomixis
Chapter 14. The Seed
Chapter 15. Embryology in Relation to Taxonomy

1. Historical Account and Present Status

The Embryology deals with the study of all the events starting from microsporogenesis, megasporogenesis, pollination and fertilization till the development of a mature embryo.

In order to trace the developmental history of embryology, one has to go as back as third century B.C. when Theophrastus, known as the father of Botany, for the first time described pollination of the date palm in his text "Enquiry into Plants" based on the account of Herodotus. The Arabs and Assyrians had some knowledge of role played by two plants for ensuring good production of dates. Till sixteen century, no progress had been made in this direction and even mention of sex in plant was considered inappropriate and obscene.

It is only in the second half of the seventeenth century, the biological science got impetus with the invention of microscope by Leeuwenhoek (1677). Grew (1682) in his "Anatomy of Plants", made first explicit mention of the stamen as the male organ of the flower and role of pollen grains in fruit production.

R.J. Camerarius (1694) had more scientific approach as his findings were based on some experiments. He found that the female plants of mulberry (*Morus indica*) and *Mercurialis annua* when kept isolated from male plants fruits developed only abortive seeds. He had a similar finding with *Ricinus* and *Zea mays*. He summarized his work in a famous treatise called "De sexu plantarum". In conclusion he said: "In the plant kingdom, the production of seed, which is the most perfect gift of nature and general means of maintenance of species, does not take place unless the anthers have previously prepared the young plant contained in the ovary". He considered anthers as male and ovary with its style as female.

J.G. Kölreuter (1761) highlighted the significance of pollination in seed set and gave detail account of role of insects in pollination. He successfully produced hybrids in *Nicotiana, Dianthus, Matthiola, Hyoscyamus*.

In the year 1824, an Italian scientist G.B. Amici observed the splitting of the adhered pollen grains on the stigmatic hairs of *Portulaca oleracea* and pushing out of a tube or "gut" but subsequent developments could not be ascertained. Influenced by the significant observation of Amici, French Botanist Brongniart (1827) examined a large number of pollinated pistils with a view to understand the interaction between the pollen and stigma and introduction of the fertilizing substance into the ovule. He was of the view that pollen tube after penetrating the stigma burst to release 'spermatic granule', which enters the placenta and then ovule.

Amici (1830) once again applied himself to the problem and re-examined the germination and growth of pollen tube of *Portulaca oleracea*, *Hibiscus syriacus* and other taxa and ruled out completely the findings of Brongiart and came out with the conclusion that pollen tube elongates bit by bit and finally comes in contact with ovule, one tube for each ovule.

In 1837, Schleiden pointed out that the development of embryo from the tip of pollen tube, which enters into the embryo sac, the latter only functions as incubator into which pollen tube was nourished to give rise to new plantlet. Amici (1842) once again came out boldly to oppose the views of Schleiden and tried to establish that the embryo did not arise from the tip of the pollen tube but from the portion of the ovule, which was already in the existence and was fertilized by the fluid in the pollen tube. His views were further consolidated with the publications of Hofmeister (1849), based on his observations on 38 species belonging to 19 genera. Schleiden and Schacht (1850) continued to hold their previous opinion. However, controversy came to an end with publication of a comprehensive review of Radlkofer (1856), in which he fully supported the detailed observations of Hofmeister. He is also being credited for his observation of tetrad formation and presenting organization of embryo sac.

Reichenbach, Hartig and several other workers noted the presence of two nuclei in the pollen grains of several species of angiosperms. Strasburger (1877) and his student Elfving (1879) extended these observations to several other families and demonstrated the wide spread occurrence of the binucleate condition in the pollen grains. In order to understand the fate of nuclei, Elfving made preparation of pollen tubes from dissected styles and was able to find out three nuclei, out of which two were known to be male gametes and one tube or vegetative nucleus. Strasburger and Elfving made the mistake of interpreting smaller cell in the pollen grain as vegetative or prothallial cell and large cell as generative cell. Nevertheless, the observation was of great importance and Strasburger later rectified the mistake in his subsequent paper, published in 1884.

We are indebted to Hofmeister (1847-1861) for our knowledge of the organization of the embryo sac. He succeeded in identifying two groups of cells at the opposite poles of the embryo sac. Those lying at the micropylar end were called as "germinal" or "embryonal" vesicle, all capable of giving rise to embryo. The cells at the chalazal end were considered to be prothallial. Despite his outstanding contribution, he failed to distinguish clearly between the egg and the synergid and regarded all of them as having the same function. He also failed to trace out the origin of embryo sac.

Strasburger (1897) for the first time demonstrated the development of embryo sac from the megaspore mother cell, derived from the nucellus and the organization of the embryo sac from eight nuclei of megaspore, which organized into egg apparatus, secondary nucleus and antipodals. Subsequently in 1884, he described the process of syngamy or the fusion of male and female gametes. Hanstein (1870) was the first to follow the developmental sequence of embryo in *Capsella* and *Alisma*.

Nawaschin (1898), while studying the fertilization process in *Lilium martagon* and *Fritellaria tenella*, showed that both male gametes are involved in fertilization, one fusing with egg (syngamy) and the other with secondary nucleus (triple fusion). Thus the concept of double fertilization came into existence, which subsequently turned out to be of universal occurrence in angiosperm.

Embryological studies gained great momentum with the invention of microtome and refinement in staining technique. By the year 1900, enough information was gathered about the development of the gametophyte and embryo and a substantial summary was given by Coulter and Chamberlain (1903) in their book, entitled "Morphology of Angiospers". After 26 years, Karl Schnarf published two valuable books, namely *"Embryologie der Angiospermen"* in 1929 and *"Vergleichende Embryologie der Angiospermen"* in 1931. These two treatises are still considered to be an outstanding compilation for reference on embryology.

Due to painstaking efforts of P. Maheshwari, the India took the lead in the field of embryology in early thirties and forties, of the twentieth century. His book "An Introduction to the Embryololgy of Angiosperms" in 1950 is a classical presentation. This was followed by an edited volume by P. Maheshwari, entitled "Recent Advances in the Embryology of Angiosperms" in 1963. Recently, two books have been published by B.M. Johri, namely "Embryology of Angiosperms" and "Experimental Embryology of Angiosperms" in 1982 and 1984, respectively. A third book was edited by Johri et al. (1992), entitled "Comparative Embryology of Angiosperms" All the three books are providing valuable information about the comparative embryology of angiosperms.

The earlier systems of classifications were based mainly on external characters of flowers, fruits and seeds. However, gradually it was realized that embryological data are of great significance in deciding taxonomical identification of a number of doubtful cases. Schnarf (1931) clearly highlighted the importance of embryology in taxonomical considerations. Two important examples to be cited are families Cyperaceae and Onagraceae, which were identified from other angiosperms on the developmental basis of male and female gametophytes, respectively.

Investigations on sperms by Cass (1973) showed that two sperms produced by male gametophytes remain together for some time. However, later large numbers of workers have reported the presence of male germ unit (MGU), which was considered to be of universal occurrence in angiosperms. With the isolation of sperms from a bursting pollen tube, the possibility of sperm manipulation got the ground and a new field of pollen biotechnology has emerged (Shivanna and Sawhney, 1997).

In 1960's, interest was diverted towards modern techniques and applied embryology, which involved the modern methods of histochemistry, biochemistry, electron microscopy, fluorescence, autoradiography, computers, and tissue cultures etc. Applying modern techniques Jensen and his colleagues studied the structure and role played by synergids during fertilization. Heslop-Harrison and his student made detailed study of pollen wall development and pollen-pistil interaction.

The beginning of the experimental embryology can be traced to the observations of Gartner (1894), Massart (1902) and Fitting (1909), who treated the ovaries of certain plants with *Lycopodium* spores, dead pollinia and aqueous extracts of pollens respectively and found some swelling in the ovary. An important aspect of the experimental embryology was the artificial culture of the excised embryos. As early as 1904, Hanning had initiated work on the embryo culture. Subsequently, Laibach (1925, 1929) utilized the technique for hybrid embryo culture, thus laid the foundation for embryo rescue of inter-specific or inter-generic hybridization. Presently, this technique is successfully exploited for the production of disease-resistant economically important plants.

The tissue culture has its beginning with the concept of totipotency of plant cells laid down by the German scientist, Haberlandt (1902). However, he himself did not succeeded in culturing plant cells but predicted that in appropriate nutrient media, the cells will divide, grow and differentiate.

Plant tissue culture has progressed steadily ever since its inception in 1902. The initial experiments related to a simple sustained prolonged culture of tissues only. The differential response of the culture tissues under variable chemical conditions provided the impetus to utilize the technique in a profitable manner. Over the years efficacy of the technique became apparent and felt that the technique could be used not only to understand the growth and differentiation but has become an integral part of biotechnology and is being routinely employed for the improvement of crops and other plants of economic importance.

It is only in 1930s, three scientists working in different countries and on different plant tissues demonstrated experimentally that cells or tissues can be cultured continuously in defined media and plantlets can be regenerated. Gautheret in 1934, succeeded in culturing cambial cells of tree species including *Salix capraea* and *Populus nigra*, and observed their proliferation for a few months on Knop's solution containing glucose and casein hydrolysate (CH). Further, he for the first time established continuous growing tissues on culture media from carrot and Jerusalem artihoke root cambium. At the same time, White (1939) has also reported such cultures from tumour tissue of the hybrid *Nicotiana glauca* x *N. langsdorfii*. Nobecourt (1939) working independently on carrot obtained the same result. In subsequent years, further research carried out on other plants consolidated the concept of totipotency. Muir (1953) isolated single cell from callus tissues of *Nicotiana tabacum* and *Tagestes erecta* in liquid medium and established their cultures. Vimla and Hilderbrandt (1965) demonstrated the totipotency of isolated single cell of tobacco that regenerated plantlets.

To ensure continuous culture of tissue and their differentiation into root/shoot or regeneration into complete plants many synthetic media have been developed by different scientists all over the world from time to time. Of the various media, the medium formulated by Murashige and Skoog (1962) is widely used. The Tissue culture got a great stride as the range of possible experiments on the control of growth of cultured embryos was greatly extended with the discoveries and isolation of auxins (1934), gibberellic acid (1939) and cytokinins (1955). Murashige and Skoog (1957) in their classical experiment on tobacco tissues showed that the organogenesis is under the controlled mechanism of exogenously supplied hormones in the medium.

Other potential uses are embryoid formation from somatic cells of carrot (Reinert, 1958; Steward *et al.*, 1958). Thus somatic embryogenesis has been established as an effective pathway for clonal multiplication.

Haploid and homozygous diploids are highly desirable in breeding programme for superior traits. Guha and Maheshwari (1964), for the first time regenerated haploids from *Datura innoxia*. Subsequently, large numbers of plants have been regenerated though anther cultures. The significant use of this technique is not only genetical but a number of new varieties of economic importance have been developed by the application of this technique.

Induction of polyploidy and other disadvantages associated with the anther cultures, attempts have been made to isolate pollen grains/microspore and to culture them in suitable media. Tuleke (1953), for the first time reported callus induction from pollen grains of *Ginkgo biloba*, a gymnosperm. However, in *Brassica* sp., an angiosperm, successful production of callus for the first time has been reported by Kameya and Hinata (1970). Nitsch (1974), developed synthetic medium and succeeded in raising haploid plants from isolated pollens of *Nicotiana* and *Datura*.

The limitation of anther/pollen culture for haploids has forced scientists to look for female gametophyte as an alternative source. In this direction, San Noeum (1979) for the first time reported successful *in vitro* ovary culture in *Hordeum vulgare*.

The technique of *in vitro* pollination of ovule was for the first time developed by Maheshwari and his students at the University of Delhi with *Papaver somniferum* (Kanta *et al*. 1962). Kanta and Maheshwari (1963) applied this technique successfully to overcome the self-incompatibility in *Torentia fournieri*, which was followed by many successful results obtained with different plants.

One of the most striking features of plant tissue culture has been the isolation and fusion of the protoplasts (Cocking, 1972). The important practical application of protoplast culture or somatic hybridization was successfully demonstrated by Carlson and his co-workers (1972) as they produced first interspecific hybrid between *Nicotiana glauca* and *N. langsdorffii*. Melchers *et al*. (1978) with the fusion of protoplasts of tomato and potato produced first intergeneric hybrid. The hybrid developed and produced flowers and fruits called '**pomato**' and the suckers '**topato**', the latter was thick, flat and full of starch (equivalent of potato).

The concept of **'synthetic seed, synseed or artificial seed'** was first introduced by Toshio Murashige in 1977. Kitto and Janick (1982) for the first time coated clumps of carrot embryoids with polyoxyethylene to develop artificial seeds. Subsequently, Redenbaugh *et al*. (1986) used more effective hydrogel, such as sodium alginate for encapsulation of somatic embryos of alfalfa and celery, which developed into plantlets.

In vitro fertilization in plant is not that easy as in animals. However, much progress has been made towards the *in vitro* fertilization in plants with the success of protoplast culture and somatic hybridization. The first successful report of fusion of isolated sperm and egg came from the laboratory of Kranz (Kranz *et al*., 1991). It took a couple of year to induce embryogenesis and plantlet formation (Kranz and Lorz, 1993).

The information about the genetics of crown gall tumour formation in dicots by *Agrobacterium tumefaciens* has opened frontiers for understanding the molecular mechanism of growth and development. The Ti plasmid of the bacterium is being successfully used as a vector for the transfer of foreign genes and regeneration of transgenic plants. In the area of genetic engineering the tissue culture is employed as a tool for the production of transgenic plants. The first transgenic plant was produced for tobacco with the help of *Agrobacterium tumefaciens* (Horsch *et al*., 1984). This field has now become one of the potential areas, which hold much promise and could be effectively utilized for qualitative and quantitative improvement of the economically important plants.

Scientific events in 1970s opened avenues for genetic engineering, which prompted scientists to undertake genetic manipulation of the embryological processes. In this reference work of Mariani *et al*. (1991) is worth mentioning, as male sterility was introduced in *Brassica napus* with the help of ***barnase gene*** isolated from bacterium *Bacillus amyloliquefaciens*. Another important milestone towards genetic engineering is the production of parthenocarpic fruits in tobacco and egg plant with the insertion of chimeric gene ***DefH9-iaaH*** (Rotino *et al.*, 1997).

Now, the attempts are there all over the world to provide information relevant to the molecular mechanism of various embryological events. One of the recent works is the application of 455bp antisense cDNA fragment of BcMF3 gene (codes for methylesterase) through *Agrobacterium tumefaciens*, which affects microsporogenesis and pollen tube growth in *Arabidopsis thaliana*, thus inducing male sterility (Le-cheng *et al.*, 2006).

2 The Flower

There are three important stages encountered in the life cycle of a plant. The first stage is embryonic phase during which all the essential organs of the vegetative plants are moulded. The second crucial stage is the vegetative phase, when the plant develops into an autonomous, photosynthetically active plant and; the third crucial stage is the flowering phase, when flower formation takes place. Flower formation signifies a transition from the vegetative to the reproductive phase of development. The shoot meristem is induced to develop sepals, petals, stamens and carpels instead of leaves. This transition can occur at a particular time in the life cycle of plant, which varies from species to species.

The flower is a specialized reproductive shoot of limited growth bearing reproductive organs, microsporophyll (stamen) and megasporophyll (carpel) or the only one, often with two accessory whorls, calyx and corolla, sometimes only one or even none at all (Fig. 2.1).

A flower is regarded as **complete** if it has all the four whorls i.e., two lower accessory whorls; **calyx** and **corolla** and two upper essential whorls; **androecium** and **gynoecium**. If any of these parts is missing, it is called **incomplete**. With reference to reproductive parts, a flower is called **perfect** when it consists of both male and female sex organs whereas in

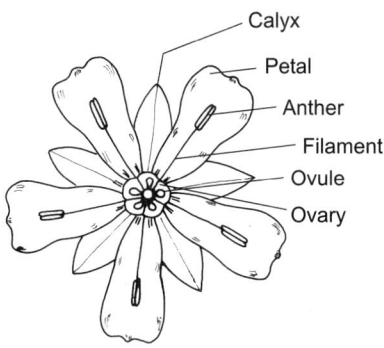

Fig. 2.1 Hypothetic view of a flower.

imperfect flower any one of the two reproductive parts is missing. A perfect flower is always **bisexual** or **hermaphrodite** e.g., members of the families Brassicaceae, Malvaceae and Solanaceae etc., while imperfect flower is always **unisexual** or **diclinous**, may be either male (**staminate**) or female (**pistillate**) e.g., members of the families Euphorbiaceae, Cucurbitaceae etc.

A plant is said to be **monoecious,** if both male and female flowers are found on the same plant e.g., *Zea mays*, *Cucurbita maxima,* while **dioecious** shows male and female flowers on two different plants e.g., *Carica papaya*, *Borassus flabellifer* etc. Mango and lichi show **polygamous** condition where all types of flowers i.e. male, female and bisexual are found on the same plant.

Further it may be noted that a flower without calyx and corolla is said to be naked or **achlamydeous**, as in betel; a flower with only one whorl is **monoclamydeous**, as in *Polygonum* and a flower with both the whorls is **diclamydeous**, as in china rose.

Flowers exhibit great variation in sizes, colours, shapes and insertion of floral parts. As with the size, the smallest flower *Wolffia microscopica* shows 0.1 mm diameter while the largest flower, *Rafflesia* sp., a total root parasite shows 1.0 m diameter.

PARTS OF FLOWER

In order to describe the structure of a typical flower, the most common ornamental china rose (*Hibiscus rosa-sinensis*) has been chosen as a type specimen to show different parts of a flower (Fig. 2.2). Floral parts like calyx, corolla, androecium and gynoecium are inseted on a condensed axis, called **thalamus**. At the base of calyx a whorl of bracteole is produces called **epicalyx** (Fig. 2.2 A, B). A flower with a stalk or pedicel is called **pedicellate,** while without pedicel it is called **sessile**.

(i) Thalamus

The thalamus, **torus** or **receptacle** is a floral axis formed by the direct prolongation of the pedicel, on which floral parts are inserted. It is normally slightly swollen, knob- like structure but sometimes it becomes conical and bears floral leaves arranged in a spiral manner e.g., *Artabotrys hexapetalous, Michelia champaca* etc. The thalamus also shows nodes and internodes just like the ordinary shoot but normally it is highly condensed bringing all the floral parts close to each other. However, there are examples where internodes between the floral parts i.e. calyx and corolla (**anthophore** e.g., *Silene*), corolla and androecium (**androphore or gonophore;** e.g., *Gynandropsis*), or androecium and gynoecium (**gynophore,** e.g., *Gynandropsis, Moringa, Pterospermum* etc.) become distinct by elongation showing the shoot nature. When both androphore and gynophore develop, they are together known as **androgynophore**.

(ii) Bracts

Bracts are special leaves from the axil of which one or more flowers arise. When a small leafy or scaly structure is present on any part of the flower-stalk (pedicel), it is called **bracteole**. Bracts vary in shape, size, colour and duration and accordingly may be of different types.

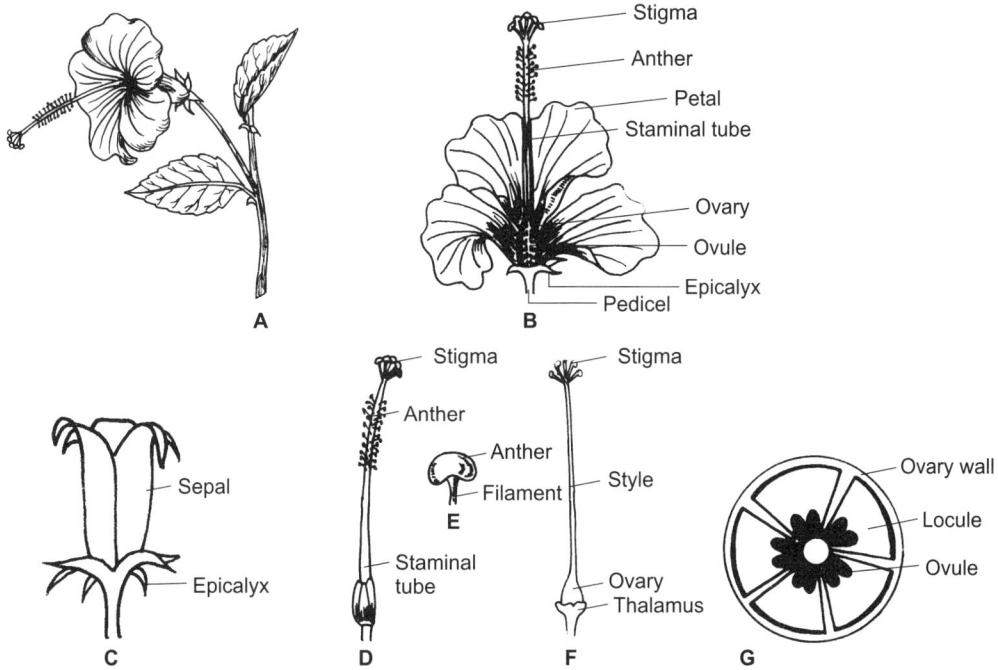

Fig. 2.2 *Hibiscus rosa-sinensis*. **A**. A flowering twig. **B**. Longitudinal view of a flower. **C**. Epicalyx and calyx. **D**. Staminal tube enclosing style **E**. A stamen. **F**. A carpel. **G**. T.S. of ovary showing ovules

One of the types is **epicalyx,** which shows one or more whorls of bracteoles developing at the base of the calyx. Epicalyx is characteristics of Malvaceae e.g., china rose (Fig. 2.2 C), cotton, lady's finger etc. Bracts give protection to the floral bud or becoming brightly coloured, as in *Bogainvillea* and invites insects for pollination.

(iii) Calyx

It is the first lowermost or the outermost accessory whorl (Fig. 2.2 C). Each unit of the calyx is called sepal, which are usually green (**sepaloid**) or rarely coloured (**petaloid**) e.g., *Sterculia roxburgii, Delonix regia* etc. Sepals may be free (**polysepalous**) or united (**gamosepalous**). Sometimes, sepal becomes large and leaf-like showing foliar nature e.g., *Mussaenda frondosa* or modified into **pappus** as in sunflower. The sepals may be caducous, deciduous, or persistent type. In addition to their protective roles, calyx sometimes attracts insects because of the showy nature.

(iv) Corolla

It is the second accessory whorl (Fig. 2.2 B). Each unit is called petal, which may either be free (**polypetalous**) or united (**gamopetalous**). Petals are usually highly coloured due to water-soluble pigments like anthocyanin, anthoxanthin or carotenoid. The petals may be caducous, deciduous or may be of persistent type. Corolla may also be **regular** or radially symmetrical (e.g., members of the family Malvaceae and Brassicaceae), **zygomorphic** or bilaterally symmetrical (e.g., members of the family Papilionaceae) or **irregular**. Since the petals are brightly coloured and sometimes provided with outgrowth or special appendages, such as pouch or sac, e.g., snapdragon (**saccate or gibbous**), prolonged into tube (**spur**) e.g., *Delphinium, Impatiens* etc. In many flowers of Asteraceae, Lamiaceae, and Rubiaceae etc. a special gland called **nectary** develops containing nectar. Sometimes transverse splitting of the corolla lobe results **corona** which may be seen in *Passiflora*, *Cuscuta* and *Nerium* etc. In addition to their protective role they are one of the important sources of entomophily.

(v) Perianth

In a few monocots, there is no distinction between calyx and corolla because only one whorl is present (**monochlamydeous**), called **perianth**. Each unit is called **tepal** which may either be free (**polyphyllous**) or united (**gamophyllous**). The perianth may be caducous, deciduous or of persistent nature.

(vi) Androecium

The androecium is the first essential or the third accessory whorl, which constitutes the male reproductive part. Its unit is **stamen** or **microsporophyll**, which shows a slender stalk or **filament**, an apical spore-containing sac called **anther** and the **connective,** which connect anther lobes and filament.

Each anther is either single lobed (**monothecous**) e.g., china rose (Fig. 2.1 E) or bilobed (**dithecous**) e.g., pea. The anther lobes show pollen chambers or **microsporangia,** which contain large number of pollen grains or **microspores,** representing the first cell generation of the male gametophyte. Stamens may either remain free or may be united (coherent). There may be different degree of cohesion of stamens and this may be called **adelphous** condition: when the filaments are united together but the anthers remaining free, it is called **monadelphous**, e.g., china rose (Fig. 2.2 D); **Diadelphous,** when the stamens are united in two bundles and the anthers remaining free, e.g., Fabaceae; **Polyadelphous**, when the filaments are united into a number of bundles but the anthers are free, e.g., cotton, castor etc; **Syngenesious**, when the anther are united together into a bundle but the filaments are free, e.g., Asteraceae and **Synandrous**, when the stamens are united throughout their length, both by the filaments and the anthers, e.g., Cucurbitaceae.

Likewise, there is different degree of adhesion of the stamens: **Epipetalous**, when they are attached to the petal wholly or partially, e.g., Malvaceae, Solanaceae and Asteraceae; **Gynandrous**, when united with the carpels either wholly or by their anthers only, e.g., *Calotropis, Aslepias* etc.

Even the length of the stamens is different in a flower. They may be **didynamous**, when out of the four stamens two are long and other two are short, e.g., Brassicaceae and ; in **tetradynamous,** of the six stamens inner four are long and the outer two are short, e.g., Lamiaceae. Sometimes the stamens are sterile, called **staminode**.

(vii) Gynoecium

It is the second essential or the fourth floral whorl constituting the female reproductive part. Its unit is **carpel** or **pistil** or **megasporophyll**, which shows three parts: ovary, style and stigma (Figs. 2.2 D, F).

The pistil may be one or many, either free (**apocarpous**) or fused (**syncarpous**). The carpels may be united either throughout their length as in syncarpous pistil or they may be united in the region of the ovary alone where style and the stigma remaining free, as in *Dianthus, Linum*; or in the region of ovary and style, stigmas remaining free, as in china rose and cotton; or in the region of the style and the stigma, ovary only remaining free as in *Catharanthus*, Nerium; or in the region of stigma (and the style partly), as in *Calotropis*. The ovary consists of large number of ovules, arising from the placentae (Fig. 2.2 G) or directly from the thalamus e.g., sunflower. Within the ovule, the embryo sac or the **microsporangium** develops which represents the first cell generation of female gametophyte. Each embryo sac shows egg apparatus, secondary nucleus and antipodals.

In addition to the floral parts being either spiral or whorled in arrangement, the level of insertion of the sepals, petals and stamens on the floral axis varies in relation to the level of ovary or ovaries. If the sepals, petals and stamens are attached to the receptacle, below the ovary, the ovary is said to be **superior** and the flower is said to be **hypogynous**, e.g., china rose. In some flowers with superior ovaries, the petals and stamens are attached to the margin of a cup-shaped extension of the receptacle, e.g., pea, bean and goldmohur etc. The ovary is said to be **half-inferior** and the flower is called **perigynous**. In still other flowers the sepals, petals and stamens grow from the top of the ovary, which is **inferior**, e.g., sunflower, guava, cucumber etc. Such flower is called **epigynous**.

The style is a long, slender structure situated above the ovary, which ends into stigma. The style may be apical, lateral or **gynobasic** depending upon whether the style has its origin apically (china rose), laterally or from the depressed center of four-lobed ovary, e.g., Lamiaceae.

The stigma may be simple or branched. The surface of the stigma is rough, papillose, hairy or with sticky secretion. The stigma provides a platform for landing of pollen grains and allows pollen of right mating type to function normally while others are discarded due to incompatibility. Sometimes, pistil becomes sterile called **pistillode.**

SEED

A mature ovule is called seed, which consists embryo (dividing zygote), endosperm (fertilized secondary nucleus) and seed coat (arising from integuments). Depending upon the number of cotyledons, the seed may be **monocot** or **dicot**. Further, depending on the presence or absence of endosperm seeds may be **endospermic (albuminous)** or **non-endospermic (exalbuminous)**. In seeds of some plants species of citrus, the embryo contains chloroplast and is green (Casadoro *et al.*, 1980). Seeds ensure the perpetuation of plant and also dissemination of plants to different areas.

2.6 Plant Embryology: Classical and Experimental

FRUIT

A mature ovary is called fruit, which may either develop after fertilization or may develop without fertilization, called **parthenocarpic** fruit. It has two parts: pericarp and seed. A fruit

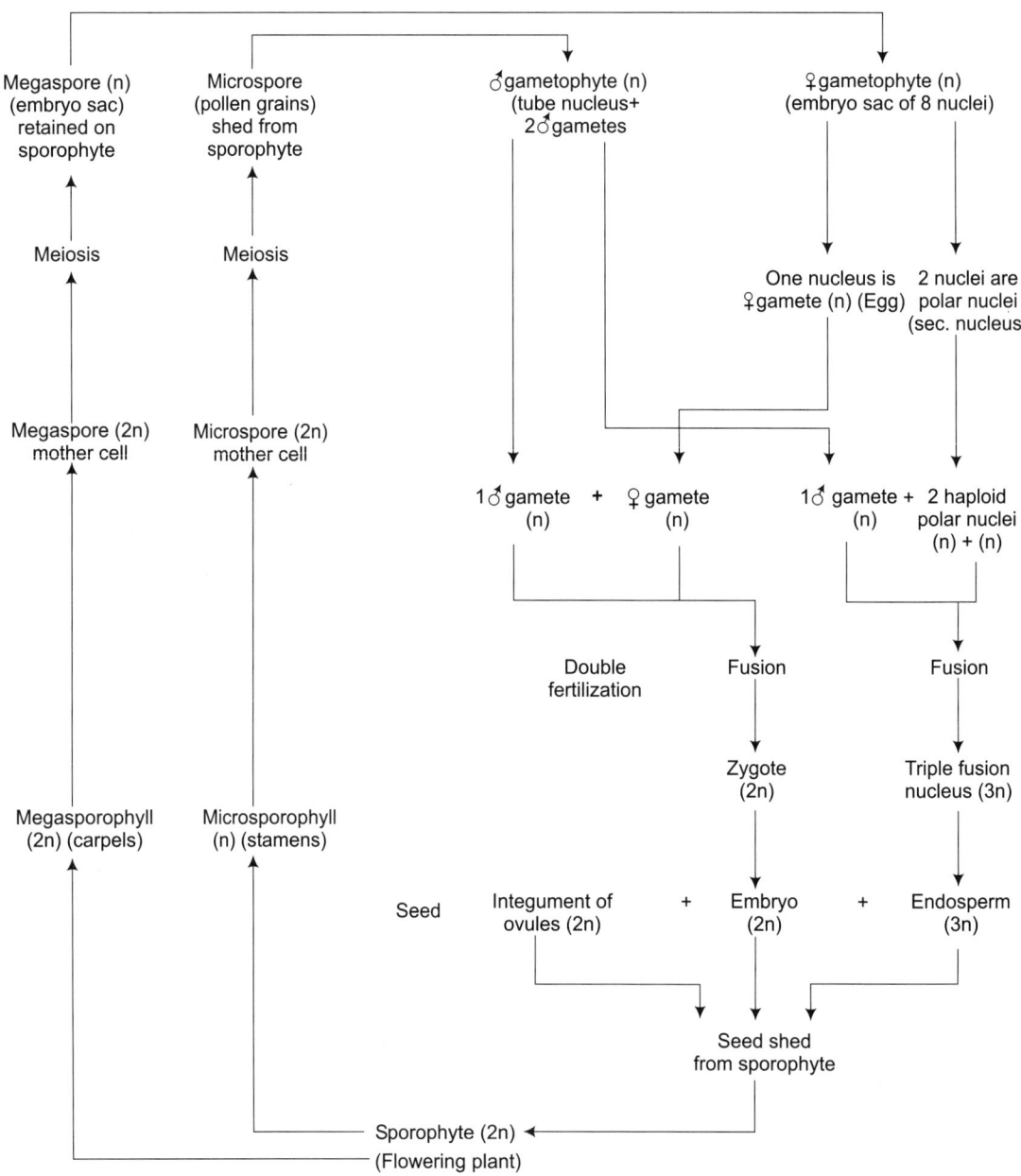

Fig. 2.3 Schematic representation showing life cycle of a flowering plant.

may be **true**, if it develops from the ovary of a flower and **false** if the parts other than ovary participate in the formation of fruit. Fruits either true or false may be classified into three types viz., simple, aggregate and multiple. In some plants, e.g., in the peanut the fruits develop only after the gynophore penetrates into the soil carrying the carpel with fertilized ovule.

Some plant produce fruits of more than one kind on the same individual. This phenomenon is called **heterocarpy.** *Aethionema heterocarpum* belonging to Brassicaceae, for instance, produce dehiscent siliqua and one-seeded indehiscent nuts in the same raceme. There are cases in which different fruits are borne above and below the ground are termed **amphicarpy,** e.g., *Trifolilium polymorphum, Amphycarpaea bracteata* (Fabaceae), *Gymnarrhena micranta* (Asteraceae). Amphicarpy is someway related to cleistogamy (Evenary *et al.*, 1977). In *Cardamine chenopodifolia* (Asteraceae) the plants which arise from seed of the subterranean fruits differ from those which develop from seeds of the aerial fruits (Cheplick, 1983).

Life cycle of a flowering plant has been shown in schematic diagram (Fig. 2.3).

3 Microsporangium

Microsporophyll or the stamen is the male reproductive part of the flower, which consists filament, anther and connective. In Ranales leaf-like stamens are found having three veins and bear the microsporangia on the abaxial surface. However, presence of a single vascular bundle is the prevailing condition in most of the angiosperms. Anther is the fertile part of microsporophyll, sometimes it is sterile called **staminode**.

A typical anther is bilobed (**dithecous**), consisting four spore sacs or the **microsporangium** (Figs. 3.1 A, B, C). However, at maturity two sporangia of each side become confluent because of the breaking of the partition wall between them. In some plants such as *Hibiscus rosa-sinensis* anther is one lobed (**monothecous**) consisting two microsporangia, which is fused at maturity. Rarely, there may be just one microsporangium per anther e.g., *Arceuthobium minutissium*. A T.S. of dithecous microsporangium reveals wall layers, microsporangia and different types of tissues (Fig. 3.2).

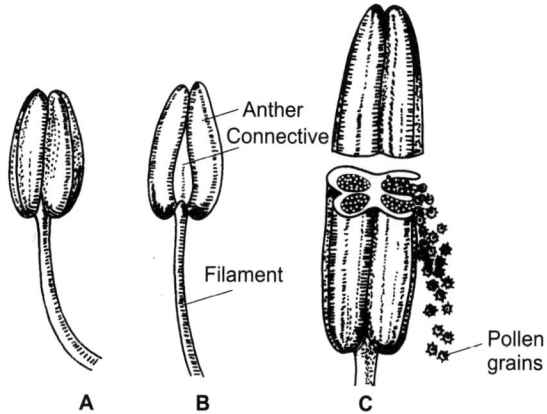

Fig. 3.1 Stamens. **A.** Ventral view. **B.** Dorsal view. **C.** T.S. of anther showing release of pollen grains.

DEVELOPMENT OF ANTHER

A cross section of a very young anther shows mass of homogenous meristematic tissues surrounded by the epidermis. It soon becomes slightly four lobed, where a row of hypodermal cells differentiated by becoming larger in size, radially elongated and show distinct nucleus. These differentiated cells form **archesporium**, which either appears as a single cell or plate of cells in cross section.

The archesporial cells divide to form a **primary parietal layer** towards the outer side and a **primary sporogenous layer** towards the inner side. The cells of the former divide by periclinal and anticlinal walls to give rise to series of concentric layers, 3-5 cells thick, constituting the wall of the anther. The primary sporogenous cells either function directly as spore mother cell or may undergo a few mitotic divisions to form large number of cells. The spore mother cells divide meiotically to form pollens or microspores, which are haploid and represent the first cell generation of male gametophyte (Fig. 3.3 A-H).

(i) Anther Wall Layers

A mature anther wall consists of four layers: epidermis, endothecium, middle layer, and tapetum.

(a) Epidermis: The epidermis is the outermost layer of the anther. It undergoes anticlinal divisions only to keep pace with the rapidly enlarging internal tissues. In mature anther they are greatly stretched and flattened. Epidermis performs protective function.

(b) Endothecium: The layer of the cells lying just below the epidermis is the endothecium. It is usually single layered but sometimes multilayered as in *Nicotiana tabacum*. Endothecium is generally incompletely differentiated, either in the bulging part of the anther or sometimes near the connective. However, in *Triticale* a complete ring of endothecium is present. The cells of the endothecium become radially elongated and from their inner tangential wall fibres run upwards ending near the outer wall of each cell. The endothecium thickening shows high content of cellulose material. Its maximum development is attained at the time

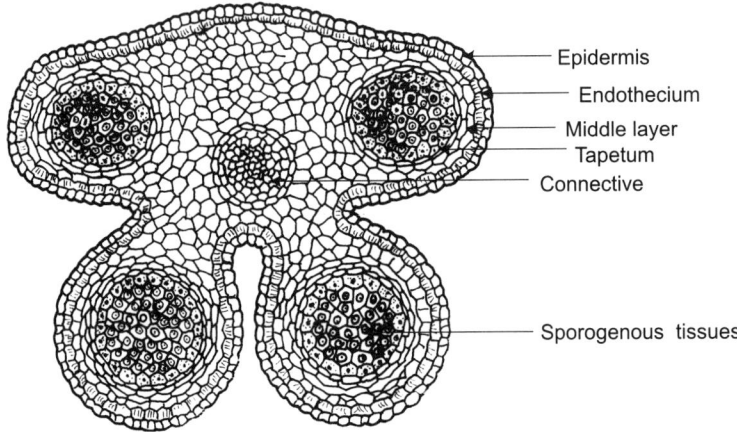

Fig. 3.2 T.S. of microsporangium showing different types of tissues.

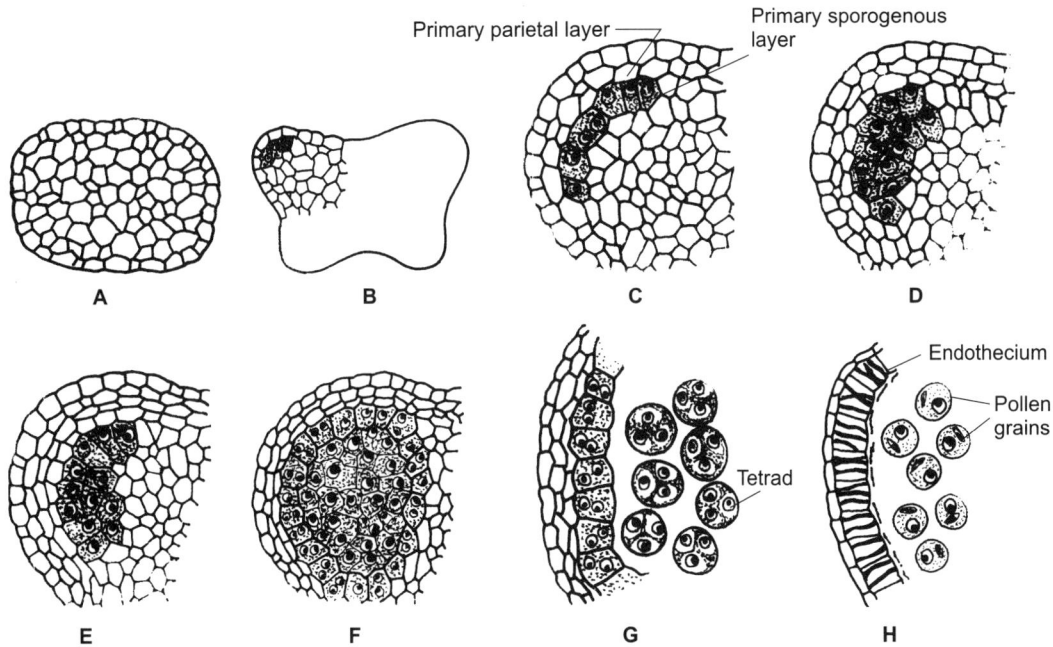

Fig. 3.3 Different developmental stages of the anther. **A.** T.S. of young anther. **B.** Differentiated archesporial cell. **C.** Differentiated primary parietal cell and primary sporogenous cells. **D-F.** Differentiation of wall layers and meiocytes. **G.** Microspore tetrad. **H.** Pollen grains and endothecium.

when the pollen grains are about to shed. However, in aquatic forms like Hydrocharitaceae, some cleistogamous forms and those plants whose anthers open by apical pore, the endothecium fails to develop fibrous thickenings. Among other exceptions like *Musa*, *Annona* and *Ipomoea* etc. the fibrous thickening is absent but the epidermis undergoes cutinization and lignification over the entire surface.

The endothecial layer is responsible for the dehiscence of the anther due to differential hygroscopic expansion of the outer and the inner tangential wall.

(c) Middle layer: Next to the endothecium, there are 1-3 layers of cells constituting middle layer or there may be several as in *Lilium*. As a rule, all of them become flattened and crushed at the time of meiotic division in the microspore mother cells; however, there are a few exceptions where outermost layer persists for a long time. Interestingly, in *Gloriosa* the outermost layer develops fibrous thickening similar to those of endothecium.

The cells are generally rich in reserve foods such as starch, thus suggesting its nutritive role in pollen development.

(d) Tapetum: The innermost wall layer or the tapetum is of considerable physiological importance because all the food materials entering into the sporogenous tissues pass through it. It attains its maximum development at tetrad stage.

Typically, the tapetum is composed of single layer of cells, which shows dense cytoplasm and distinct nucleus; however, in *Nicotiana* and *Costus* it is composed of several layers.

Tapetum is usually derived from primary parietal layer forming homogenous layer of cells. However, in a few species the tapetal cells are **dimorphic** in nature as it shows two different origins (primary parietal cells and connective cells). Tapetal cells of different origin show different sizes and different structures e.g., *Alectra thomsoni*. However, there are certain examples, where tapetum has its origin from the peripheral cells of the sporogenous tissues.

Nuclear Division of the Tapetum: The nucleus of the tapetum initially shows uninucleate condition but subsequently become multinucleate (2, 4, 8 or 16) due to free nuclear divisions. The tapetal cells also become polyploid, either due to **endomitosis**, **restitution nuclei** or by the formation of **Polytene**.

(i) **Endomitosis:** In this type longitudinal splitting and separation of chromosomes take place within the same nucleus without the formation of spindle resulting in the formation of polyploidy nucleus (Fig. 3.4 A-D). Examples are *Spinacia*, *Cucurbita*.

(ii) **Restitution nuclei:** In this type normal mitosis occurs up to the early anaphase but the two sets of chromosomes are finally incorporated into a common nuclear membrane to form a restitution nucleus (Fig. 3.5 A-C).

(iii) **Polyteny:** If the two chromatids replicates longitudinally but fail to separate, thus a single chromosome will consists many chromatids within the same nucleus (Fig. 3.6 A-B).

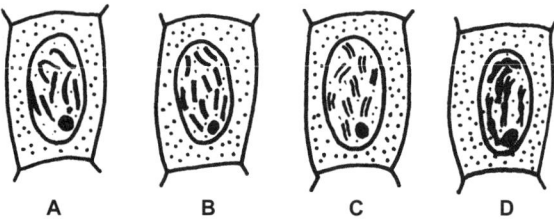

Fig. 3.4 Endomitosis in tepetal cells. **A.** Endoprophase. **B.** Endometaphase. **C.** Endoanaphase. **D.** Endotelophase.

Types of Tapetum: Based on behaviour there are two types of tapetum; (i) **Amoeboidal** and (ii) **Secretary or Glandular**.

(i) Amoeboidal Tapetum: It is also known as **invasive** or **periplasmodium** tapetum. This type of tapetum is more prevalent in monocotyledons than dicotyledons. In amoeboidal tapetum, there is breakdown of the inner tangential cell wall followed by the enlargement

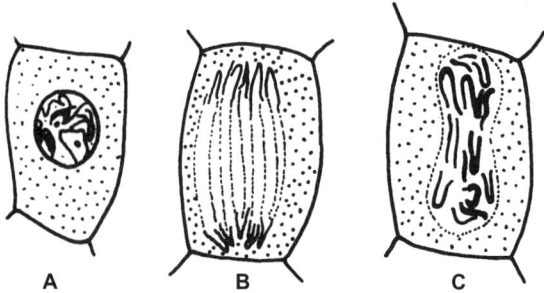

Fig. 3.5 Formation of restitution nucleus. **A.** Prophase. **B.** Anaphase. **C.** Both the sets of chromosomes are surrounded by a common nuclear membrane.

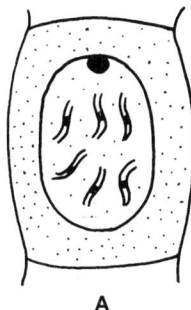 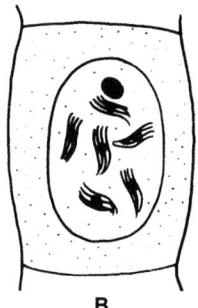

Fig. 3.6 Formation of polytene. **A.** Diploid cell with single chromatid chromosome. **B.** Tetraploid cell, each chromosome consists 4 chromatids.

and movement of the protoplasm into the anther sac, where they fuse to form periplasmodium. The formation of periplasmodium occurs at different developmental stages ranging from meiotic prophase to tetrad stage of pollen development. Clausen (1927), therefore, classified this type of tapetum into following four sub-types:

(a) *Sagittaria type*: The tapetal cells lose their walls by the time the microspore tetrads have been formed and their protoplasts begin to project inward as soon as the microspores have separated. Examples: *Sagittaria, Alisma, Limnocharis, Hydrocharis*.

(b) *Butomus type*: In this type the formation of the periplasmodium occurs a little earlier, when the microspores are still grouped in tetrads. Examples: *Butomus, Stratiotes, Ouvirandra*.

(c) *Sparganium type*: Here also the fusion of the protoplasts begins at the tetrad stage but the tapetal cells are multinucleate. Examples: *Sparganium, Typha, Tradescantia*.

(d) *Triglochin type*: In a few types the tapetum begins its activity while the microspore mother cells are still undergoing the meiotic division. The tapetal protoplasts and the nuclei protrude into the space between the mother cells. Examples: *Triglochin, Potamogeton* and several members of the Araceae.

However, the ultrastructural detailed study of amoeboidal tapetum has been shown in *Arum italicum* by Pacini and Juniper (1983). The anther wall shows single cell thick epidermis and endothecium; three-cell thick middle layer and two-cell thick tapetum. Development of microspores in relation to events in tapetum can be described under following stages:

(i) In the premeiotic phase tapetal cells have inter-connected plasmodemata. The interconnections are absent in the tangential walls facing either to middle layer or microspore mother cells (Fig. 3.7A).

(ii) During meiosis the innermost middle layer becomes flattened and the tapetum cells starts dissolving. At leptotene the plasmodesmatal connections widen to form cytomictic channels (Fig. 3.7 B); by pachytene the tangential wall between the tapetal cells disappear (Fig. 3.7 C) and; by anaphase most of the radial walls between the tapetal cells also disappear and large number of microtubules are seen towards the inner tangential wall (Fig. 3.7 D,E).

(iii) By the time tetrads are formed, the inner tangential wall of the tapetum disappear and the amoeboidal tapetum begins to envelop tetrads and some microtubules are

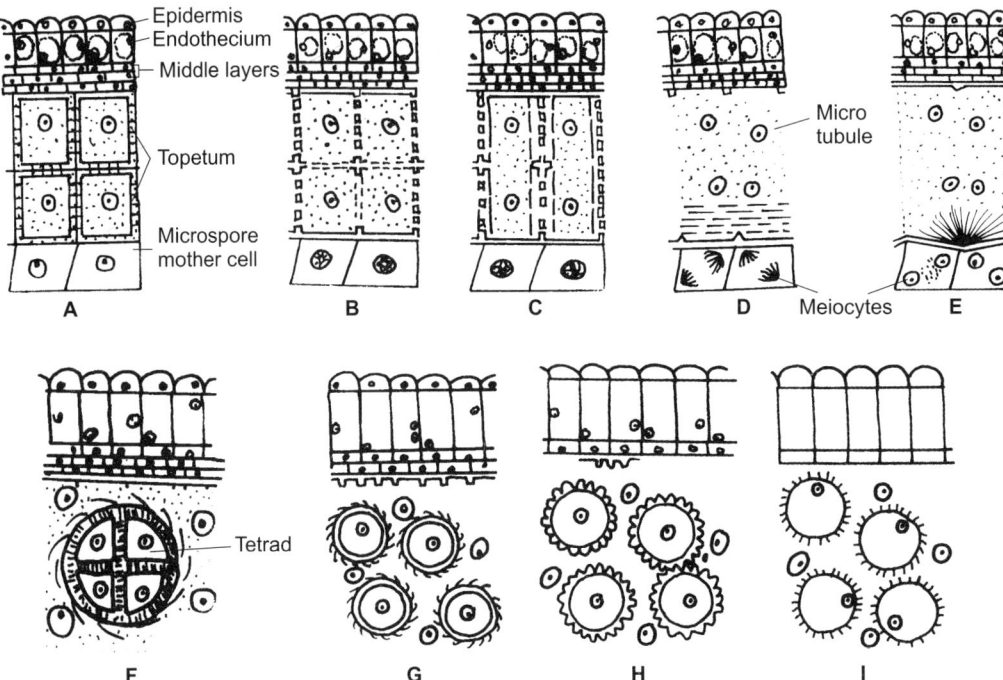

Fig. 3.7 *Arum italicum*. Amoeboidal tapetum. **A.** Premeiotic phase. **B.** Leptotene. **C.** Pachytene. **D.** Anaphase-I. **E.** Telophase-I. **F.** Tetrad. **G.** Microspore surrounded by microtubules. **H.** Microspores at later stage. **I.** Uninucleate pollen grains with pollen wall (After Pacini and Juniper, 1983).

seen running parallel to the tetrad wall (Fig. 3.7 F). By this stage inner row of the middle layer shows degeneration of their nuclei.

(iv) With the release of microspores from tetrad the outer wall of the tapetum also disappear and the tapetal contents seem to be differentiated into two zones. The inner zone surrounding the microspores contains polyribosomes microtubules, vesicles and a few dilated ER cisternae. The outer zone on the other hand contains a few small vacuoles, nuclei, ribosomes, polyribosomes, mitochondria and plastids (Fig. 3.7 G).

(v) At the mid-uninucleate stage the tapetal plasma membrane retract from the exine surface forming roughly cone-shaped space (Fig. 3.7 H). This phenomenon also occurs at the periphery of tapetum but with less precision and is discontinuous. These spaces are soon filled with material forming spines on the exine (Fig. 3.7 I). At this late uninucleate stage the tapetal cytoplasm is no longer differentiated into two zones. The number of polyribosomes and ribosomes are reduced and the microtubules disappear completely.

(vi) Towards the end of the pollen development, the periplasmodium degenerates, dries up, becomes coated on the surface of the pollen wall as **pollenkitt** material (an oily layer).

(ii) Secretory Tapetum: Secretory tapetum is more commonly observed in the dicotyledons. Unlike the amoeboidal tapetum, the cells of the secretory tapetum remains intact in their original position, however, break down occurs at the maturation of the pollen grains. The tapetal cells secrete and liberate substances to anther chamber and contribute in the development of pollens. A few important steps of tapetal behaviour and pollen development have been described as follow:

(a) At the stage of sporogenous tissues, tapetal cells possess mitochondria, plastids, a number of **Pro-Orbicular** or **Pro-Ubisch** bodies and dictyosomes with only a few associated vesicles. At tetrad stage, the numbers of Pro-Ubisch bodies increase, although distributed throughout the cytoplasm but appear to be more concentrated in the region of cell facing the anther cavity.

(b) Before the onset of meiosis the tapetal cell wall becomes thick and the cytoplasm appears to be denser due to increased number of ribosomes and pro-Ubisch bodies. Soon after the separation of the microspores from one another, the pro-Ubisch bodies migrate to the cell surface, where they get coated with **sporopollenin** (derived from oxidative polymerization of carotenoid) and are now called **Ubisch** bodies. They are released into the anther cavity, where they are involved in the external thickening or sculpturing of the pollen exine.

(c) By the time pollens have separated tapetal cell wall becomes considerably thinner and finally disappears; cytoplasm is completely disorganized.

However, Cimiapolini *et al.* (1993) recorded twelve stages of tapetum-pollen development in *Cucurbita pepo*.

The tapetum is a highly active layer of cells surrounding the sporogenous tissues and helps in the nutrition of the developing microspores. It is associated with enzymatic activity, i.e. callase synthesis, which brings about callose degradation and release of microspores from the tetrads. Tapetum is involved in synthesis and the release of the pollenkitt and tryphine materials, which are associated with pollen wall formation. Morphological and cytological studies on anther development in male fertile and male sterile lines have revealed the formation of non- viable pollens in sterile lines is either associated with non- functioning or the abnormal functioning especially callase activity due to its early release (prophase-I) as shown in *Petunia*.

SPOROGENOUS TISSUES

Sporogenous tissues directly functions as microspore mother cells or may undergo a few mitotic divisions before functioning as mother cell.

Usually there are four groups of sporogenous tissues (tetrasporangiate), two in each anther lobe; however, there are some plants where fewer than four groups of sporogenous tissues are found. In Malvaceae, there are two groups while *Vallisneria* shows all gradations from unilocular to tetralocular conditions. In plants like *Korthalsella*, there are six groups due to synandrium (fusion of three stamens).

Although all the sporogenous tissues are capable of giving rise to microspores but there are some examples, where some of them frequently degenerate and become nutritive in function, thus providing nutrition to the microspores as in *Ophiopogon*, *Haloptelia* and *Zostera* etc.

(i) Syncytium Formation and Meiosis

In the young anther before the onset of meiosis, the different cell layers i.e. wall, tapetum and sporogenous tissues exhibit plasmodesmata connections, showing free flow of nutrients amongst the cells of different layers (Fig. 3.8 A). However, at the beginning of meiosis, the connections between different layers get severed although the cells of the same type are still connected.

An interesting structural change occurs in the sporogenous cells soon after the meiosis starts. Each sporogenous cells develop additional walls made up of callose (a polysaccharide made up of B-1, 3 glucans) on the inner side of the cellulosic cell wall, which finally disintegrates. Plasmodesmatal connection between sporogenous cells is replaced by cytoplasmic bridges, through which cytoplasm and the organelles make free movement between meiocytes (Fig. 3.8 B). They attain their maximum development in zygotene-pachytene stage. Thus, meiocytes or the microspore mother cells together constitute a functional identity called **Syncytium**. This condition establishes synchrony amongst the meiocytes of a microsporsangium.

At the end of meiotic prophase the callose wall completely surrounds microspore mother cell and cytoplasm connections are cut off. After the completion of meiosis additional callose wall is laid down between microspores (Fig. 3.8 C). Microspores are released with dissolution of callose deposition (Fig. 3.8 D-E).

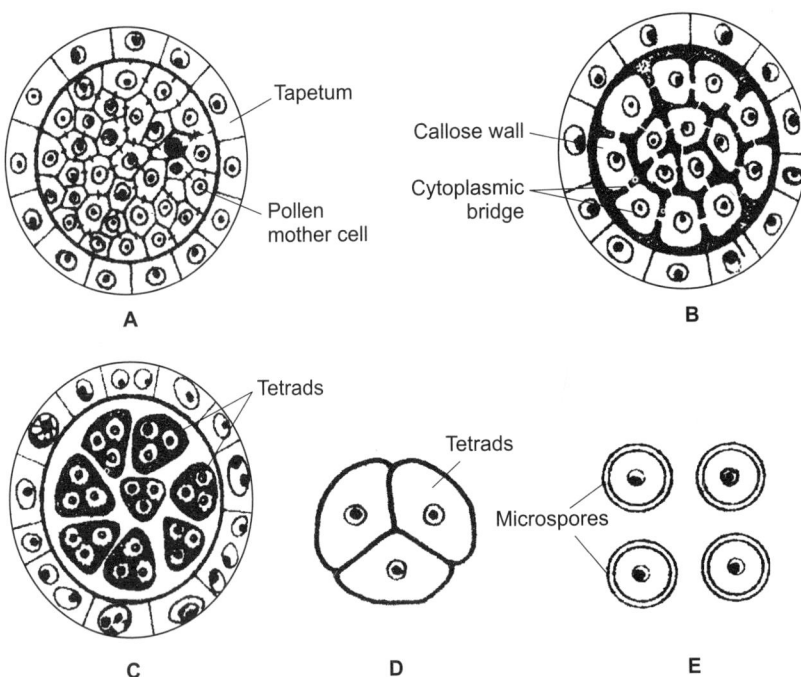

Fig. 3.8 Diagrams to show cellular inter-connections in developing anthers. **A.** Preleptotene stage; plasmodesmatal connections between PMCs and between tapetal cells and the PMCs. **B.** Meiotic Prophase -I. The PMCs are now enclosed in thick callose wall and are inter-connected by cytoplasm channels between PMCs. **C.** Tetrad stage. **D.** Isolated tetrad. **E.** Pollen grains.

(ii) Cytokinesis

The wall formation during cytokinesis is of the two types: (a) Successive and (b) Simultaneous, resulting in the formation of microspore tetrads.

   (a) **Successive type:** In the successive type, the two nuclei formed after the first nuclear division is separated by the laying down of the wall, thus to form a dyad. The nucleus of the dyad soon undergoes a division followed by second wall (Fig. 3.9 A).

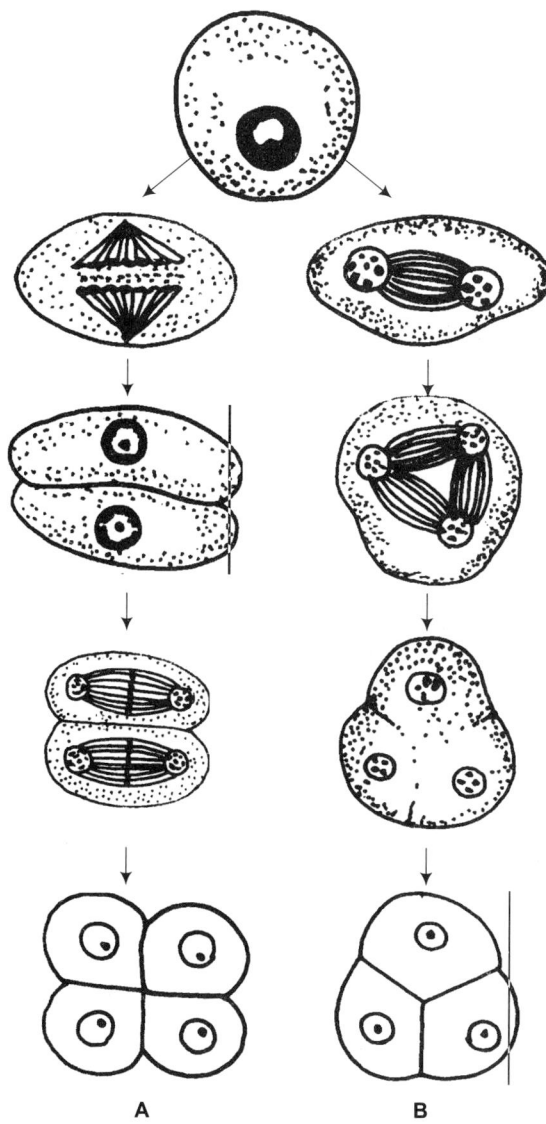

Fig. 3.9 Microsporogenesis. **A.** Successive type cell division. **B.** Simultaneous type cell division.

(b) **Simultaneous type:** This type of the cytokinesis is of common occurrence. Here the first nuclear division is not followed by wall formation immediately but the walls are laid down simultaneously i.e. only after the formation of all the four nuclei (Fig. 3.9 B).

(iii) Microspore Tetrad and Polyspory

The arrangement of microspores in tetrad is variable and shows five types (Fig. 3.10): (i) **Tetrahedral** (ii) **Isobilateral** (iii) **Decussate** (iv) **Linear** and (v) **T-shaped**.

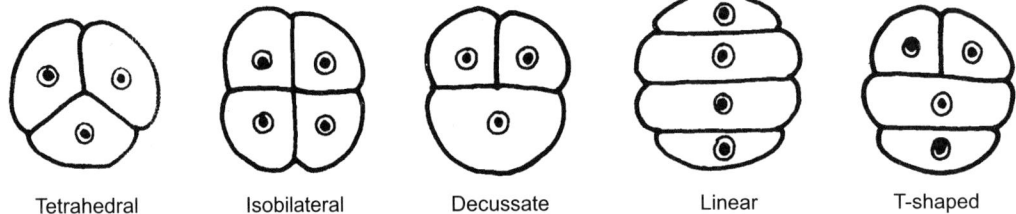

Fig. 3.10 Different types of microspore tetrads.

The first two types are most common in angiosperms. In *Musa, Agave* and *Laurus* etc. two or three types of disposition may be found, however, in the *Aristolochia elegans* all the five types of the tetrad have been reported. Interestingly in *Zostera* the microspore mother cell is elongated, which divides in a plane parallel to the longitudinal axis to form four filiform microspores.

Occasionally, there are cases where fewer than four spores result from the division of microspore mother cell or may form more than four spores called **polyspory**. Mahabale and Chennaveeraiah (1957) have reported polyspory in *Hyphaene*, which is often due to the divisions of the members of the tetrads (Fig. 3.11). Tetrad with as many as eleven microspores has been reported in *Cuscuta reflexa*.

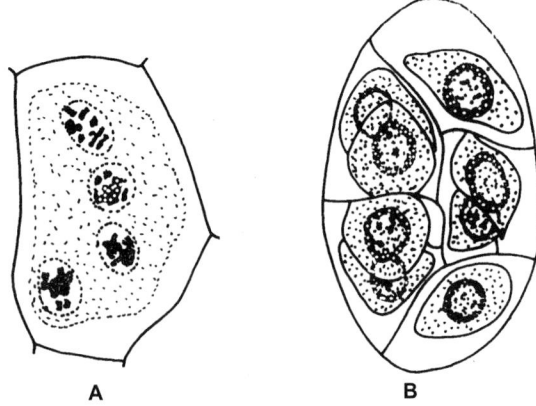

Fig. 3.11 Polyspory in *Hyphaene indica*. **A**. Microspore mother cell after meiosis. **B**. Octant (After Mahabale and Chennaveeraiah, 1957).

(iv) Compound Pollen Grains and Pollinium

Usually microspores soon separate from one another; however, in some plants they adhere to form compound pollen grains e.g., *Drimys, Annona,* and *Drosera* etc. In the tribes of *Ophrydeae* and *Neotieae* this tendency is carried further and the compound grains are themselves held together by means of viscin threads in small units called **massulae** (Fig. 3.12A, B).

In certain plants, e.g., *Acacia* and other species of Mimosaceae the tetrads stuck together in groups, **polyads** (Fig. 3.12 C) which contain as many as 64 pollen grains; however, in Asclepiadaceae and Orchidaceae all the microspores in a sporangium remain glued together to form a single mass called **pollinium** (Fig. 3.12 D).

(v) Release of Pollen Grains

Before the liberation of pollen grains, outer wall or the exine is laid down. After the development of the exine, the callose wall dissolves due to secretion of an enzyme, callase from the tapetal cells. The development of the male gametophyte begins after the liberation of individual microspores. In majority of angiosperms the pollen grains or microspores are shed from anthers in two-celled stage but in some at three-celled stage. The pollen representing the male gametophyte enters into next stage i.e. production of male gametes.

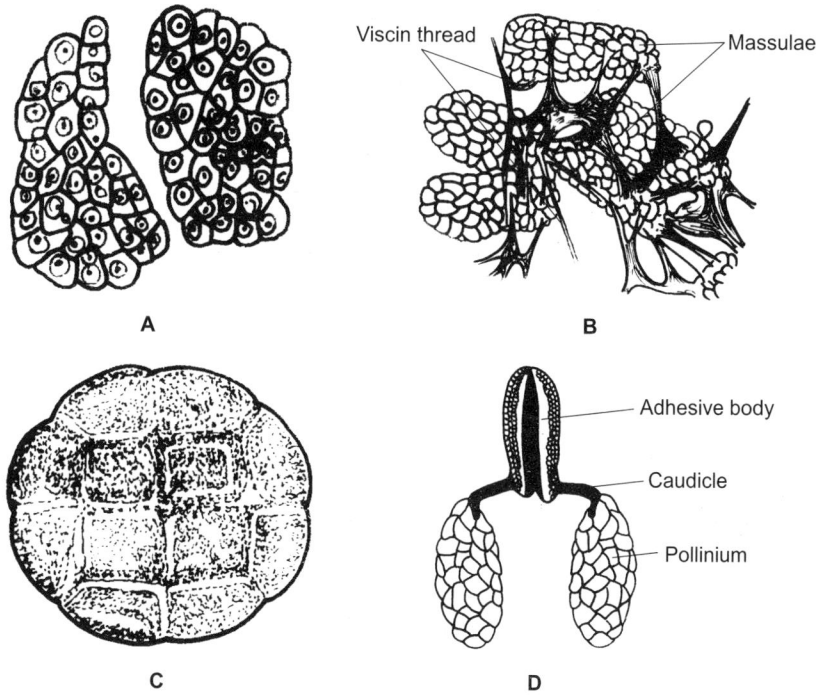

Fig. 3.12 **A.** Massulae. **B.** A group of massulae showing the connecting viscin thread. **C.** A polyad. **D.** Pair of pollinia in *Calotropis*.

4 Male Gametophyte

The microspores or the pollen grains represent the first cell generation of the male gametophyte. The further development of the male gametophyte or microsporogenesis gives rise to a large **vegetative cell** and a small **generative cell** (Fig. 4.1 A-D). The latter undergoes another mitotic division to form two male gametes (sperms) either in the pollen grain (Fig. 4.1 G-H) or in pollen tube (Fig. 4.1 I-J). Along with these changes the wall synthesis is also followed. The detail account of the male gametophyte can be studied under four heads:

(1) Microspore
(2) Vegetative and generative cells
(3) Division of the generative cell
(4) The pollen wall
(5) Unusual type of pollen development.

MICROSPORE

The newly formed pollen grains have very dense cytoplasm with distinct nucleus. Soon after its release from tetrad, it undergoes rapid increased in volume. This is followed by vacuolation, which pushes the nucleus into thin peripheral cytoplasmic film lining the wall. The nucleus of the pollen may divide or may remain in resting period varying from a few days to several weeks depending upon the species. The pollen grains have reserve food materials in the form of carbohydrates, proteins or lipids. Interestingly the nature of the food materials has some relations with the type of pollinations, as in entomophilous pollens are rich in lipids, while anemophilous have abundant starch.

VEGETATIVE AND GENERATIVE CELL

The first division of the microspore gives rise to two unequal cells: the larger one is vegetative cell, which develops into pollen tube, while the smaller one is the generative cell, which with the subsequent mitotic division forms two male gametes or sperms.

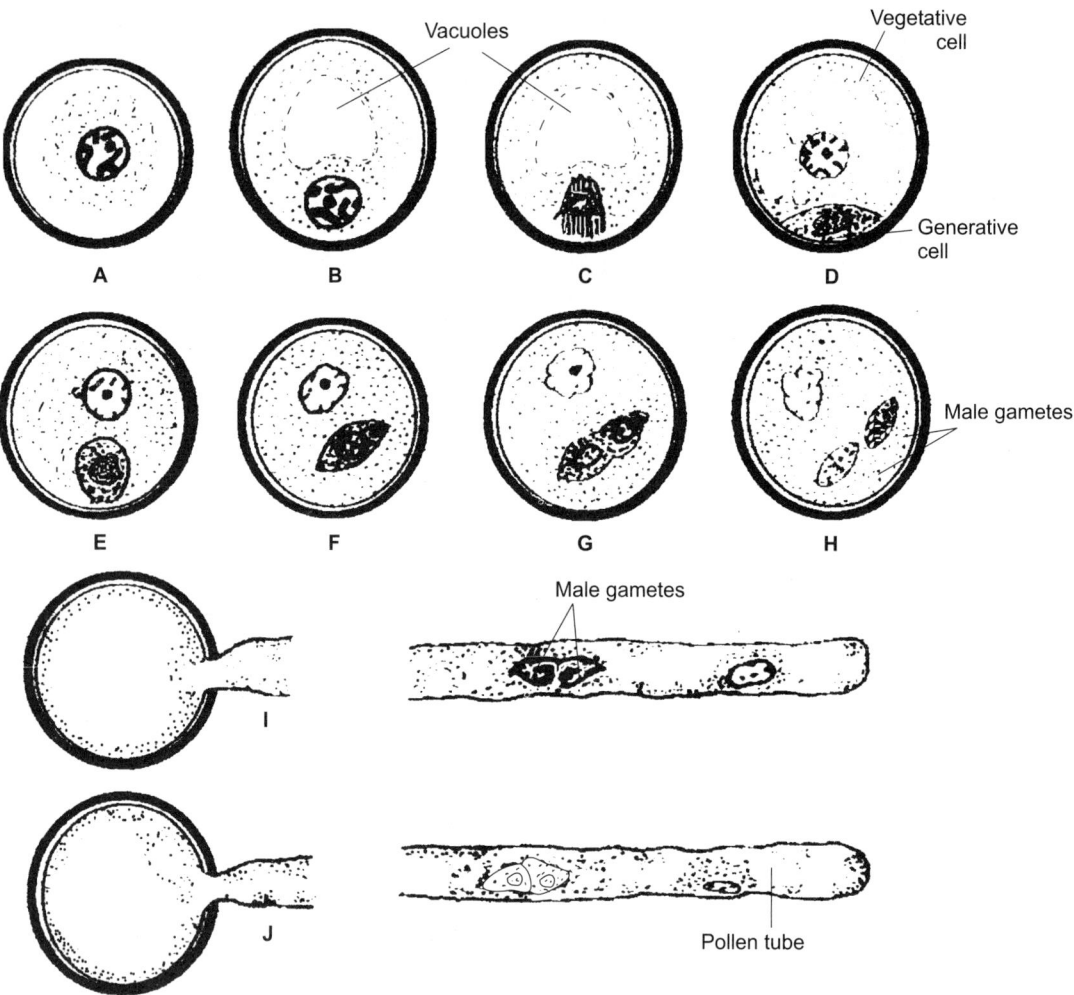

Fig. 4.1 Diagrams to illustrate important stages in the development of male gametophyte. **A**. Newly formed microspore. **B**. Old stage showing vacuole formation. **C**. Dividing microspore nucleus. **D**. Two-celled stage showing vegetative and generative cell. **E**. Generative cell losing contact with wall. **F**. Generative cell lying free in the cytoplasm of vegetative cell. **G-H**. Division of the generative cell in pollen grain. **I-J**. Division of the generative cell in pollen tube (After Maheshwari, 1949).

Geitler (1935) observed that the metaphase spindle usually shows pronounced asymmetry, the wall-ward pole being blunt and the free pole acute (Fig. 4.2 A-C). The cause of asymmetry has been attributed to differences in the time of the development of the spindle poles: generative pole develops more slowly than the vegetative because the smaller amount of the cytoplasm is associated with the former. Symmetrical spindles have been observed only occasionally e.g., *Asclepias, Anthericum,* and *Adoxa* etc.

The divisions of the microspores within the same chamber do not show any synchronization as in the microspore mother cells. However, the microspores of the same

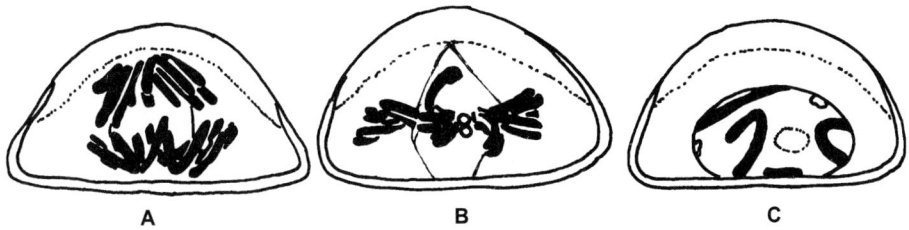

Fig. 4.2 *Allium cernuum*, first division of microspore. **A.** Prophase. **B.** Metaphase **C.** End of Anaphase (After Maheshwari, 1950).

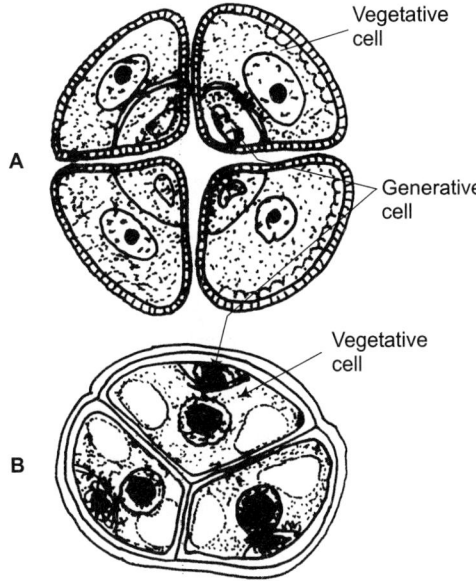

Fig. 4.3 Microspore showing vegetative and generative cell. **A.** Pollen tetrad showing generative cell cut off toward inner side of each microspore. **B.** Pollen tetrad showing generative cell cut off toward outer side of each microspore.

tetrad show same stage of development but a complete synchronization is observed in **massulae** and **pollinia**, where microspores are united together as in Mimosaceae, Asclepiadaceae and Orchidaceae.

The generative cell is either cut off inner side (Fig. 4.3A) or on the outer side (Fig. 4.3B) or on the radial wall or in a corner. The generative cell soon loses contact with the wall of the microspore and come to lie freely in the vegetative cell.

Immediately after detachment, the generative cell is spherical. However, variation in shapes, like elliptical, lenticular, spindle-shaped or sometimes vermiform have also been observed in different species. The generative cell contains all usual cell organelles except plastids with a few exceptions. The nucleus is smaller but contains higher amount of DNA.

DIVISION OF THE GENERATIVE CELL

The generative cell either divides when the pollens are still within the anther or it may takes place after the pollen discharge. In the former case the pollen are shed at 3-celled stage as in *Beta* and *Hordeum*. In latter condition, the pollens are discharged at 2- celled stage but the subsequent development takes place as follow:

 (i) It may occur on the stigma as in *Haplotelia integrifolia*.
 (ii) It may occur in the pollen tube, which is the most common type.
 (iii) It may occur after the pollen tube has reached the embryo sac as in *Euphorbia terracina*.

Cytokinesis of the generative cell resulting in the bipartition may takes place either by the usual cell plate formation, as in *Asclepias* and *Potulaca* or by the process of furrowing as in *Juncus*. Witmer (1937) has observed both cell plate and furrowing in *Vallisneria*.

The two male gametes or sperms thus formed are discrete cells and their cytoplasmic sheath persists throughout their course in the pollen tube.

Dimorphic Sperms and Male Germ Unit

Normally, both the sperms are of the same types, however, recent studies on the structure of the pollen grains of *Plumbago zeylanica*, *Brassica campestris* and *Zea mays* have revealed that the sperms are markedly dimorphic i.e. one is larger than the other and the differences have also been observed in the cell organelles. Smaller sperm cell shows the predominance of plastids, while most of the mitochondria are located in the larger cell. The two sperms are directly linked to each other by common transverse cell wall perforated with plasmodesmata. Further, in the *P. zeylanica*, one of the sperms is connected to the vegetative nucleus by the cytoplasmic extension, thus forming a 3-celled **Male Germ Unit** (**MGU**; Fig. 4.4). This organization determines the future course of sperms. The smaller free sperm unites with egg

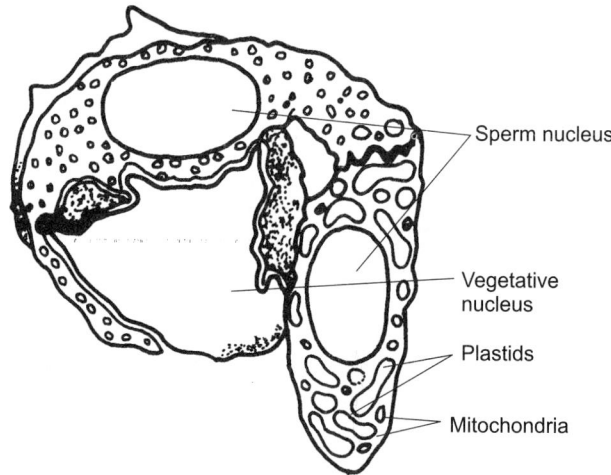

Fig. 4.4 Male Germ Unit (**MGU**) of *Plumbago zeylanica* showing vegetative nucleus and sperm nucleus (After Russell, 1984).

while the larger sperm, which shows association with the vegetative nucleus, participates in triple fusion (Russell, 1987).

Thus the occurrence of **MGU,** whether formed before or after pollen germination appears to be the universal feature amongst the angiosperms.

POLLEN WALL

Pollen grains are rather uniform in their wall architecture: (i) **tectate exine** and (ii) **pilate exine** (Fig. 4.5A-B). A mature pollen wall shows two distinct layers-outer **exine** and inner **intine.** The former is resistant to acetolysis, while latter is an acid degradable. The exine is composed of sporopollelin, a highly resistant material believed to be produced by the oxidative polymerization of carotenoid pigments (Brook and Shaw, 1971). The exine is sculptured with varying degree of elaboration. The sculptured part of the exine is made up of radially oriented rod-like baculae (columella) which may remain open (pilate grain) or covered by a roof or tectum (tectate grain). The exine covers the entire pollen grain except germinal apertures (germ pores) where it is absent or highly reduced. The baulae (in pilate grain) may be enlarged to form knob-like structure and stand free, or may be fused together to give a raised wall often disposed in a reticulate pattern. In a tectate grain the void between baculae open to the outside through perforations termed **micropores**. The exine is composed of an outer ektexine (sexine and nexine) and an inner endexine. The endexine is often well developed in dicots and is absent in monocots. In order to avoid confusion diagrammatic representation of the old and new terminology has been presented (Fig. 4.6).

The intine is composed of pecto-cellulose materials. In some taxa, particularly Poaceae a middle layer is rich in pectic polysaccharides, termed as **Z**-layer, is distinguishable. This layer is thickened at germ pore and is termed "Zwischenkorper" (Heslop-Harrison, 1977).

(i) Development of Pollen Wall

During microsporogenesis, after meiosis four haploid spores are enclosed in a common callose wall and the microspores lack wall of their own. The individual spore is also

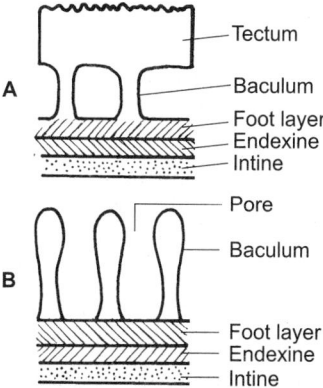

Fig. 4.5 Pollen wall. **A.** Tectate exine. **B.** Pilate exine.

4.6 Plant Embryology: Classical and Experimental

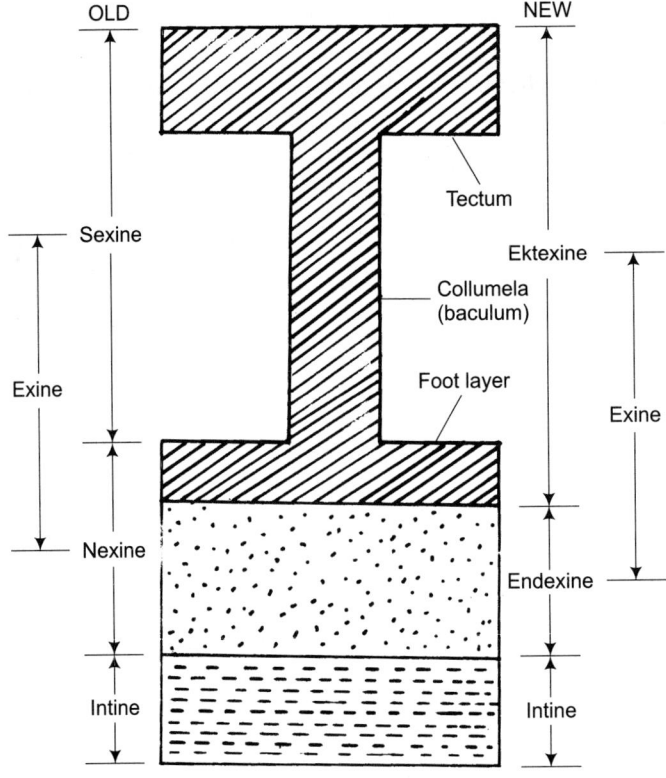

Fig. 4.6 Diagram showing old and new terminology

separated from one another by means of callose partition. The pollen wall is partly contributed by the cytoplasm of microspore and partly by tapetum (Fig. 4.7).

The first layer of the wall to be formed is **primexine**, which is cellulosic in nature and is laid down between callose and plasmalemma, when the spores are in tetrad stage. The primexine becomes gradually thick and forms the ektexine. Later with the digestion of the callose, the spores are free. In this free state pollen grains synthesize the **intine** (pecto-cellulosic) and the innermost layer of the exine (**endexine**). The sporopollenin, which is secreted by both the cytoplasm of the spores and the tapetal cells of the anther, contribute in the thickening of the exine.

At certain regions in the cytoplasm the endoplasmic reticulum are closely applied to the plasmalemma. The cellulosic **primexine** is discontinuous at these regions which mark the position of future germ pore. The transition from primexine to exine takes place in the following manner (Fig. 4.8):

(a) The primexine increase in the thickness creating additional gaps where convoluted lamellae are deposited forming columns, called **pro-bacula**.
(b) Now the cytoplasm starts synthesizing sporopollenin and pro-bacula is now called **bacula**.

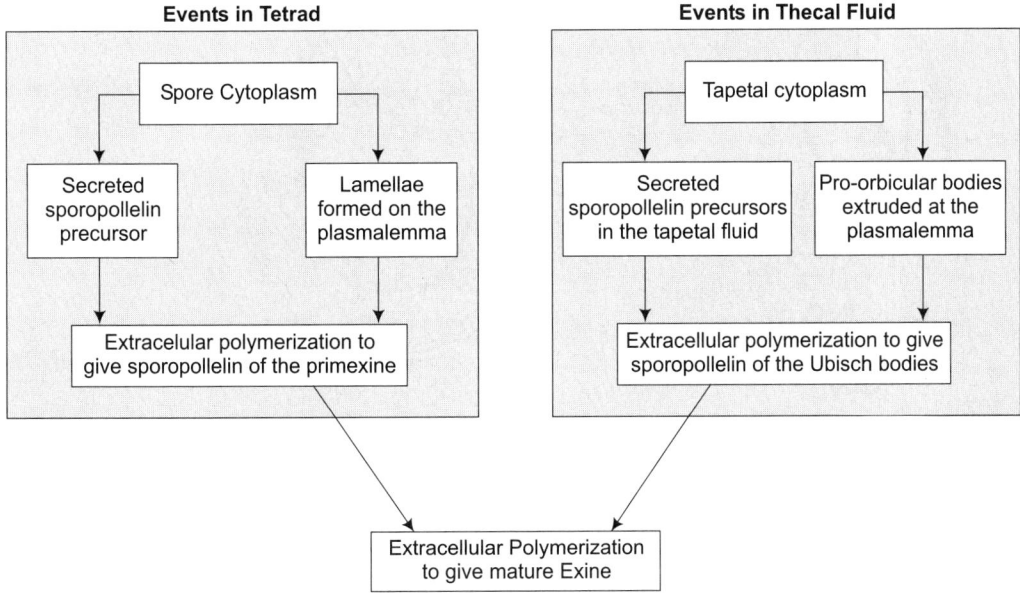

Fig. 4.7 Schematic representation of the deposition of sporopollelin in the tapetal orbicules or Ubisch-tapetal bodies and exine development in *Lilium* (After Heslop-Harrison, 1971).

(c) Later the lower end of the bacula spreads sideways forming **foot-layer**.

(d) The top of the bacula-column also spreads sideways in all the directions forming **tectum**. The tip of the bacula sometimes enlarges to form **knobs**.

(e) Further, the deposition of the sporopollelin is continued and the whole pollen grain wall expands radially and laterally as the pollen grain enlarges.

(f) The callose wall is gradually digested and the pollen grains lie free in the pollen sac.

(g) In this stage, pollen grains synthesize the innermost layer of exine, called **endexine**.

(h) The final stage in the development is the formation of the cellulosic **intine**. This process is associated with an increase activity of the dictyosomes and randomly distributed microtubules.

(ii) Processes Occurring in the Tapetum

(a) During maturation of the pollen grains, structural changes occur in the tapetum. Simultaneous with maturation of the tetrads, tapetal cells of the anther enlarge and begin to form specific spherical bodies, called **Pro-Ubisch bodies (pro-orbicules)**.

(b) These bodies migrate to the cell surface, where they are coated with sporopollelin and are released into the anther cavity.

(c) The sporopollelin coated Ubisch bodies are involved in the external thickening of the exine.

The whole event of wall formation takes partly when the spores are at tetrad stage and partly when spores are at free stage and the wall material is contributed by both microspore and tapetum (Fig. 4.9).

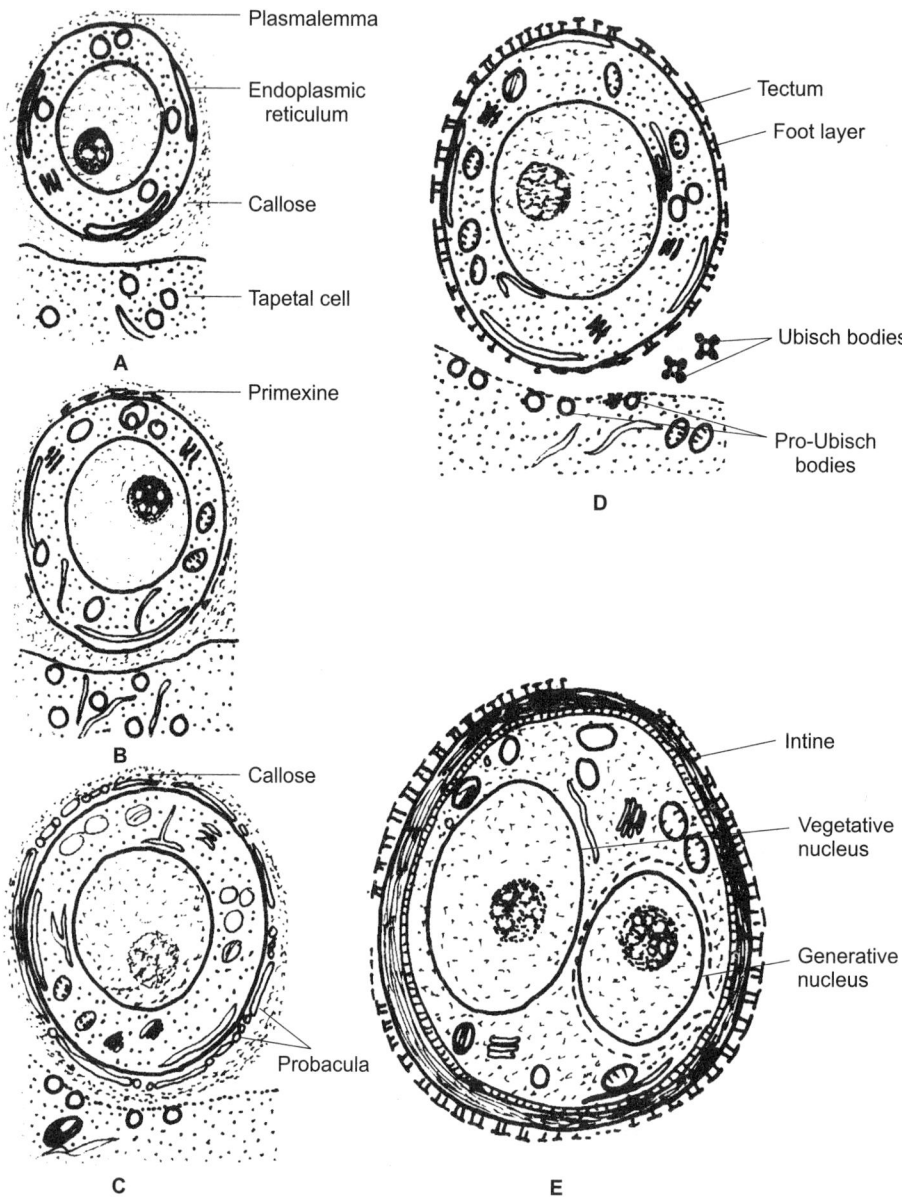

Fig. 4.8 Formation of Ubisch bodies and development of pollen wall (After Echlin, 1968).

A mature pollen grain contains large amount of starch and in certain species fatty substances which are apparently absorbed from the tapetum. In many plants the starch disappears from the pollen grains during the ripening of the anther while in others the starch disintegrates only in the pollen tube. It is assumed that there is connection between the disintegration of the starch and the high osmotic pressure of the pollen tubes which is higher than that of the cells of the style through which tube passes.

Male Gametophyte **4.9**

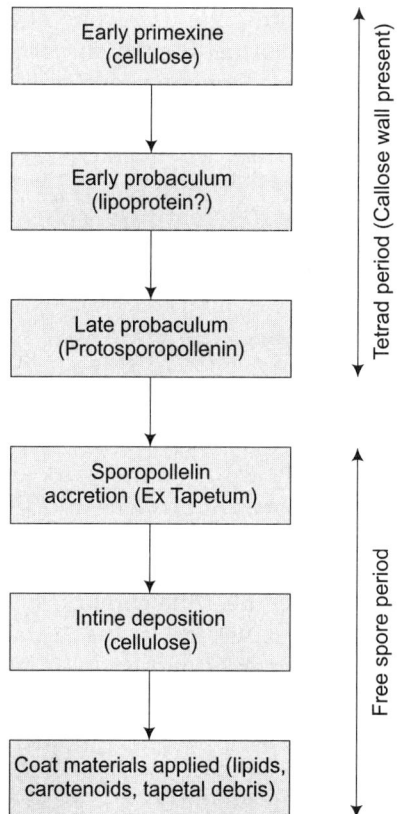

Fig. 4.9 Scheme in the events of pollen wall formation (After Heslop-Harrison, 1968)

The chemical analysis of mature pollen grains shows the following composition:

Proteins	7.0 – 26.0%
Carbohydrates	24.0 – 48.0%
Fats	0.9 – 14.5%
Ash	0.9 – 5.4%
Water	7.0 – 16.0%

(iii) Pollenkitt and Tryphine

The pollenkitt is a well-established instance of the surface transfer of material from the tapetum to the pollen surface. The pollenkitt is made up of lipoid material, flavanoids, carotenoids and degeneration products of the tapetal proteins. The enzymes involved in the biosynthesis of flavanoids are localized predominantly in the tapetal cells (Rittscher and Wiermann, 1983). Lipoidal material of pollenkitt is synthesized in plastids which disintegrate prior to tapetal degeneration. Pollenkitt is quite conspicuous in insects pollinated pollens. The function of the pollenkitt is not clear; it is supposed to act as an insect attractant, help in pollen dispersal and in protection against the damaging effect of UV rays. Deposition of flavanoids appears to be essential for normal functioning of pollen. Pollen grains of maize, which lack flavanoids are incapable of achieving fertilization (Coe et al., 1983). Lipoidal components of the pollenkitt probably have an important role in controlling water loss from the germ pore and exine of the microspore following pollen dispersal. The pollenkitt also has an important role in controlling the breeding behavior of the species.

Tryphine is often distinguished as different from pollenkitt. The former is a complex mixture of hydrophilic substances derived from the breakdown of tapetal cells, whereas the latter is principally the hydrophobic lipids containing species-specific carotenoids (Dickinson, 1973).

(iv) Pollen Wall Proteins

Tsinger and Petrovskaya-Baranova (1961) were the first to report the presence of proteins in the wall of pollen grains. It is largely through the work of J. Heslop-Harrison and his associates that the details of the pollen wall proteins and their significance have become apparent. Both layers of pollen wall, exine and intine, contains a considerable amount of protein (Knox et al., 1975). The entine proteins are present in the form of radially arranged tubules. They are generally concentrated near the germ pore. The exine proteins are present in the cavities of bacula (in tectate grains), or on the surface depression (in non-tectate grains). A part of these proteins are enzymes, mostly hydrolytic enzymes, such as esterase, acid phosphatase, amylase and ribonuclease, the proteins responsible for pollen allergy are also present in the pollen wall.

In many members of the Iridaceae, Cannaceae, Zingiberaceae and Lauraceae and aquatic plants, there is progressive reduction in the thickness of the exine or it may be completely absent. Irrespective of the nature of exine, pollen wall proteins are invariably present either in the exine or intine.

(v) Origin of Pollen Wall Proteins

Heslop-Harrison and his associates have worked out the details of the origin of wall proteins in several taxa (Fig. 4.10 A-D). As the deposition of the intine progresses, following the release of microspores from the tetrad, the plasmalemma of the pollens cytoplasm puts out radially oriented tubules into the developing intine. Eventually these tubules with their protein inclusions become cut off from the plasmalemma and are sealed off from the cell surface by the deposition of further layer of intine free from tubules. In aperturate pollens, intine proteins are largely concentrated in the region of the germ pore, whereas in non-aperturate pollens proteins are distributed throughout the intine.

Exine proteins, on the other hand, originate in cells of the adjoing tapetum, a sporophytic tissue. During meiosis in the microspore mother cells, proteins and lipids accumulate in the tapetal cells; the former in single membrane-bound vesicles and apparently derived from the rough endoplasmic reticulum, and the latter in the plastids. When the tapetal cells break down towards the end of pollen development, these proteins and lipids are released into the thecal cavity and eventually become deposited in the surface depression of the exine. In tectate grains the proteins fractions passes through the micropores of the tectum and

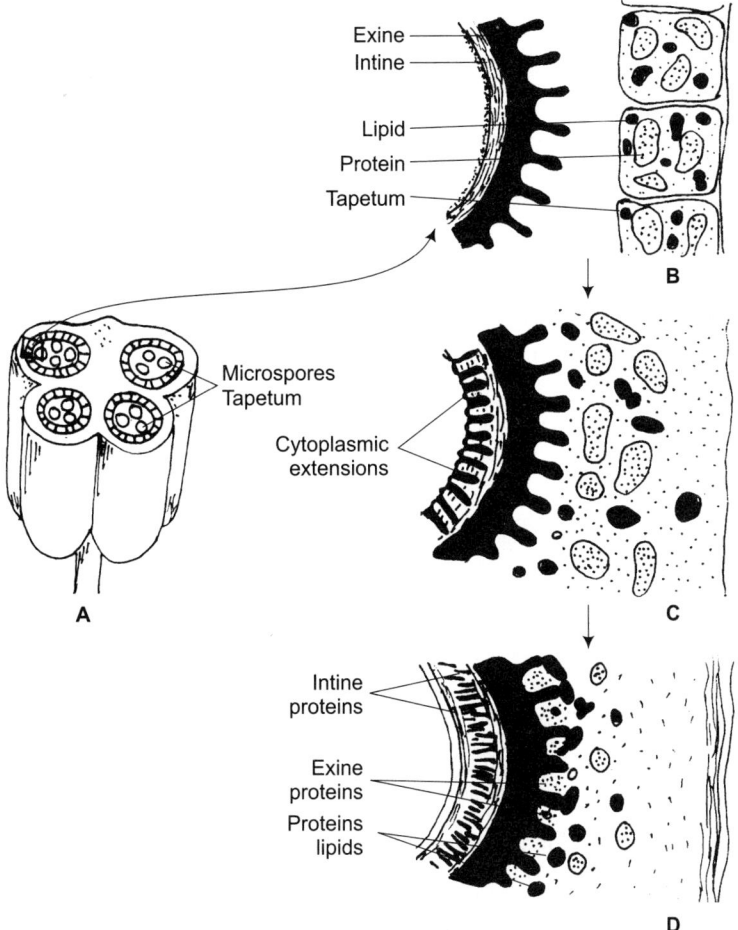

Fig. 4.10 Origin of intine and exine proteins (After Shivanna, 1977).

accumulate in the space between the baculae, while the lipids remain on the surface of the tectum.

Intine proteins are, therefore, the product of the pollen cytoplasm (the male gametophyte) and the exine proteins of the tapetum (sporophytic tissues). The Studies on the enzymes revealed that acid phosphatase activity was largely confined to pollen grains and was rather low in the tapetum. Esterase activity was insignificant in the pollen but in the tapetum it increased steadily until the breakdown of the tapetum.

UNUSUAL TYPE OF POLLEN DEVELOPMENT

Apart from the normal pollen development, there are a few examples showing abnormal pollen development.

(i) Development of Pollens in Cyperaceae

The members of Cyperaceae exhibit that out of four nuclei formed after meiosis, only one functions normal to develop into functional pollen grains. The remaining three are non-functional and is pushed to one side, where they gradually degenerate (Fig. 4.11 A-C).

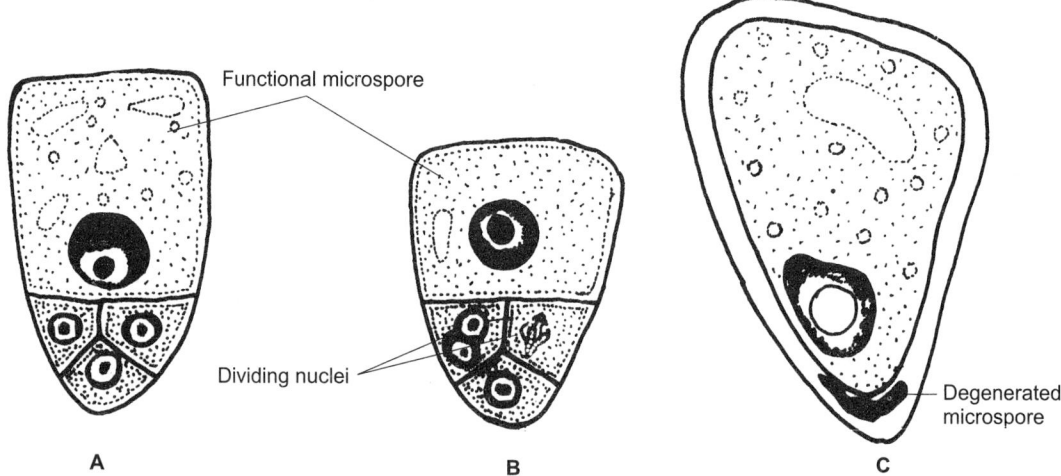

Fig. 4.11 Pollen development in *Cyperus*. **A**. Microspore tetrad showing one large functional microspore and three non-functional microspores. **B**. Two of the non-functional microspore nuclei have divided. **C**. The non-functional microspores have degenerated (After Khanna, 1965)

(ii) Embryo Sac-like Pollen Grains

In 1898, Nemec noted that in the pentaloid anther of *Hyacinthus orientalis*, the pollen grains sometimes form 8-nucleated structure showing a surprising resemblance to embryo sac (Fig. 4.12 A-D). He believed that they arose as a result of degeneration of the generative nucleus and the three divisions of the vegetative nucleus. This has been called as "**Nemec-Phenomenon**" after Nemec and has been confirmed by other scientists as well. The pollen embryo sac consists 3-celled egg apparatus and a central cell. Some pollen embryo sacs have also been recorded with larger or lesser number of nuclei.

Stow (1930, 1934) concluded that all pollen grains are potentially capable of assuming either male or female forms. Under normal condition "**Male potency**" is dominant over the "**Female potency**" but under abnormal conditions, where there is release of "**necrohormone**" from the dead pollen grains, the female potency gets expressed resulting in the formation of pollen-embryo. Naithani (1937), however, held different view and attributed the pollen-embryo sac formation to the effect of temperature.

(iii) **Pollen sterility:** Gamete sterility, failure of the formation of functional gametes, is quite common in flowering plants. Gamete sterility is categorized into two kinds: (i) non-selective and (ii) selective; The main difference is that the former operates in both male and female gametes, in the latter it operates in one of the sexes, generally the male. Male sterility results in the arrest of pollen development which may occur at any time from the initiation of anther

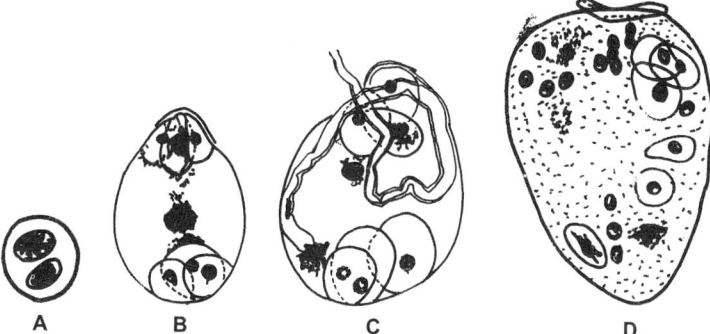

Fig. 4.12 Formation of pollen embryo sac in *Hyacinthus orientalis*. **A.** Normal pollen showing vegetative and generative cell. **B.** Pollen embryo sac. **C.** Pollen embryo sac affected by pollen tube from pollen grain of another variety. **D.** Fertilized pollen embryo sac; smaller are presumed to be product of division of triple fusion (After Stow, 1934).

to the maturation of pollens. Selective male sterility has important practical applications and extensive work has been carried out particularly on cytoplasmic sterility (Laser and Lersten, 1972).

Genic male sterility is controlled by a single gene, in most taxa and is manifested under homozygous recessive conditions. Cytoplasmic Male Sterility (CMS) is inherited through the female parent. Factors involved in male sterility are as follows:

(a) Many reports indicate that the male sterility is caused by the viral infection.

(b) Involvement of cell organelles, particularly mitochondria and plastids. This is largely because they are autonomous system, possess the genetic material DNA and are generally transmitted through the egg.

(c) Tapetal development is normal until about the time of meiosis. Later the tapetal cells in sterile anthers show many irregularities. They enlarge and more vacuolated; the cells undergo cytoplasmic disorganization. Often the tapetal cells become highly vacuolated and enlarge to fill most of the locule. The breakdown of the radial walls of these cells results in the formation of intra-tapetal syncytia. Also unlike in the normal anthers, the inner primary tangential wall persists and become spongy; the orbicular layer is deposited below this primary wall. In some anthers tapetal cells enlarge and their vacuoles contain amorphous mass. Thus, tapetum irregularities or abnormalities cause failure of nutrient supply to developing pollens or mis-timing of callase activity.

(d) Some investigations suggest significant variation in the amount and types of amino acids in sterile and fertile pollens, for example, analysis of amino acids in pollens of seven diploid (fertile) and seven triploid (sterile) of apple revealed more proline and less histidine in fertile pollens compared to sterile ones (Tupy, 1963).

(e) Environmental factors significantly affect male sterility in both male sterile and fertile lines. The degree of male sterility caused by genic or cytoplasmic factors varies, depending on the environmental factors; light and temperature play an important role. In cotton homozygous sterile plants produce 100% sterile anthers at

32°C. Around 38°C the heterozygotes also become fully sterile. When the maximum temperature exceeds 38°C, even fertile plants produce more and more sterile anthers (Meyer, 1969).

(f) Male sterility can be induced by the application of different kinds of chemicals: (i) colchicines, (ii) ethyl methane sulphonate (EMS), (iii) gamma rays irradiation, (iv) gametocidal chemicals, such as Mendok and Dalapon, Ethephon, Gibberellic acids, Maleic hydrazide etc.

(g) Failure of anther dehiscence-normal pollens are formed but not released.

(h) Male sterility can be induced through genetic engineering. Introduction of **BARNASE** gene isolated from *Bacillus amyloliquefaciens* encodes an RNAase which brings about male sterility. However, introduction of another gene called **BARSTAR**, can restore the male fertility by inactivating the proteins produced by **BARNASE** gene. Therefore, transgenic plants expressing both **Barstar** and **Barnase** genes are fully male fertile.

5. Megasporangium

The megasporium or the ovules located within the ovary, arise from the placentae, consists of nucellus and integument. It is attached to the placentae on the inner wall of the ovary by means of stalk called funiculus (Fig. 5.1A). A mature ovule shows an opening called micropyle and the basal region of the ovule where the funiculus is attached is called hilum. Within the nucellus a female gametophyte or the embryo sac is present (Fig. 5.1B).

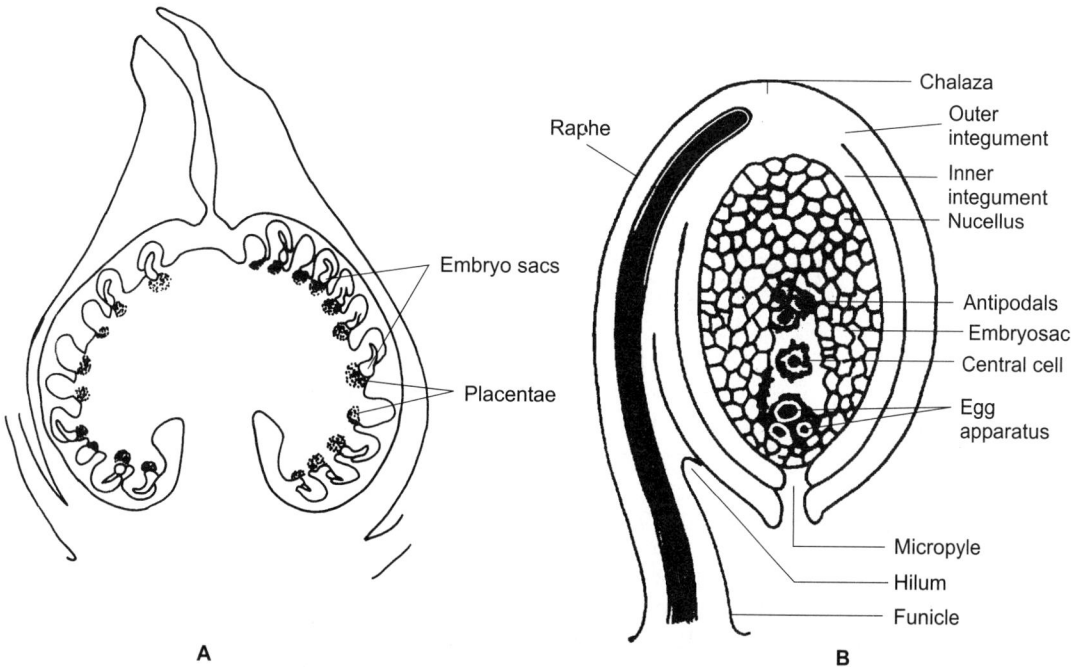

Fig. 5.1 **A**. L.S. of ovary showing the origin of ovule. **B**. Ovule showing different parts.

TYPES OF OVULE

Mature ovules are classified into six types based on the position of micropyle with respect to funiculus, which are as follow:

(i) Orhotropous (Atropous) Type

The micropyle and the funiculus lie in the same line as in the Polygonaceae, Urticaceae, Cystaceae and Piperaceae (Fig. 5.2A).

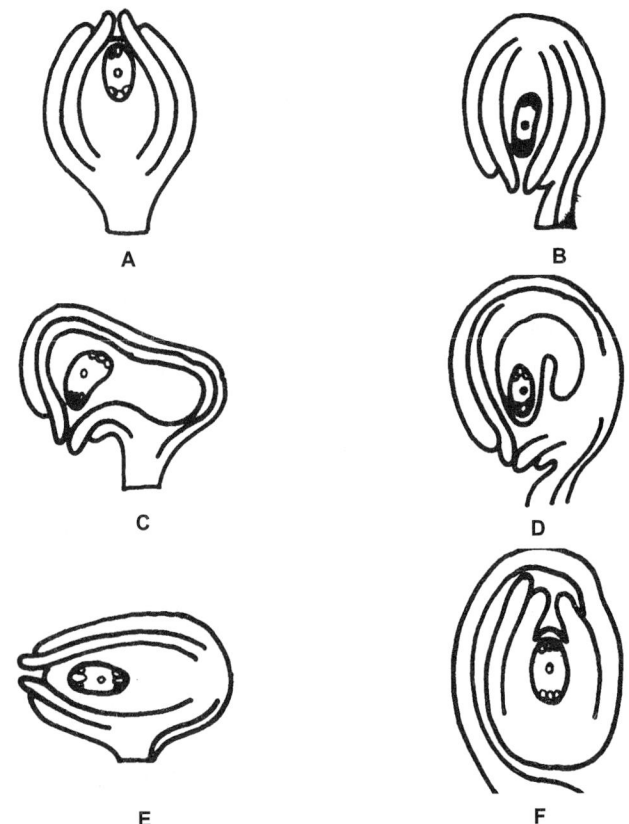

Fig. 5.2 Different types of ovules. **A.** Orthotropous. **B.** Anatropous. **C.** Campylotropous. **D.** Hemianatropous. **E.** Amphitropous. **F.** Circinotropous.

(ii) Anatropous Type

In this type the body of the ovule becomes completely inverted so that the micropyle and the hilum comes to lie very close to each other (Fig. 5.2B). This type is universal almost in all the members of Sympetalae and is also found in several other families belonging to Dicotyledons and Monocotyledons. According to Davis (1962), 82% of the angiospermic families show this type of ovule.

(iii) Campylotropous Type

In this type, the ovule is curved as in some members of Capparidaceae, Chenopodiaceae, Resedaceae and Leguminosae (Fig. 5.2C).

(iv) Amphitropous Type

When the curvature is more pronounced and also affects the embryo sac, so that the latter becomes bend like a horse shoe as in Alimaceae, Butomaceae, Centospermales (Fig. 5.1D).

(v) Hemianatropous (Hmitropous) Type

In this type the nucellus and integument is at right angle to the funiculus e.g., *Ranunculus, Nothoscordum and Tulbaghia* (Fig. 5.2E).

(vi) Circinotropous Type

Initially the ovule of orthotropous type but with continuous unilateral growth the ovule becomes anatropous and subsequently the micropyle again points upward in fully mature ovule as in Plumbaginaceae and Cactaceae (Fig. 5.2F).

PARTS OF OVULE

(i) Integuments

Normally ovule has either one or two integuments. The number is constant in most families and only in rare cases unitegmic and bitegmic ovules occur in the same family. In Sympetalae unitegmic condition is universal except in Plumbaginales and Primulales. In Archichlamymydeae and Monocotyledons most genera have bitegmic conditions. In some plants there is also a third integument, called **aril** e.g., *Almus, Asphodelus* and *Trianthema* (Fig. 5.3).

A special structure called **caruncle** is found in several members of Euphorbiaceae, which arise by the proliferation of the integumentary cells at micropylar end (Fig. 5.4A-C).

There are reports of the occurrence of stomata and chlorophyll in the outer integument. However, a complete absence of integument is known only in some members of Loranthaceae and Balanophoraceae.

(ii) Micropyle

The micropyle, which is constituted either by the inner integument as in Centrospermales and Plumbaginales or by both inner and outer inteuments as in Pontederiaceae. Less frequently as in Podostemaceae, Rhamnaceae and and Euphorbiaceae it may be formed by outer integument alone. The passage formed by the outer integument is called **exostome** while the **endostome** is the passage formed by inner integument.

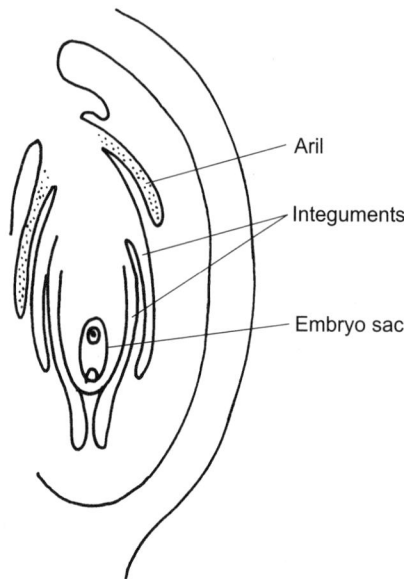

Fig. 5.3 Ovule showing third integument (aril).

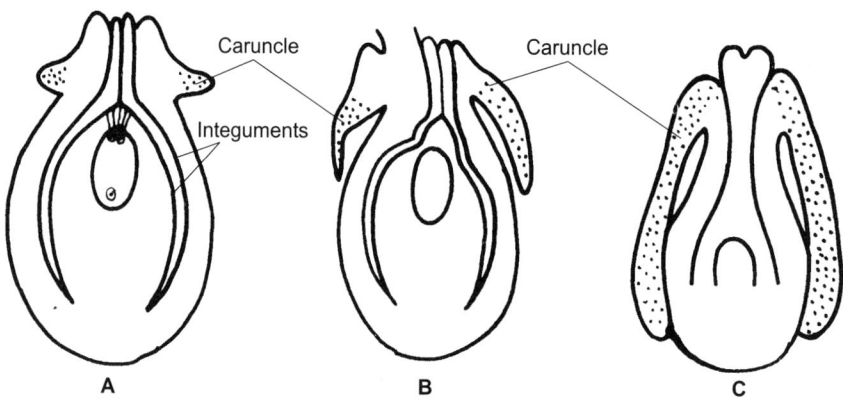

Fig. 5.4 Schematic diagram of the development of caruncle.

(iii) Nucellus

Nucellus is the major portion of the ovule into which embryo sac develops. During the early development of ovule, at the stage of appearance of integumentary primordia, either one or many hypodermal cells situated at the apex of the ovule primordium are differentiated from the neighbouring tissues by their increased size, bigger nuclei and denser cytoplasm. Such differentiated cells are called **primary archesporial cells** from which nucellus develop (Fig. 5.5). Depending upon the mode of development of nucellus, the ovule is of two types: (a) **Crassinucellate** and (b) **Tenuinucellate** (Fig. 5.5).

Megasporangium **5.5**

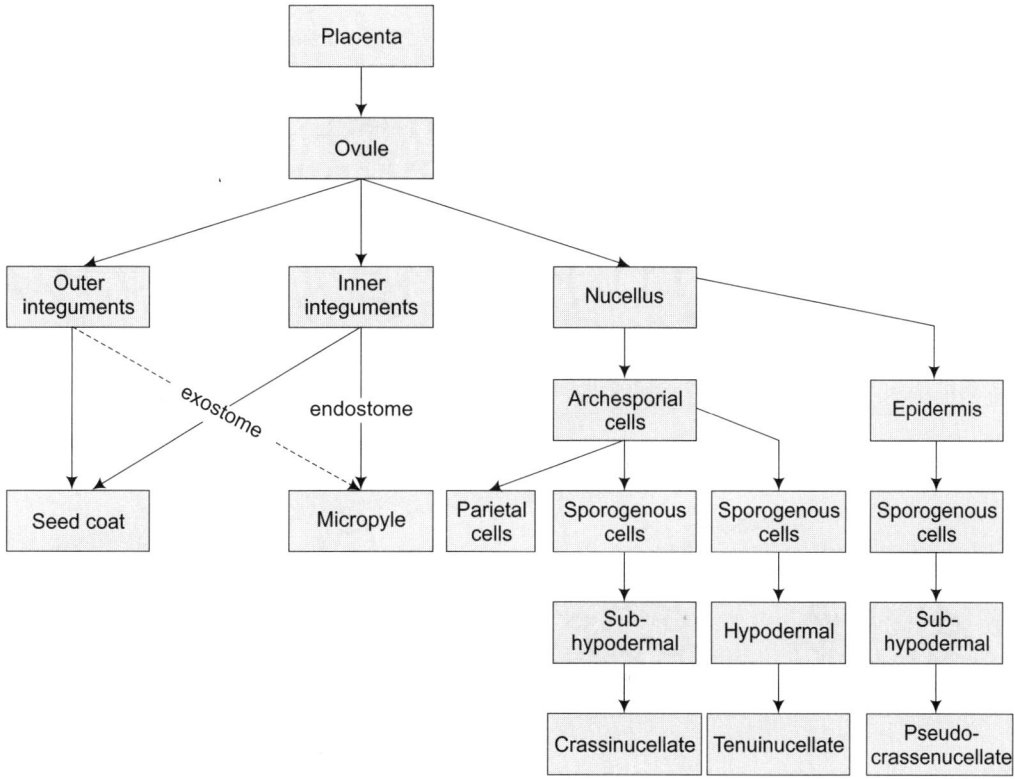

Fig. 5.5 Schematic diagram of the developmental stages of ovule

(a) **Crassinucellate:** In this type, the primary archesporial cell divides periclinally into an outer smaller **primary parietal cell** and an inner bigger **primary sporogenous cell.** The former divides to form one or several layers of parietal tissues, while latter undergoes a few divisions or may directly functions as **megaspore mother cell.** Thus, megaspore mother cell is separated from the nucellar epidermis by one or several layers of cells (Fig. 5.6A-D). However, in *Anemone* the nucellar epidermis also contributes in peripheral nucellus.

(b) **Tenuinucellate:** In the second type, parietal cells are absent and the megaspore mother cell lies directly below the nucellar epidermis (Fig. 5.7A-D). Teninucellate ovules are supposed to be more advanced type.

Davis (1966) proposed a special term **pseudo-crassinucellate** should be given for all those ovules where division in the nucellar epidermis is responsible for the sub-hypodermal nature of the sporogenous tissues.

According to Davis (1966), of the 314 families for which information is available 179 families show Crassinucellate ovule, 105 families bear Tenuinucellate ovules and 11 families possess Pseudo-crassinucellate ovules.

Several members of Salicaceae, Nyctaginaceae, Euphorbiaceae, Polygonaceae and Cucurbitaceae are characterized by having a beak-shaped nucellus, which reaches out into the micropyle (Fig. 5.8).

5.6 Plant Embryology: Classical and Experimental

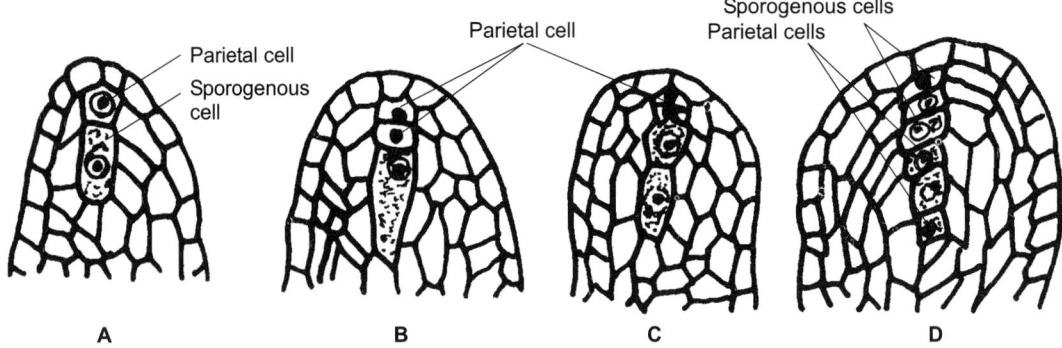

Fig. 5.6 Megasporogenesis in crassinucellate ovule of *Myriophyllum intermedium* (After Bawa, 1969).

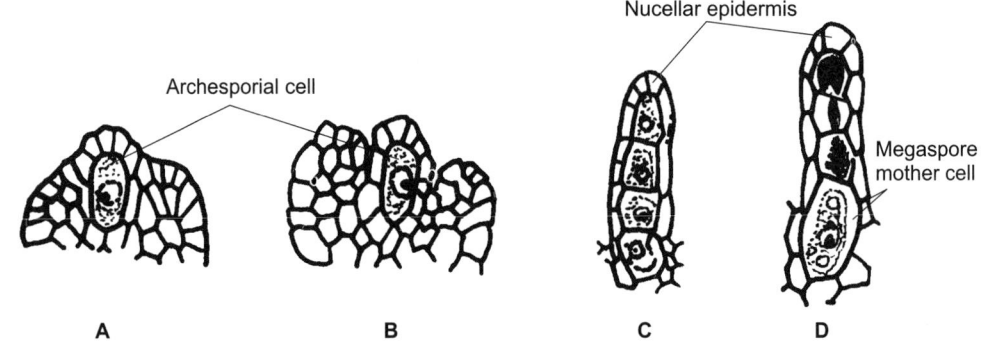

Fig. 5.7 Megasporogenesis in tenuinucellate ovule of *Elytraria acaulis* (After Johri and Singh, 1959).

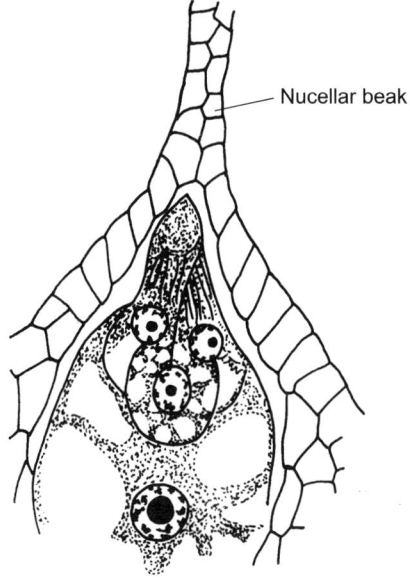

Fig. 5.8 Ovule showing formation of nucellar beak in *Polygonum persicaria*.

As the embryo sac matures, the nucellar cells gradually become used up. In some plants it persists in mature seeds as nutritive tissue called **perisperm**, e.g., Piperaceae, Amaranthaceae, Zingiberaceae, and Cannaceae.

(iv) Endothelium or Integumentary Tapetum

In those plants, where the nucellus is disorganized, especially tenuinucellate ovule, the embryo sac comes in direct contact with the integument. The cells of this layer now frequently become radially elongated and sometimes become binucleate and form endothelium or integumentary tapetum (Fig. 5.9).

Endothelium is a nutritive layer; however, there are reports, which states that it contains diastase and other enzymes. At later stage when the embryo is reaching to maturity the inner surface of the endothelium takes up the nutritive function.

(v) Hypostase

Van Tieghem (1901) first reported this type of nucellar tissues lying directly below the embryo sac, called hypostase (Fig. 5.10). The cells are poor in the cytoplasmic content but have partially cutinized or suberized wall composed of highly refractive materials. Hypostase occurs in many families, namely Crosomataceae, Umbelliferae, and Elaeagnaceae.

Various functions have been attributed to hypostase; like a sort of barrier preventing the growth of embryo sac into the base of the ovule, stabilizes the water balance; food transport; glandular tissues responsible for the secretion of hormones or enzymes for the growth of embryo sac.

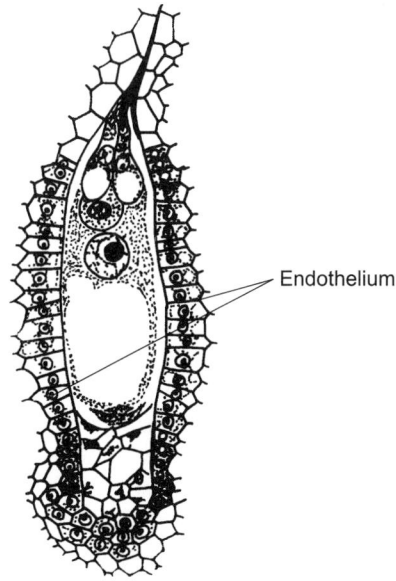

Fig. 5.9 *Labelia trigona* ovule showing endothelium (After Kausik, 1935).

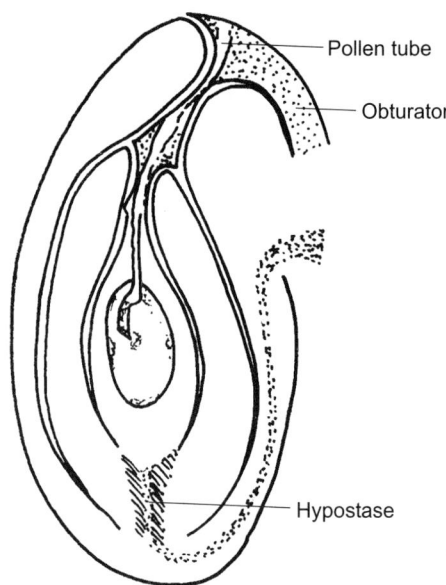

Fig. 5.10 *Acalypha indica* ovule showing obturator and hypostase. The obturator showing pollen tube. (After Johri and Kapil, 1953).

(vi) Epistase

Van Tieghem also reported the occasional presence of similar well-marked tissues in the micropylar part of the ovule, called epistase (Fig. 5.11). Usually it originates from the apical cells of the nucellar epidermis, which shows well-marked radial elongation and becomes some what thickened or suberized. Sometimes, epistase forms nucellar cap which persists and could be seen even in the advanced stage of embryo development. Presence of epistase has been reported in several taxa, e.g., *Costalia, Costus, Nicolaia*.

(vii) Obturator

In several families the ovule is characterized by the presence of a placental outgrowth, called obturator which grows into the space between the nucellus and the integument or also between the ovule or the ovary wall so as to help in bringing about a connection between the stylar tissue and the nucellus (Fig. 5.10).

The obturator is devoid of vascular tissues. It is presumed to guide the growth of pollen tube to the micropyle. This tissue is either composed of thin walled compact tissues or made up of loose hairy outgrowths. The cells of the obturator show dense cytoplasm containing large number of endoplasmic reticulum, dictyosome and vesicles. The cells of the obturator produce surface exudates and provide mechanical and chemical guidance to the growing pollen tube. After fertilization its cells shrinks and disappear.

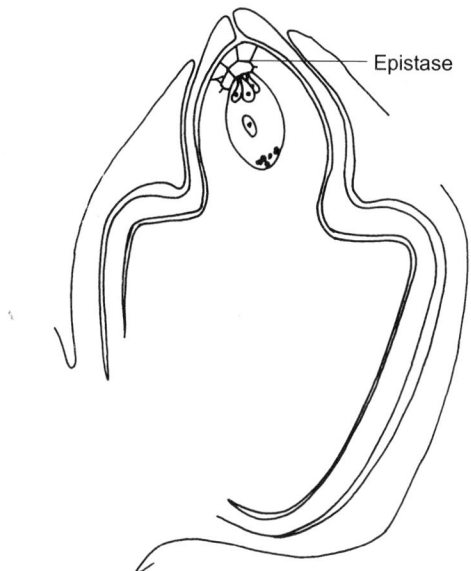

Fig. 5.11 Ovule showing epistase (After Chikkanaiah, 1962).

(viii) Vascular Supply

The vascular bundles enter into the ovule through the chalazal end and may branch to supply integument, nucellus, embryo sac or even sporogenous tissues.

(ix) Archesporium

The archesporial tissue is of hypodermal origin. In general one cell of the nucellus situated directly below the nucellus epidermis becomes more conspicuous than the others owing to its larger size, denser cytoplasm and more conspicuous nucleus. This is the **primary archesporial cell**.

The primary archesporial cell may divide to form primary parietal cell and primary sporogenous cell or it may functions directly as megaspore mother cell in crassinucellate and tenuinucellate ovules, respectively. The primary parietal cell may remain undivided or it may undergo periclinal and anticlinal divisions to form variable number of wall layers. The primary sporogenous cell functions as megaspore mother cell without undergoing any further divisions.

The outline presented above is subject to many variations. In *Hydrilla*, there are sometimes two or three archesporial cells in single row, while *Scurrula* and *Dendrophthoe* show massive archesporium.

MEGASPOROGENESIS

The megaspore mother cell undergoes usual meiotic division to form tetrad. The first division is always transverse, giving rise to dyad cells. Typically the second division is also transverse, resulting in linear tetrad of 4-megaspore. Anticlinal division either in the micropylar dyad cell or the chalazal dyad cell results in T-shaped or ⊥-shaped tetrad.

However, isobilateral or tetrahedral arrangement of megaspore is also reported (Fig. 5.12A-D). The genus *Musa*, is of special interest, because all 4-type of arrangements occur in the same species.

Normally, it is the chalazal megaspore of the tetrad, which functions and gives rise to embryo sac, while remaining three degenerate and disappear. However, in *Langsdorffia* and *Balanophora* the micropylar megaspore is functional giving rise to embryo sac, while other three soon degenerate. A very interesting case is observed in *Culcitium reflexum*, where both micropylar and chalazal megaspores are functional. In *Rosa*, second megaspore sometimes becomes functional, while in *Aristotelia* belonging to Elaeocarpaceae; the third megaspore from the micropylar end is functional. Rarely, as in *Gloriosa, Ostreya, Poa* and *Casuarina*, all the four or any of the megaspore may become functional.

SOME ABNORMAL AND REDUCED OVULES

The young ovule of *Crinum* consists of megaspore mother cell covered by a few layers of cells derived from the nucellar epidermis. In post-fertilization stages the endosperm absorbs not only the whole of the nucellar tissues and the integument but also the pericarp with the result that the seed consists of only the endosperm and the embryo. The outer layers of the endosperm behave like a phellogen and produce a few layers of cork tissues. Chloroplasts develop in the peripheral cells of the endosperm lying below the phellogen.

In some members of the Olacaceae the ovules show no clear differentiation into nucellus and integuments. In *Anacolosa frutescens* the nucellus is represented by a single epidermal cell situated above the archesporial cell while the cells lying laterally divide periclinally and symbolize the integument.

The Loranthaceae have gone a step further. Here no ovules are present as such and the placental-ovular complex (placental column bearing the ovules) is known as **marmelon or**

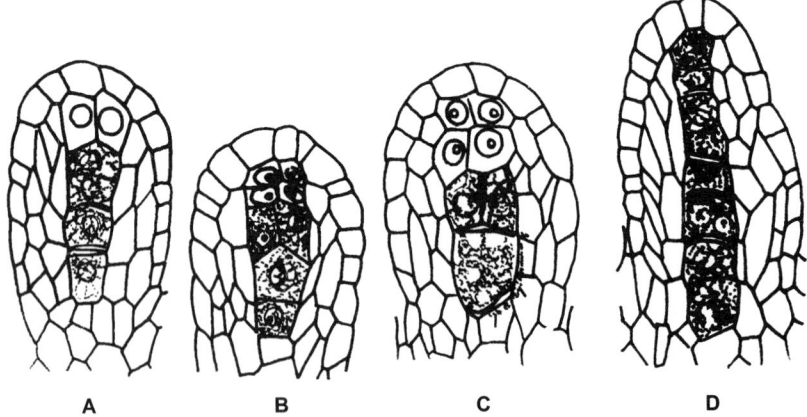

Fig. 5.12 Megaspore tetrad in *Urginea indica*. **A.** Linear tetrad. **B.** T-shaped tetrad. **C.** Tetrad showing decussate arrangement. **D.** Two tetrad lying in the same row (After Maheshwari, 1950).

placenta. The integuments have disappeared and there is gradual reduction in the lobing and differentiation of the marmelon. In *Helixanthera* and *Moquiniella* the marmelon is absent and the archesporial cells differentiate at the base of the ovarian cavity.

Mention may also be made of the Ranunculaceae which are usually considered as members possessing uniovulate carpels. However, some of them like *Adonis annua, Clematis gauriana* have 2-4 accessory sterile ovules in addition to the single fertile, seed producing ovule.

A peculiar condition occur in *Rosularia, Sedum, Laurus, Potentilla* and some members of Rubiaceae, the megaspore give out lateral tubes, which grows upwards as haustorial processes and draw nutrition from the adjacent tissues (Fig. 5.13 A-B).

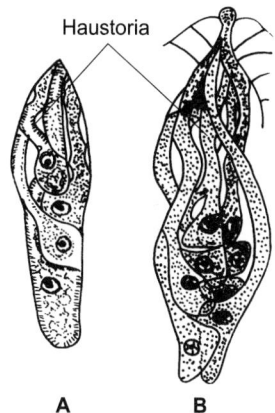

Fig. 5.13 Megaspore haustoria.
A. *Sedum sempervivoides.*
B. *Rosularia paluda*
(After Mauritzon, 1933).

6 Female Gametophyte

The female gametophyte or megasporophyte represents the embryo sac, which is usually 7-celled structure, consisting three distinct groups of cells; **Egg apparatus**, **Central cell** and **Antipodal apparatus**. Egg apparatus is situated towards micropylar end and shows a central egg flanked on the either side by **synergids**. The central cell situated in the centre, consists of two-fused nucleus or secondary nucleus, coming from the two polar ends. The antipodals consists of three cells situated towards chalazal end. All the cellular structures within the embryo sac are haploid except secondary nucleus, which shows different stages of polyploidy (usually diploid).

DEVELOPMENT OF EMBRYO SAC OR MEGASPOROGENESIS

The primary sporogenous tissues differentiated in the nucellus undergo a few divisions or may directly functions as megaspore mother cell and enters into meiotic division to form four megaspores. Out of the four, only one situated either at the micropylar or the chalazal end survives and the rest degenerates. The single megaspore develops into embryo sac or female gametophyte (Fig. 6.1 A-I).

Depending upon the number of megaspore nuclei taking part in the development of female gametophyte of angiosperms, it may be classified into three types which may further be divided into different sub-types (Fig. 6.2):

(1) Monosporic embryo sac
(2) Bisporic embryo sac
(3) Tetrasporic embryo sac

(1) Monosporic Embryo Sac

In this type only one nucleus takes part in the development of the female gametophyte. In the monosporic type there are two sub-types, namely (i) *Polygonum* type and (ii) *Oenothera* type.

6.2 Plant Embryology: Classical and Experimental

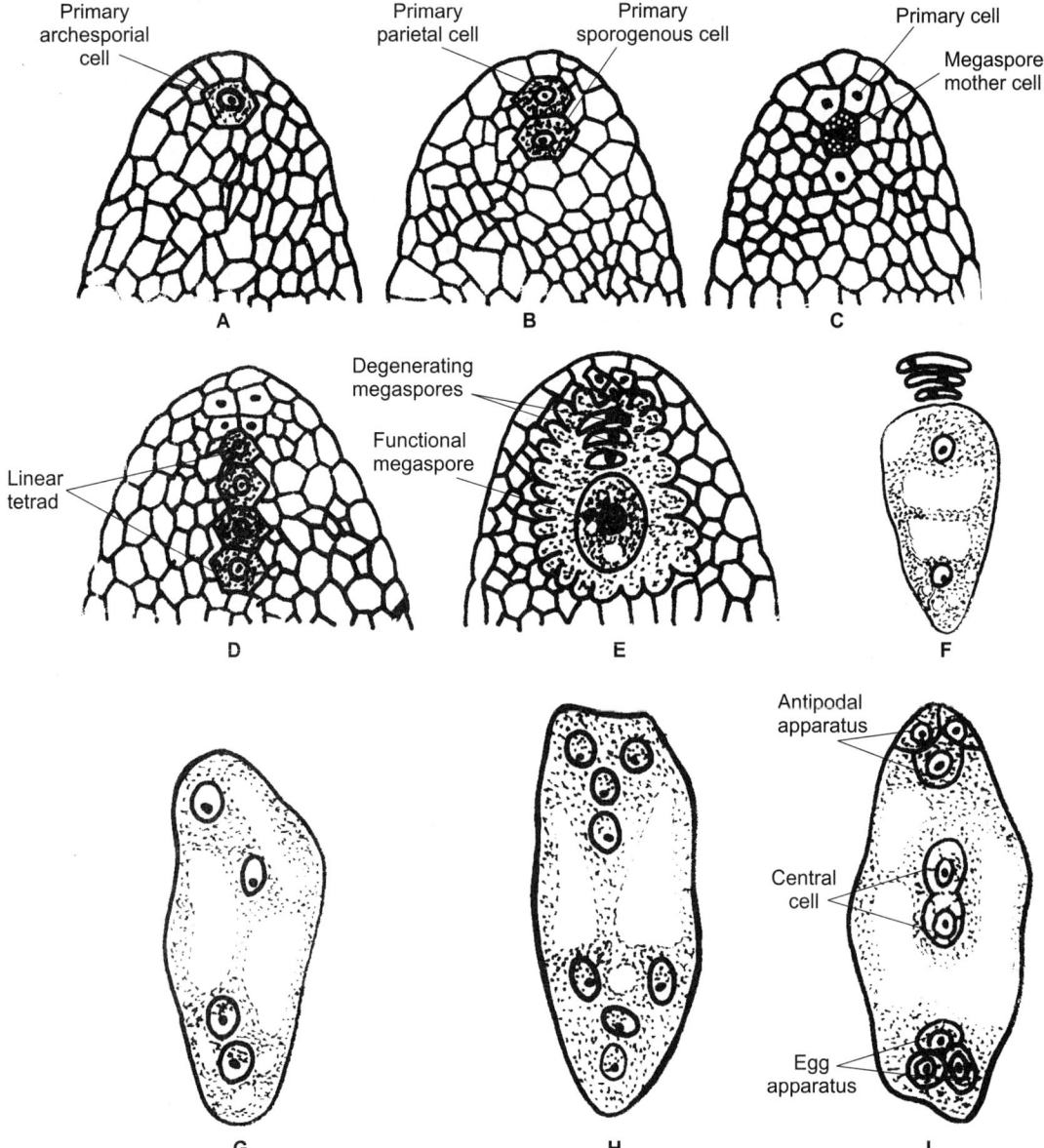

Fig. 6.1 Diagram showing sequential steps of development of embryo sacs (megasporogenesis).

(i) *Polygonum Type*: The development of the embryo sac begins with the elongation of functional megaspore (haploid) accompanied by increases vacuolation, which is located at the chalazal end of the megasporagium or ovule. The centrally situated nucleus divides mitotically to form two nuclei, which are gradually pushed towards two poles due to vacuole formation. Both the nuclei at polar ends divide twice to form four nuclei at each pole. Thus, 8- nucleate stage of embryo sac is now followed by cellular organization. Out of the 4 nuclei at the micropylar end, three organized to form egg apparatus, while the remaining one forms

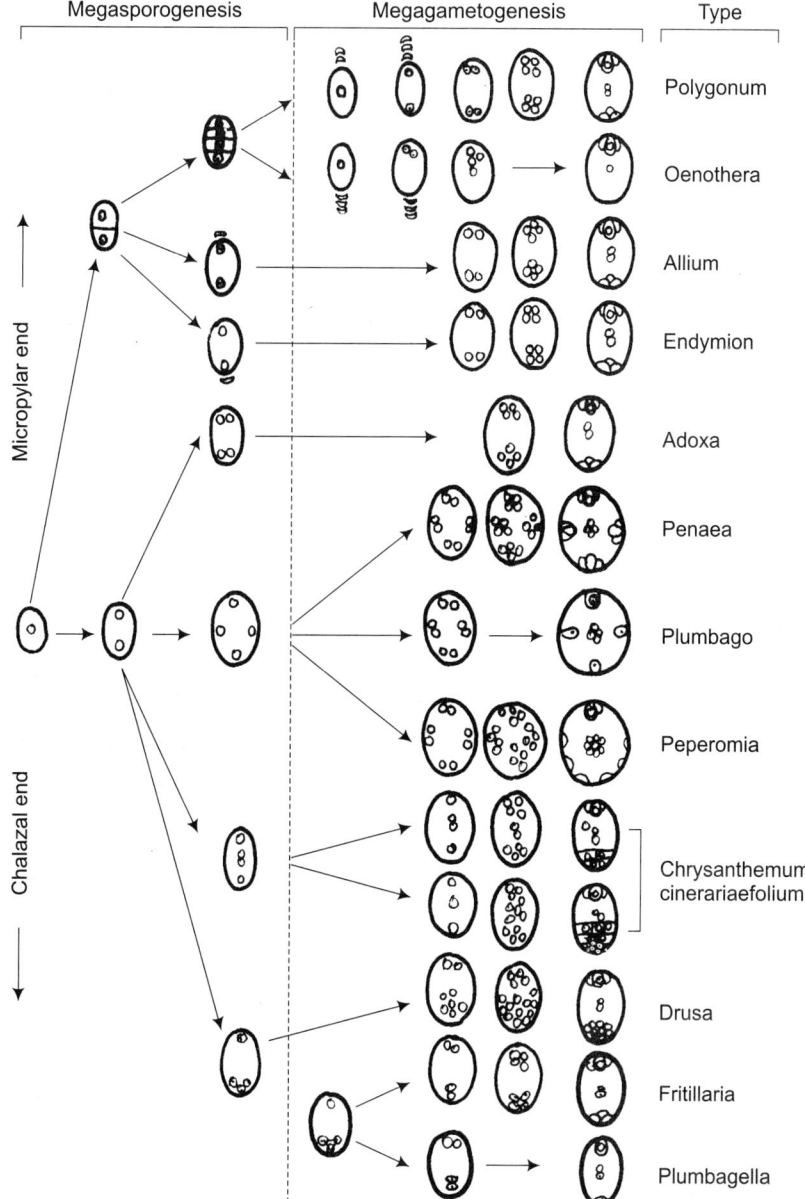

Fig. 6.2 Diagrammatic representation showing the origin and development of different types of embryo sacs.

upper polar nucleus. Similarly, out of four nuclei at the chalazal end, three nuclei organized to form three antipodal cells and the remaining one forms the lower polar nucleus. Thus two polar nuclei, each from either pole fuses in the centre to form secondary nucleus or binucleate central cell.

In some cases there is reduction in the number of nuclei due to suppression in the nuclear division of chalazal nuclei. In a few examples such as *Chaeoichris, Elatine, Thesium,* and *Geodorum,* 6-nucleate embryo sacs are found, whereas *Orchis morio* shows 5-nucleate embryo sac.

However, there are cases where embryo sac with more than 8 nuclei are also found either due to: (i) fusion of two embryo sacs e.g., *Elatine hydropiper, Muda* var. IR–53, (ii) migration of the nuclei of the nucellar cells into the embryo sac e.g., *Hedychiyum garsnerianum, Pandanus* sp., or (iii) occurrence of the secondary divisions in some of the first formed 8-nuclei e.g., *Crepis capillaris, Umbilicus intermedius* and *Nicotiana.*

(ii) Oenothera Type: This type shows 4- nucleate embryo sac, which has for the first time been reported by Hofmeister (1847, 1849) in a few members of Onagraceae. Later, Greet (1908) has given a detailed account of the developmental stages in *Oenothera lamarkiana.*

The embryo sac is usually formed by the micropylar megaspore of the tetrad, which only goes two nuclear divisions instead of usual three occurring in the *Polygonum* type of embryo sac. Since the third division is omitted and all the nuclei are situated in the micropylar part of the developing embryo sac, there is neither a lower polar nucleus nor antipodal cells. *Schisandra chinensis*, however, is the only example outside the family showing 4- nucleate embryo sac (Yoshida, 1962; Swamy, 1962), where the functional; megaspore is the chalazal in origin.

Rarely, more than 4 nuclei may be seen in an embryo sac due to incorporation of adjacent megaspore and its contents or due to further divisions of the nuclei of embryo sac e.g., *Anogra palida* (Johansen, 1931). However, sacs with fewer than 4 nuclei have also been found in *Harmannia tetraptera* (Johansen, 1929), *Jussaea repens* (Khan, 1929).

(2) Bisporic Embryo Sac

The bisporic embryo sac arises from one of the two dyad cells formed after meiosis I. Since no wall is laid down after meiosis II and both the megaspore nuclei formed in the functional dyad cell take part in the development of the embryo sac. Bisporic embryo sacs are of two types: (i) *Allium* type and (ii) *Endymion* type.

(i) Allium Type: This type of bisporic embryo sac was first described in *Allium fistulosum* (Strasburger, 1879) and has since then been reported in several species of this genus. It is showing 8- nucleate embryo sac.

The two nuclei of the lower dyad cell (chalazal end) divides to form 4 and then 8 nuclei, which undergo cellular organization to form usual embryo sac with 3-celled egg apparatus, central cell with two polar nuclei and 3-celled antipodals.

Allium type has been reported in several plants belonging to diverse families *viz.,* Podostemonaceae, Butomacae, Alismaceae, Amaryllidaceae, Orchidaceae etc.

However, reduction in the number of nuclei is found as in *Machaerocarpus californicus* a member of family Alismaceae, where 6-nucleate embryo sac is found in the embryo sac because the antipodal nuclei fail to divide beyond two nuclei.

(ii) Endymion Type: This type also consists of 8-nucleate embryo sac like *Allium* type but it derived from the micropylar dyad cell as in *Scilla* species.

(3) Tetrasporic Embryo Sac

In this type of embryo sac, the meiotic divisions I and II of megaspore mother cell is not followed by any wall formation, so that the megaspore at the end of the meiosis show 4 nuclei (**coenomegaspore**). Here all the 4 nuclei take part in the development of embryo sac. A tetrasporic embryo sac is more interesting because all the 4 contributing nuclei are genetically different and further arrangement and mitotic divisions give rise to seven types of embryo sacs: (i) *Peperomia* type, (ii) *Penaea* type, (iii) *Drusa* type, (iv) *Fritillaria* type, (v) *Plumbagella* type, (vi) *Plumbago* type and (vii) *Adoxa* type. The basic arrangement of the 4 nuclei in coenomegaspore before the beginning of the mitotic divisions is mainly of the following 4 types:

(A) 1 + 1 + 1 + 1 arrangement;
   (i) Peperomia type
   (ii) Penaea type
   (iii) Plumbago type
(B) 1 + 3 arrangement;
   (i) Drusa type
   (ii) Fritillaria type
   (iii) Plumbagella type
(C) 2 + 2 arrangement;
   (i) Adoxa type

(i) *Peperomia* Type

This type has for the first been time reported by Campbell (1899) and Johnson (1900) in *Peperomia pellucida*. Each of the 4 megaspore nuclei, which are arranged tetrahedrally (1 + 1 + 1 + 1) divide twice, resulting in a total of 16 nuclei which become more or less uniformly distributed in the rather thick layer of cytoplasm lying at the periphery of the of the embryo sac. According to Johnson, 2-nucleus at the micropylar end now organized to form the egg and a synergid, 8 nuclei fuse to form secondary nucleus in the centre and the remaining 6 are cut off to form antipodals at the chalazal end.

Certain variations have been marked in this type. The variations are mainly either towards increase or decrease in the number of nuclei, which fuse to form secondary nucleus and also in the number of nuclei which are left over to form antipodals. In *Peperomia hispidula* for example 14 nuclei fuse to form secondary nucleus, while the embryo sac of *Gunnera* species show 3 nuclei at the micropylar end which form egg apparatus whereas 7 nuclei fuse in the centre to form secondary nucleus.

(ii) *Penaea* Type

Stephen (1909) for the first time described this type of embryo sac in three genera of the Penaeaceae, viz., *Penaea*, *Brachysiphon* and *Sarcocolla*. Here the 4 tetrahedrally arranged nuclei undergo division to form 16 nuclei, with four distinct quarters, one at each end of the embryo sac and two at the sides. Now three nuclei of each quarter is cut off as cells, while the fourth remaining and moves to centre to form secondary nucleus. The egg cell of the micropylar triad alone is functional.

Embryo sac of this type have since then been described in several members of the families Malpighiaceae, Euphorbiaceae.

Special mention may be made of *Acalypha indica* (Maheshwari and Johri, 1941), which does not entirely fit into the described Penaea type as mature embryo sac shows four groups of two each at the periphery and 8 free nuclei in the centre. Apart from this type many more variations have been found in the embryo sacs of the ovules.

A few cases are reported in which fewer than 16 nuclei have been found and others with more than this number.

(iii) *Plumbago* Type

This type has been reported for the first time by Haupt (1934) in *Plumbago capensis*. The 4-nucleate stage shows crosswise arrangement of 4 nuclei, which undergo further division resulting in 8 free nuclei arranged in four pairs. One nucleus of the micropylar end is now cut off to form egg cell, while 4 nuclei one from each group of the remaining 7 nuclei undergo slight increase in size and gradually approach one another to form tetraploid secondary nucleus. The remaining three nuclei degenerate at their original place but occasionally 1, 2 or all 3 of them cut off at the periphery to form cells, which may persist and assume an egg-like appearance. This embryo sac is most conspicuous by the absence of synergid.

(iv) *Drusa* Type

This type of embryo sac was first recorded by Hakkansson (1923) in *Drusa oppositifolia*, a member of the family Umbelliferae. After the meiotic division four nuclei are arranged in 1 + 3 fashion, 3 nuclei at the chalazal end while one nucleus remains at the micropylar end. The 1 + 3 arrangement is followed by 2 + 6 and then 4 + 12 stage. Out of the 4 nuclei at the micropylar end three organize to form egg apparatus and an upper polar nucleus, while 12 chalazal nuclei form 11 antipodal cells and one lower polar nucleus, which fuse with the upper polar nucleus to form a central secondary nucleus.

Variation, especially in the number of nuclei towards the reduction has been found in *Ulmus americanum*. Shattuck (1905) reported 8-nucleate embryo sac of the *Adoxa* type. Several embryo sacs have been reported, where 10, 12 or 14 nuclei have been seen in the embryo sac.

(v) *Fritillaria* Type

Eight-nucleate *Fritillaria* type embryo sac has been for the first time described by Treub and Mellink (1880) in *Lilium bulbifera*, which was later supported by many investigators.

However, detailed developmental stages were studied by Bombacioni (1928) in *Fritillaria* and *Lilium*. He observed that the formation of 4-nucleate megaspore does not divide directly into 8-nucleate stage but by secondary 4-nucleate stage in which the 2 chalazal nuclei are much larger than the micropylar due to fusion. At first there is 1 + 3 arrangement of the megaspore nuclei similar to *Drusa* type. But the three nuclei at the chalazal fuse to form triploid nucleus, which represents second 2-nucleate stage and is soon followed by division, which results a second 4-nucleate stage with 2 haploid micropylar nuclei and two triploid chalazal nuclei. One more division occurs, resulting in 8-nucleate stages. The sequence is, therefore, as follow: megaspore mother cell, primary 2-nucleate stage, primary 4-nucleate stage, secondary 2-nucleate stage, secondary 4-nucleate stage and last the 8-nucleate stage. At this stage, organization of the nuclei results in micropylar egg apparatus with three haploid cells, three triploid antipodals and a tetraploid secondary nucleus formed by the fusion of two polar nuclei (haploid and triploid)

However, there are a few examples with 7-nucleate embryo sac in *Gagea* (Ramanov, 1936); 6-nucleate embryo sac in *Statica* (Fagerlind, 1939) or 5-nucleate embryo sac in *Tulipa maximovicii* (Ramanov, 1939) due to reduction in the nuclear division at the chalazal end.

(vi) *Plumbagella* Type

This type has so far been reported only in *Plumbagella micrantha* (Fagerlind, 1938; Boyes, 1939). The 4 megaspores take up 1 + 3 arrangement like *Fritillaria* type. The three chalazal nuclei fuse to form triploid nucleus. This result in a secondary 2-nucleate stage followed by a secondary 4-nucleate one, in which two micropylar nuclei are haploid and the chalazal are triploid. There is no further division. The nuclei nearest to micropylar end organize into an egg; the triploid nearest the chalazal end forms the antipodal cell and the remaining two nuclei, one haploid and one triploid fuse to form a tetraploid secondary nucleus.

(vii) *Adoxa* Type

This type has been for the first time described by Jonson (1879-1880) in *Adoxa moschaetillina*. In *Adoxa* type the 4 nuclei are arranged in 2 + 2 fashion. The 4 megaspores nuclei undergo one more division to form 8-nucleate embryo sac. Reorganisation results 3-celled micropylar egg apparatus, 3-celled antipodals and two polar nuclei in the centre fuse to form secondary nucleus, thus showing resemblance with mature *Polygonum* type of embryo sac.

ABERRANT AND UNCLASSIFIED TYPE

In addition to the above fairly distinct and well-established types of tetrasporic embryo sac development, there are a few variant types and therefore are unclassified type. A few important types are as follow:

(i) ***Limnanthes douglasii:*** In this plant, development of the embryo sac has been investigated by three different workers, namely Stenar (1925), Eysel (1937) and Fagerlind (1939). According to Fagerlind (1939), the embryo sac shows 6-nucleate embryo sac. The micropylar nuclei form 3-celled egg apparatus and an upper polar nucleus; the remaining two nuclei form lower polar nucleus and a antipodal.

(ii) ***Balsamita vulgaris:*** In this plant, Fagerlind (1939) described 8-nucleate embryo sac developing from the micropylar nucleus of 4-nucleate stage of the archesporial cells with 1 + 3 arrangement. Of the four only micropylar nucleus functions while the other three degenerate. Interestingly, there is formation of tubular outgrowth into which surviving nucleus moves and divides to form 8–nucleate embryo sac.

(iii) ***Chrysanthemum cinerariaefolium:*** Martenoli (1939) has described a peculiar mode of development of the embryo sac forming two types of embryo sac depending either upon the free or fusion stage of two central megaspore nuclei of 4-nucleate stage (1 + 2 + 1 arrangement). In the former case 12-nucleate embryo sac is formed (3–celled egg apparatus, 7-antipodals and two central polar nuclei), whereas in the latter case it is 9 or 10-nucleate type (3-celled egg apparatus, 4 antipodals and 3 central nuclei).

Usually one type of embryo sac is formed in the same species; however, there are reports where the same species shows different types of embryo sac or even the same individual. Kapil and Prakash (1966), for example in *Delosperma cooperi* (Aizoaceae), recorded 14% ovules of *Polygonum* type embryo sac, 18% *Endymion* type and 68% of tetrasporic type (47% *Drusa* type, 37% *Penaea* type and 18% *Adoxa* type).

Of the various factors, genotype is of paramount importance; however, there are reports that environmental factors especially the temperature to some extent influence the type of embryo sac (Hjelmqvist and Grazi, 1965).

MATURE EMBRYO SAC

It has been reported that in the majority of the angiosperms, the mature embryo sac is of the *Polygonum* type, with 3-celled egg apparatus, 3-celled antipodal and 2-nucleate central cell (Fig. 6.3). Besides, *Allium*, *Endymion*, *Adoxa* and *Fritillaria* types also show similarity with *Polygonum* type when fully organized. Except for *Plumbago* and *Plumbagella* types, the egg is always associated with two synergids (rarely, one in *Peperomia* type). The antipodals are always present; however, their number and ploidy are variable. In some cases, embryo sacs are found where usual polarity and organization are absent. In *Saccharum officinarum*, *Woodfordia floribunda*, *Crinum asiaticum* etc., the embryo sac show egg apparatus at the chalazal end and antipodals at the micropylar end. In certain other plants two egg apparatus are seen, one at each end. However, in *Poa alpine*, there are two groups of antipodals, one at micropylar end and the other at chalazal end.

The structure and the functions of different parts of the embryo sac are as follow:

(1) Egg

The egg represents the female gamete. It shows common wall with the two synergids and the central cell. The nucleus and most of the cytoplasm lie in the lower part of the cell while a

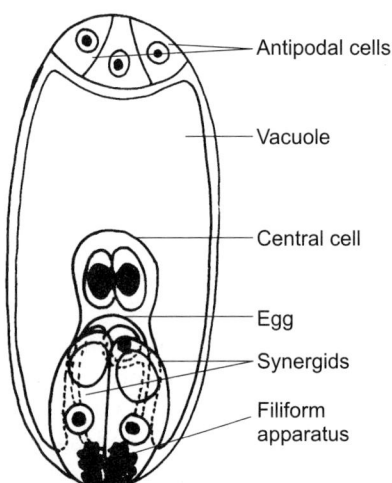

Fig. 6.3 Diagram of embryo sac (After Jensen, 1973).

vacuole, facing the micropylar end, is present in the upper part of the egg. The young egg cell shows aggregation of all the organelles except plastids. The cytoplasm is rich in starch content as reserve food material, which is consumed during fertilization or early embryogeny. In *Plumbago capensis*, many finger-like projections arise at the micropylar end of the egg (Cass, 1972), similar to filiform apparatus of synergids. In the absence of synergids in this plant, the egg itself absorbs nutrition from the nucellus, in addition to its gametic function. In *Crotaderia jubata* (Philipson, 1981), rarely the egg may develop into haustorium.

Ultrastructure profile of the egg indicates that the metabolic activity of the egg is rather low. The synergids are metabolically more active.

(2) Synergids

The synergids are elongated cells present at the micropylar end of the embryo sac. They are usually two in number, sometimes one or may be absent altogether. The nucleus lies lower side towards the micropylar end, while the upper part of the cell contains a large or many small vacuoles. The lower part is occupied by finger-like projections, called filiform apparatus. The form of filiform apparatus is variable; it may be spherical, wedge-shaped or elongated. Recent microscopy work has revealed that each projection of the **filiform apparatus** has a core of tightly packed microfibrils (possibly cellulosic) enclosed by non-fibrillar sheath. They are rich in polysaccharides. The cytoplasm shows the concentration of cell organelles near the filiform apparatus. Similarly, lipids, RNA and proteins are abundant and concentrated in this region. A few plastids are also reported (Fig. 6.4).

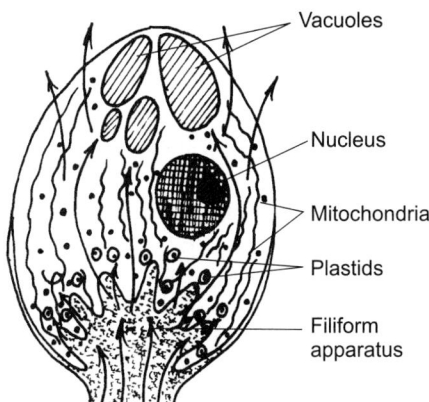

Fig. 6.4 Diagrammatic representation of synergid. Arrow indicates the direction of flow of nutrients (After Jensen, 1965).

Usually, synergids are ephemeral structures, which degenerate and disappear soon after the fertilization or even before it. In some cases, however, one or both persists for sometimes and show sign of considerable activity.

In *Quinchamalium chilense* (Fig. 6.5), a member of Santalaceae, the synergids and antipodals are extensively elongated to form a haustorial structure (Johri and Agrawal, 1965). Following functions have been attributed to synergids:

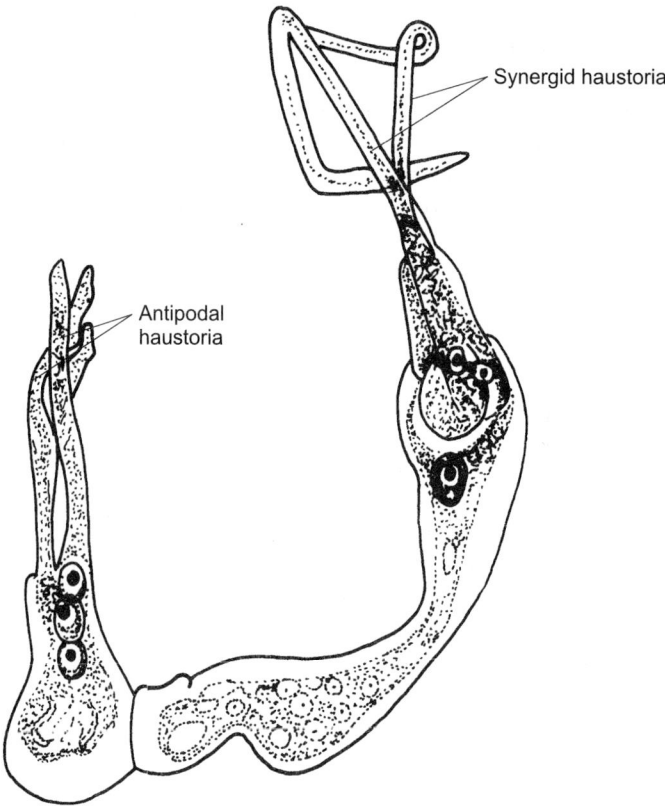

Fig. 6.5 Mature embryo sac of *Quinchamalium chilense* showing synergid and antipodal haustoria (After Johri and Agarwal, 1965)

(i) They play an active role in directing the pollen tube growth by secreting some chemotropically active substances.
(ii) Filiform apparatus helps in absorption and transportation of materials into the embryo sac from the nucellus.
(iii) The degenerating synergids form the seat for pollen tube discharge in the embryo sac.

(3) Antipodal Cells

Although short-lived, the antipodal cells show considerable increase in size and number. Usually, the number is three but may be 10-12 or even 300 cells have been reported in *Sasa paniculata*, a member of Bambusae or may be absent altogether as in *Oenothera* type. The antipodal cells of the some members of Ranunculaceae, e.g., *Aconitum napellus* (Fig. 6.6), become greatly enlarged and assume glandular appearance. Antipodals may have two or more nuclei e.g., *Fritillaria, Plumbagella, Phyllis* etc. Antipodals also shows high degree of polyploidy e.g., *Caltha palustris* (4n or 8n).

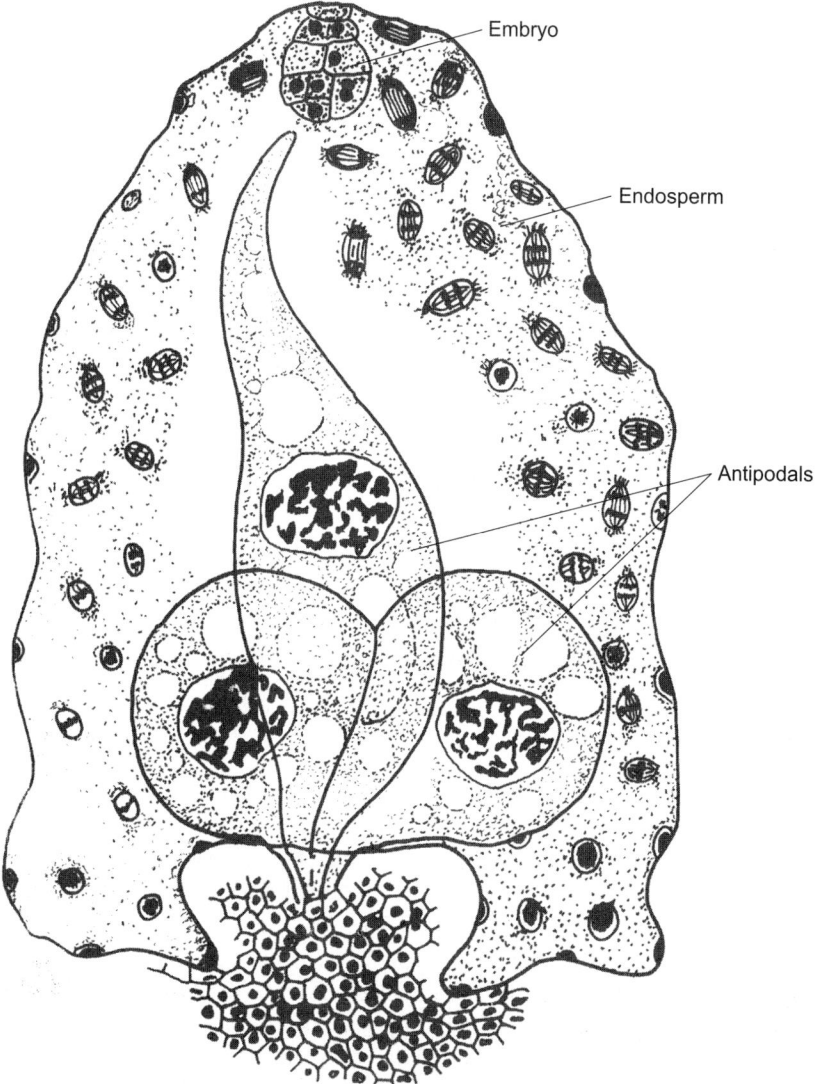

Fig. 6.6 Embryo sac of *Aconitum napellus* showing three large antipodal cells; young embryo at micropylar end and endosperm nuclei in different stages of division (After Maheshwari, 1950).

Haustorial behaviour of antipodals is known in many plants, for example *Grindelia, Haplopappus, Putoria* and *Galum etc.*

Often a nutritive role has been suggested for the persistent antipodal cells. This suggestion has further been supported with the finding of wall projections in antipodal cell of maize and poppy.

(4) Central Cell

The central portion of the embryo sac containing polar nuclei, which eventually gives rise to endosperm and therefore, been called **"endosperm mother cell"**. The fusion of the polar nuclei may occur either before or during or sometimes after the entry of the pollen tube inside the embryo sac and results diploid secondary nucleus. The cytoplasm of the central cell is rich with all cell organelles, suggesting its active state.

STORAGE MATERIALS

There are several reports of the occurrence of starch in the embryo sac. The stage at which the starch makes its appearance in the embryo sac varies in different plants. However, in the majority of plants, the starch appears when the embryo sac is mature and reaches a maximum, shortly after fertilization, later gradually decreasing in post-fertilization stage. A few cases are there, where starch occurs even in egg apparatus and antipodals. In *Sonneratia* certain oil bodies of unknown nature have been found. In *Aspidistra* large raphides have been reported in the mature stage.

The possible pathway for the nutrients into the embryo sac is from the chalazal end. This fact is supported by morphology of ovules, vascular supply and the presence of special tissue **hypostase**. Moreover, the biochemical studies by Ryczkowski (1971) suggest that in unfertilized ovules, the concentration of amino acids is highest at the chalazal end.

EMBRYO SAC HAUSTORIA

Apart from the general absorbing nature of the embryo sac, special haustorial structures sometimes develop (synergids and antipodals) and draw food materials from ovular tissues and carpellary tissues. In *Phaseolus, Melilotus, Torenia* (Fig. 6.7A) and *Gallium* etc. the embryo sac ruptures the nucellar epidermis and grow beyond it and come to lie in the micropyle.

However, in other plants it is a downward growth, which is more striking as in several members of Centrosperales, the embryo sac pushes forward at the chalazal end and digests its way through the nucellus. This structure is called, **'Caecum'** and is very effective haustorial organ e.g., *Digera arvensis* (Fig. 6.7B).

EVOLUTIONARY TENDENCIES

There is hardly any doubt that the Polygonum type of the embryo sac is the most primitive. It is easy to derive the Oenothera type of the embryo sac in which micropylar megaspore functions, there is no polarity at 2-nucleate stage and mitotic division is suppressed. The Allium and Endymion types can be derived from the Polygonum type by the absence of cell plate formation after meiosis II and suppression of one mitotic division during megagametogenesis. The Podostemaceae show a reduced Allium type where instead of two mitotic divisions there is only one.

If there is no cell plate formation after meiosis I and II and only a single mitotic division occurs, tetrasporic embryo sacs are formed. The failure of cell plate formation during

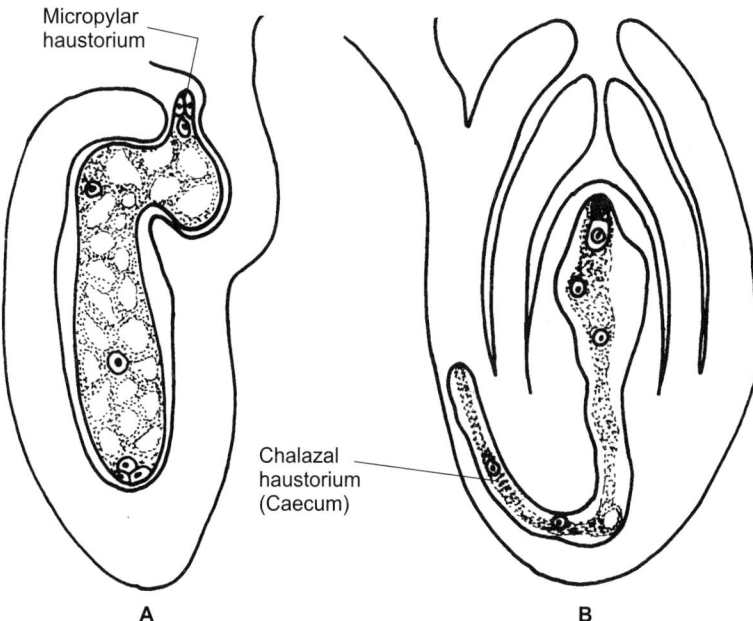

Fig. 6.7 A. *Torenia hirsuta*, showing micropylar haustorium. B. *Digera arvensis* showing "caecum" as chalazal haustorial organ.

meiosis, distinctive arrangement in the coenomegaspore (2 + 2; 1 + 1 + 1 + 1; 1 + 2 + 1 and 1 + 3) accompanied by the fusion of the chalazal nuclei in some cases and suppression of one or two mitotic divisions during gametogenesis lead to the various types of embryo sac. As a result of reduction, the Polygonum type gives rise to the Allium and then to the Adoxa type. Oenothera is a reduced Polygonum type, Plumbago a reduced Penaea type and Plumbagella a reduced Fritillaria type. Other types of embryo sac met in the angiosperms are due to an aberrant behaviour during sporogenesis or gametogenesis or both and are easily derived from one of the basic types.

Schnarf (1936) also assumed the monosporic 8-nucleate embryo sac as the most primitive. This idea is supported by the fact that this type is of wide occurrence in the angiosperms and that in the pteridophytes and gymnosperms the female gametophyte is monosporic. Moreover, all the other types can easily be derived from it. Hjelmqvist and Grazi (1964) were able to secure four different patters of behaviour in the same species by temperature alteration; hence he concluded that too much attention should not be attached to types of embryo sac.

Several theories have been put forward to explain the morphological nature of embryo sac. According to Chamberlain (1935), the angiospermic embryo sacs have been derived from that of gymnosperms. Hofmeister (1858) and Strasburger (1900) advanced the '**Gnetalean Theory**', which suggests that all the nuclei of the embryo sac are potential eggs. The theory proposed by Porsch (1907) considers that the embryo sac consists of archegonia without any prothallial cells. Schuroff (1928) described the micropylar quartet as the remnants of two archegonia and the chalazal quartet as entirely prothallial.

7. Pollination

Pollination is defined as the process of transfer of pollen grains from anther to the stigma of the same flower or of different flower. If the pollen grains are transferred from an anther to stigma of the same flower the process is called **self-pollination or autogamy**. If the pollen grains are transferred to the stigma of different flower either on the same plant or different plant, it is called **cross-pollination or allogamy**.

Taking into account the genetic constitution of the flowers to be pollinated, allogamy may be of two types: Geitonogamy, Xenogamy. If the pollen grains are transferred to different flower on the same plant (of same genetical origin), the pollination is called **Geitonogamy**. On the other hand, the transfer of pollen grains on stigma of a flower situated on different plant of the same species (of different genetic origin), or different species, the pollination is called **Xenogamy** (Figs. 7.1; 7.2).

SELF-POLLINATION OR AUTOGAMY

Self-pollination occurs usually in the bisexual flowers. In some species of plant like groundnut (*Arachis hypogea*) self-pollination occurs regularly, while in sunflower (*Helianthus*), Marigold (*Tagetes*), *Tridax* and many other genera belonging to family Asteraceae, self-pollination takes place only when cross-pollination fails. Self-pollinated species are believed to be originated from cross-pollinated species.

Contrivances for Self-pollination

For self-pollination following two types of adaptations or contrivances are found:

1. **Homogamy (Bisexuality):** This is the condition in which the anthers and the stigma of a hermaphrodite flower mature at the same time. Under this condition some of the pollen grains may reach the stigma of the same flower either through the agency of insects or wind or by the sudden bursting of anther, thus affecting self-pollination.

Fig. 7.1 Diagram showing types of pollination.

2. **Cleistogamy:** There are many plants for e.g., *Commelina benghalensis, Viola, Impatiens, Oxalis* and *Salvia* etc., where the flowers are closed called cleistogamous and self-pollination is a rule in them. In *Commelina benghalensis,* in addition to closed flowers normal brightly coloured, blue or violet, flowers are borne aerially, called **chasmogamous** flower. Here, the pollination is by means of insects (Fig. 7.3).

CROSS-POLLINATION OR ALLOGAMY

Depending upon the agency, which is instrumental for the transfer of pollens from the anthers to stigma, various terms are used to indicate the type of pollination (Fig. 7.4):

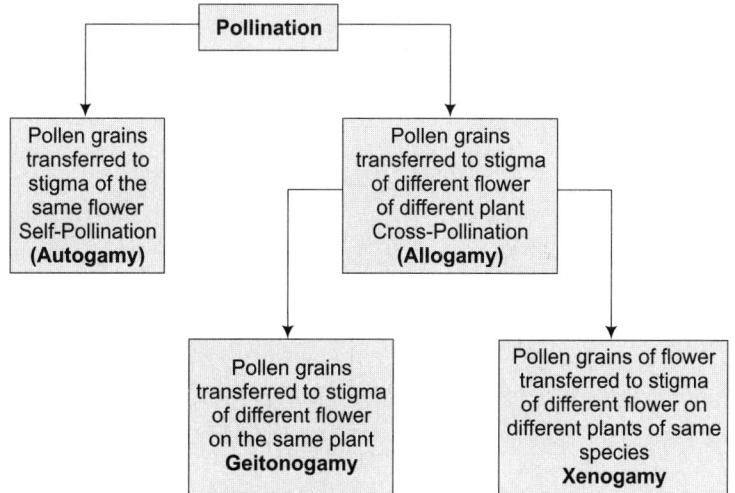

Fig. 7.2 Summary of self and cross- pollination.

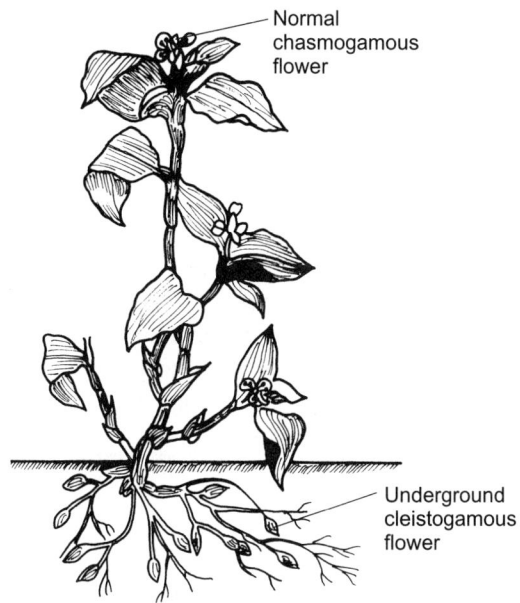

Fig. 7.3 Cleistogamous flowers of *Commelina*.

1. Anemophily or pollination by wind.
2. Hydrophily or pollination by water.
3. Zoophily or pollination by animals;
   (i) Entomophily or pollination by insects.
   (ii) Ornithophily or pollination by birds.

(iii) Chiropterophily or pollination by bats.
(iv) Malacophily or pollination by snails.
(v) Myrmecophily or pollination by ants.

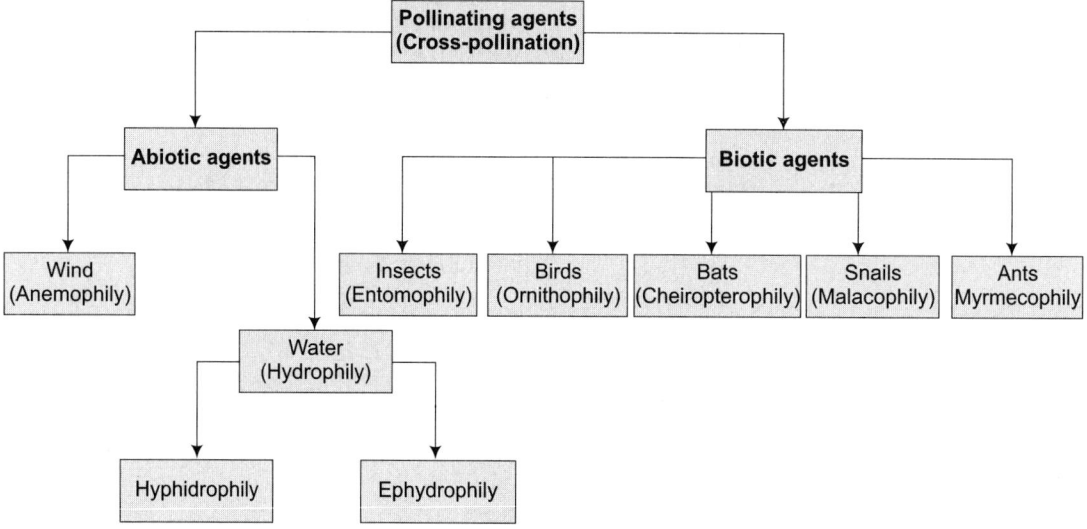

Fig. 7.4 Summary of cross pollination.

1. Anemophily

Plants with inconspicuous flowers without attractive bright colours, scent, and nectar are most likely anemophilous or wind pollinated. Pollens, in anemophilous plants, are produced in large quantity, as there is considerable wastage during transit. The sulphur rain in the hills with pine forests is an example, where large number of yellow powdery pollens covers the hilltop. They are light in weight and dry or sometimes provided with structures like air sacs or wings as in the pollens of *Pinus* trees. Stigma are comparatively large and protruding, sometimes branched and often feathery. Examples are offered by grasses, bamboos, cereals, millets, sugarcane, sedges, pines and several palms (Fig. 7.5).

2. Hydrophily

The transfer of pollen grains from anthers to stigma through the agency of water is called hydrophily. Hydrophilous flowers, as a rule is small and inconspicuous. The hydrophily may be **hyphydrophily** when pollination takes place under water as in *Ceratophyllum, Najas* and *Zostera* etc. or it may be **ephydrophily**, where the pollination takes place on the surface of water e.g., *Elodea, Hydrilla* and *Vallisneria* etc. In the submerged dioecious aquatic plant, known as *Vallisneria spiralis* belonging to family Hydrocharitaceae, the male flowers break off from the spadix and float on the surface of water, where they eventually open to liberate the pollen grains (Fig. 7.6 A, C). The female flowers, which are solitary, at maturity float on the surface of water due to uncoiling of the stalks (Fig. 7.6 B, C). The trifid stigma receives the

Pollination **7.5**

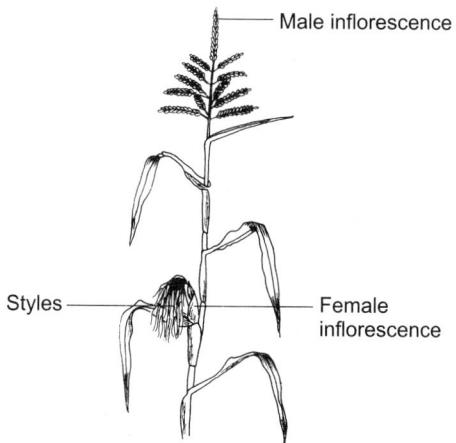

Fig. 7.5 Maize cob showing long silky styles and stigma for catching pollen grains.

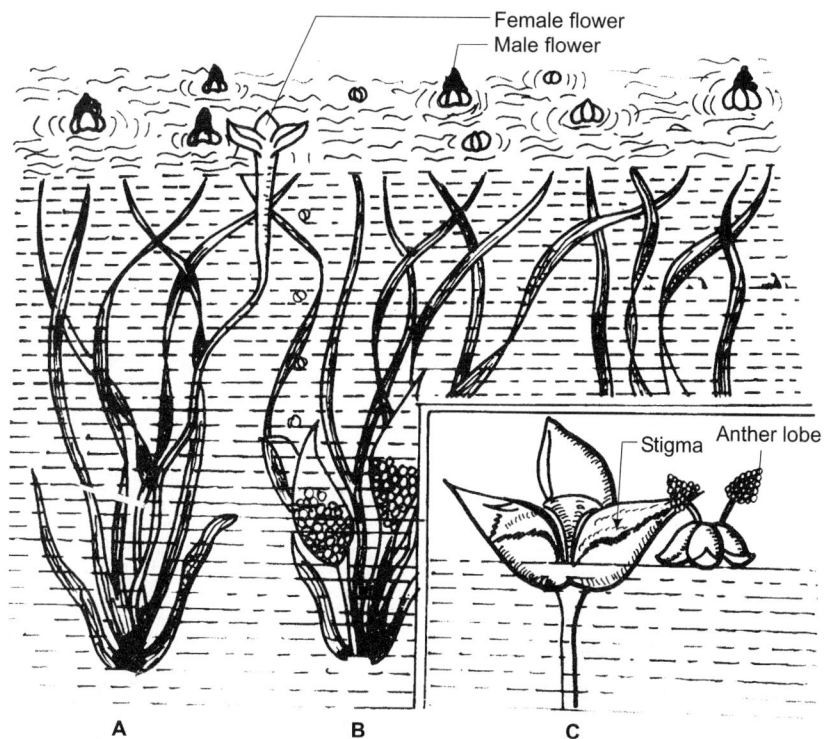

Fig. 7.6 Epihydrophily in *Vallisneria*. **A**. Female plant. **B**. Male Plant. **C**. Female and Male flowers.

pollen from the male flowers, thus causing hydrophily. After pollination the stalk of the female flower coils to bring the female flower to their original submerged position, where they mature into fruits.

3. Entomophily

Pollination by insects is of general occurrence. In order to attract insects flowers become large in size and brightly coloured. Sometimes, the flowers themselves are inconspicuous, therefore, other parts contribute in making them attractive for e.g., in *Mussaenda* sepals modify into large, white or brightly coloured structure. Likewise, in *Bougainvillaea* the bracts become large and colourful. However, in the members of Asteraceae, large number of small flowers clustered together on the thalamus to form a large, colourful, showy capitulum inflorescence to facilitate entomophily. Another important feature associated with entomophilous flowers is presence of nectar gland at the base of flowers or in a special structure called spur, which serve as a food for insects. Third significant feature that flowers possess is scent which invites insects as in *Nyctanthes*, *Cestrum* and *Quisqualis* etc. However, *Amorphophallus* and *Arum* emit stinking smell, which attracts carrion-flies and pollination is achieved.

The pollens are large and thick-walled, sticky with projecting spikes, which facilitate to stick on to the body of insects. The stigma is also large and sticky.

A very interesting case of cross-pollination by insect is seen in sage (*Salvia*), where there are two stamens in the flower with the two anther lobes each, widely separated by the elongated curved connective which plays freely on the filament. The upper lobe is fertile and the lower lobe is sterile. In the natural position the connective is upright. When insect enters the tube of the corolla, it pushes the lower sterile anther lobe of each stamen; the connective swing round with the result that the upper fertile lobe comes down and strikes the back of the insect and dusts it with pollen grains (Fig. 7.7). The flower is **protandrous**, the stigma matures later and bends down and gets in touch with the back of another insects already carrying pollens from different flowers, thus cross-pollination is achieved.

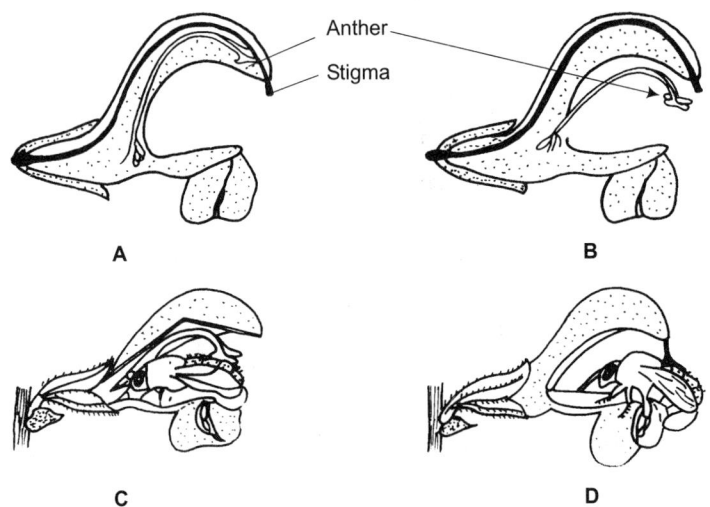

Fig. 7.7 Pollination in *Salvia*. **A, B**. Flower showing immature pistil and mature stamens. **C**. A bee with pollens glued on the back. **D**. Stigma lowered down to pick up pollens from the back of insect.

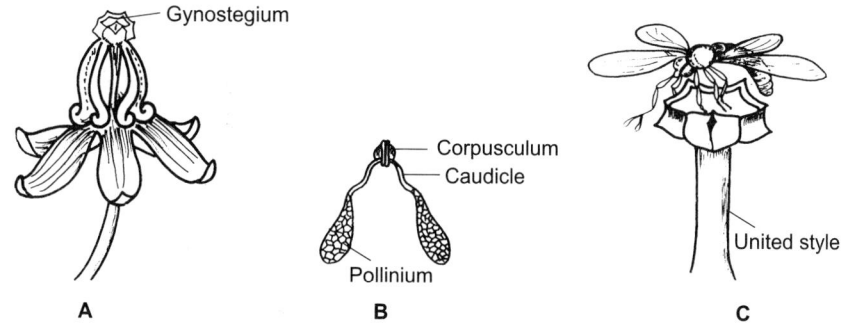

Fig. 7.8 Pollination in *Calotropis*. **A.** A flower of *Calotropis*. **B.** A pair of pollinia. **C.** A bee with pollinia attached to leg of insect.

Madar (*Calotropis*) is another interesting example of insect-pollination, where a gynostegium is formed due to fusion of stigma and androecium (Fig. 7.8 A). The pollens are glued together to form a sac called pollinium (Fig. 7.8 B). The pollinia are attached to a glandular adhesive disc at the stigmatic angle (translator mechanism). The sticky disc gets attached to the legs of visiting bees and the pollinia are pulled out, when the bees move away (Fig. 7.8 C). The bee when sits on to another flower, the pollination with the sticky pollinia is brought about.

4. Ornithophily

Pollination with the help of birds is called ornithophily. Some of the important bird pollinators are sun-birds, leaf-birds, flower-peckers and humming-birds (Fig. 7.9). Bird pollinated flowers show some adaptations to facilitate the process of pollination. They possess large, tubular or disc type of flowers, which are brightly coloured, scented and plenty of nectar. Flowers, such as trumpet vine, *Campsis radicans* and *Hibiscus* produce large quantity of nectar. Some plant species, such as *Bombax ceiba*, *B. insigne*, *Erythrina variegata* and *E. stricta* are visited by 50 different kinds of bird species (Subramanya and Radhamani, 1993).

5. Cheiropterophily

In some plants bats bring about pollination. *Eidolon helvum* and *Micropteris pusillus* are some common bats, which are attracted by large, colourful, nectar-rich, scented flowers. In *Macuna gigantean*, the flowers open after dusk and emit bad smell; at the same time pollens are released by the bursting of anther, which get stuck on the body of attracted bats, thus pollination is brought during their visit to different flowers. Other species which are pollinated by bats are *Adansonia digitata*, *Bombax ceiba*, *Anacardium occidentalis*, *Oroxylum indicum* and *Careya arborea* etc.

CONTRIVANCES FOR CROSS-POLLINATION

1. Dicliny or Unisexuality

Unisexuality is a condition where a flower is either male or female. This prepares flower for cross-pollination. Depending upon the unisexuality there are two types of plants:

Fig. 7.9 Pollination by bird (ornithophily).

monoecious, where male and female flowers are borne on the same plant e.g., maize, castor, cucumber and gourd etc., and **dioecious**, where male and female flowers are borne on two separate plants e.g., palmyra palm, papaya, and mulberry etc.

2. Self-sterility or Self-incompatibility

In all those plants, where pollens from a flower is incapable of bringing about the fertilization in the same flower are said to be self-sterile. In certain members of family Orchidaceae, the pollens have a toxic effect on the stigma of the same flower, while it is effective on the stigma of different flower. In *Petunia axillaries*, pollen tube germinates on the stigma of the same flower but the pollen tube growth is inhibited in the mid of the style. Pollens of *Passiflora* and *Malva* cannot germinate on the stigma of the same flower.

3. Male Sterility

Male sterility refers to the absence of functional pollen grains in otherwise hermaphrodite flowers. Male sterility is not very common in nature but is of great value in experimental population. Male sterility is of two types: genetical and cytoplasmic.

4. Dichogamy

In many bisexual flowers the anthers and stigma often mature at different times. This condition is known as dichogamy. There are two conditions of dichogamy:
- (i) **Protogyny:** When the gynoecium matures earlier than the anthers of the same flower. Here the stigma is pollinated by the pollens of other flowers due to immature

condition of their own anther. Examples are *Ficus* sp., *Mirabilis*, *Annona*, and *Magnolia* etc.

(ii) **Protandry:** When the anthers mature earlier than the stigma of the same flower. Here, the pollen can not germinate on the stigma of the same flower; rather they are transmitted to mature stigma of different flower. Examples are *Helianthus* sp., *Hibiscus* sp., and *Oxalis* etc. Protandry is more common than protogyny.

5. Heterostyly

There are some plant which bear flowers of two different forms: one form bears long stamens and short style (short-styled or thrum-eyed), and the other form bears short stamen and long style (long-styled or pin-eyed). This is known as **dimorphic heterostyly** for e.g., *Primula vulgaris*, *Woodfordia* etc. Stigma in short-styled flower is at the level of anther in the long-styled flower and vice-versa (Fig. 7.10 A, B). In such cases, cross-pollination is effective only when it takes place between stamen and style of the same length borne by different flowers (legitimate pollination). Some flowers, such as *Oxalis*, *Lythrum* show **trimorphic heterostyly**, where stamens and styles are of three different lengths borne by three different forms of flowers (Fig. 7.11 A, B, C). Pollens from flowers of one type can effectively pollinate stigma of only the other two types of flowers and not of its own type.

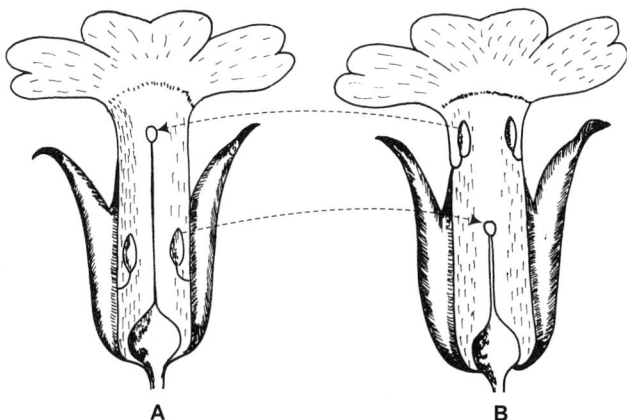

Fig. 7.10 Distyly. **A.** Flower with long style and short stamen (Pin flower). **B.** A flower with short style and long stamen (Thrum flower). Arrows indicate legitimate pollination.

6. Herkogamy

In certain flowers there is certain physical barrier, which prevents self- pollination and prepares ground for cross- pollination. In *Calotropis* (Asclepiadaceae), *Zeuxine* (Orchidaceae), the pollen grains are glued together to form pollinia, which are attached to a disc thus are kept away from the stigma. When the insects visit a flower, the pollinia get attached by their corpusculum to the legs of insects and carried to stigma of other flowers and pollination is brought about accidently.

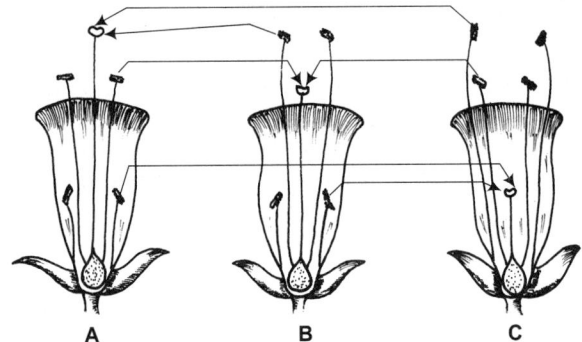

Fig. 7.11 Tristyly. **A.** Flower with long style and normal and short stamen **B.** Flower with mid-style and long and short stamen. **C.** Flower with short style and normal and long stamen. Arrows indicate the legitimate pollination.

In *Gloriosa* belonging to Liliaceae, the anthers are extrorse and the pollens, which are shed outwards, cannot reach the stigma of the same flower. An interesting case is also found in bleeding heart (*Clerodendron thomsonae*), a garden climber, bearing flowers with white calyx and red corolla where the stamens stand erect and the style bends down on one side. After the anthers burst or the pollens are removed by the insects, the filaments roll up on one side of the flower and it is the style now which stands erect and takes the same original position where the stamens was. In this position the bifid stigma directly receives the pollen grains floating in the air or from the body of he visiting insects (Fig. 7.12).

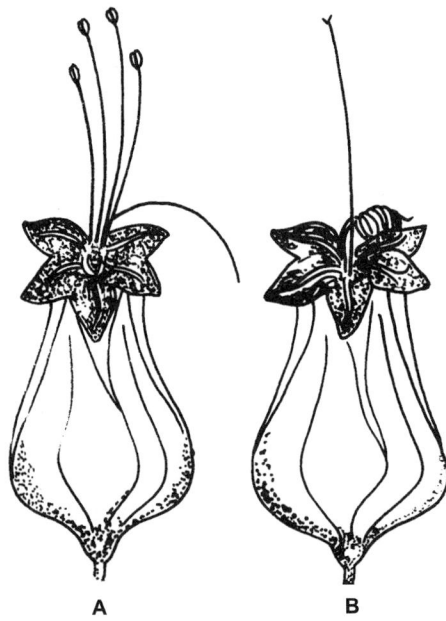

Fig. 7.12 Pollination in *Clerodendron thomsonae*. **A.** Stamens standing erect and the style bend down. **B.** Style standing erect and filaments roll up in close coils.

ADVANTAGES AND DISADVANTAGES OF SELF-AND CROSS-POLLINATION

In bisexual flowers self-pollination is a general phenomenon and pollination is almost ensured. For self-pollination plants need not to depend on external agencies to carry pollens, there is minimum wastage of pollen grains. The only disadvantage associated with self-pollination is the production of weaker progeny due to continuous selfing.

In Cross-pollination there is quantitative and qualitative improvement in the progeny, such as seeds are healthy and more viable; the fruit size is large and there is overall increase in yield; offspring are healthy; plants are better adapted in nature; the possibility of production of new varieties. The disadvantages associated with cross-pollination are dependence upon the external agencies and large-scale wastage of pollen grains.

8 Fertilization

Fertilization can be defined as fusion of the male gamete with female gamete, where plasmogamy is followed by karyogamy. Before fertilization to be brought about, pollen grains need to be transferred on to the stigma (pollination) of a flower through various agencies, like air (anemophily), water (hydrophily), insects (entomophily), animals (zoophily) etc. in angiosperms. The pollens are shed from the microsporangium either at two-celled stage or three-celled stage, which are lodged on the stigma. It is on the stigma where pollen grains germinate through the formation of pollen tube which travels through the style before reaching to the ovule. The pollen tube ultimately penetrates the ovule and releases two male gametes inside the ovule. Out of the two male gametes one unites with egg and the other unites with the secondary nucleus, thus double fertilization is brought about (Fig. 8.1). Various structures and events leading to fertilization have been described in details under the following headings.

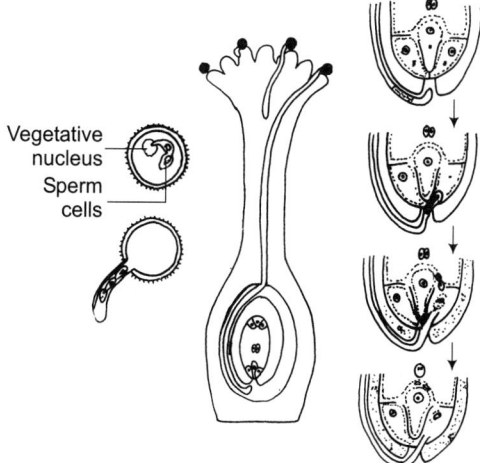

Fig. 8.1 Schematic diagram representing sequence of events involved in fertilization by tricellular pollen (After Knox and Singh, 1987).

STRUCTURE OF STIGMA

After landing on the stigma, pollen grains germinate and produce a pollen tube that carries the male gametes. The stigma has been classified into two main types depending on presence or absence of stigmatic exudates at the time of pollination: (i) dry stigma and (ii) wet stigma. These two basic types have been classified further (Table 8.1). The groupings are based on disposition of receptive cells and papillate and non-papillate nature of cells.

The stigma shows a columnar tissue divisible into two distinct zones: an upper epidermis forming the secretary zone and lower 1-3 layers, constituting storage zone. Many of the epidermal cells develop into stigmatic papillae which show much variation in morphology. In some members of Boraginaceae the stigma papillae are capitate with heavily cutinized non-receptive heads. The receptive surface is confined to the junction of the base of the papillae. The stigmatic papillae are rich in cell organelles. They may be highly vacuolated and multinucleate. The papillar part shows numerous mitochondria, ER, ribosomes and distinguishable into three parts, namely a terminal papillar part, a wide central part and a narrow basal part (Wilms, 1980). A significant outcome of the recent investigations has been demonstration of the presence of extra-cellular proteins on the surface of the stigma and in the style, irrespective of the morphological variations of the stigma and style.

Table 8.1 Classification of Angiosperm Stigma Types

1. **Dry stigma** (without apparent fluid secretion)
    Group-I Plumose, with receptive cells dispersed on multiseriate branches (Poaceae)
    Group-II Receptive cells concentrated in distinct ridges, zones or heads
    A. Surface non-papillate (Acanthaceae)
    B. Surface distinctly papillate
        i. Papillae unicellular (Brassicaceae, Asteraceae)
        ii. Papillae multicellular
            (a) Papillae uniseriate (Amaranthaceae)
            (b) Papillae multiseriate (Bromelliaceae, Oxalidaceae)
2. **Wet stigma** (surface secretion present during receptive period)
    Group-III Receptive surface with low to medium papillae; secretion fluid flooding interstices (some Rosaceae, some Liliaceae).
    Group-IV Receptive surface non-papillate, cells often necrotic at maturity; usually with more surface fluid than group III (Apiaceae).

Dry Stigma

Extra-cellular proteins in the dry stigma are present in the form of an extra-cuticular hydrated layer, the pellicle. In the dry stigma the cuticle show discontinuities and the components of the pellicle which originate from the cytoplasm of the papillae are extruded through these discontinuities (Fig. 8.2A). In *Crocus* the cuticle of the stigmatic papillae is chambered and the stigmatic secretion accumulated in these chambers (Fig. 8.2B). The third innermost layer underlying the cuticle is thick multicellular pectocellulosic wall. The cell wall of the grasses is more elaborate and comprises many layers (Fig. 8.2C). A thin mucilaginous layer is present between the pellicle and cuticle. The pectocellulosic wall is differentiated into several layers. The wall layer just below the cuticle is rich in pectic

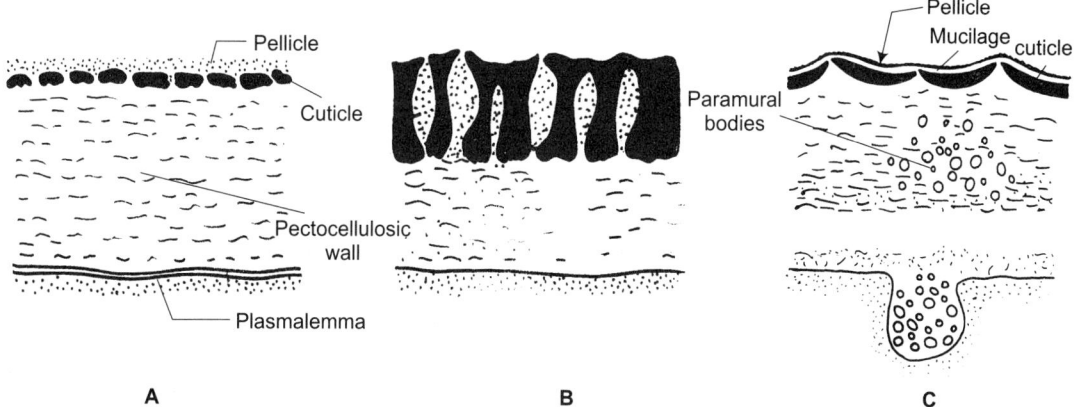

Fig. 8.2 Diagrammatic representation of the structure of the wall of stigmatic papilla in dry stigma. **A.** Typical dry stigma. **B.** Stigmatic papilla of *Crocus* to show chambered cuticle with cavities containing secretion material. **C.** In *Poaceae*, pectocellulosic wall showing paramular bodies
(After Heslop-Harrison, 1980).

polysaccharides. It also contains protein components. Paramular bodies are abundant in the papillae and appear to be involved in the secretion of surface components and the protein-polysaccharides constituent of the pectocellulosic wall and intercellular matrix (J. and Y. Heslop-Harrision, 1980). Besides proteins, the surface components also contain lipids and phenolic compounds. In a few taxa a lectin, concanavalin (con A) has been demonstrated to bind to the pellicle. The recognition and rejection reactions of pollen grains occur on the stigma surface in SSI system, (Sporophytic Self-incompatibility) placing barrier for pollen germination or penetration of pollen tube into the stigma.

The stigmatic surface of *Gladiolus* (Clarke *et al.* 1979) has been subjected to detailed analysis. A complex mixture of proteins, glycoproteins, and glycolipids are present in the pellicle. The component with carbohydrate amounts to 23% and contains the monosaccharides; galactose, arabinose, glucose and rhamnose. All the mannose is associated with a fraction that binds to con A.

Wet Stigma

The amount of exudates on the stigma and its components varies from species to species. Lipids, phenolic compounds, carbohydrates, amino acids and proteins are generally present in the exudates. The polysacharide composition of *Lilium longiflorum* is made up of arabinose (26%), rhamnose (6%), galactose (55%), glucuronic acid (11%) (Aspinall and Rosell, 1978).

The origin of exudates has been studied in many taxa with solid styles (Fig. 8.3 A-D). The stigma is secretory and exudate originates from the epidermal and subjacent cell layers. The exudates secreted by the sub-epidermal cells accumulate below the pellicle-cuticle layer. Because of the enlargement of intercellular spaces by continued secretion the epidermal cells loosen and the exudates from the subjacent intercellular spaces reaches the sub-cuticular zone. Eventually, the cuticle-pellicle layer is disrupted releasing the exudates and its flakes lie on the surface.

8.4 Plant Embryology: Classical and Experimental

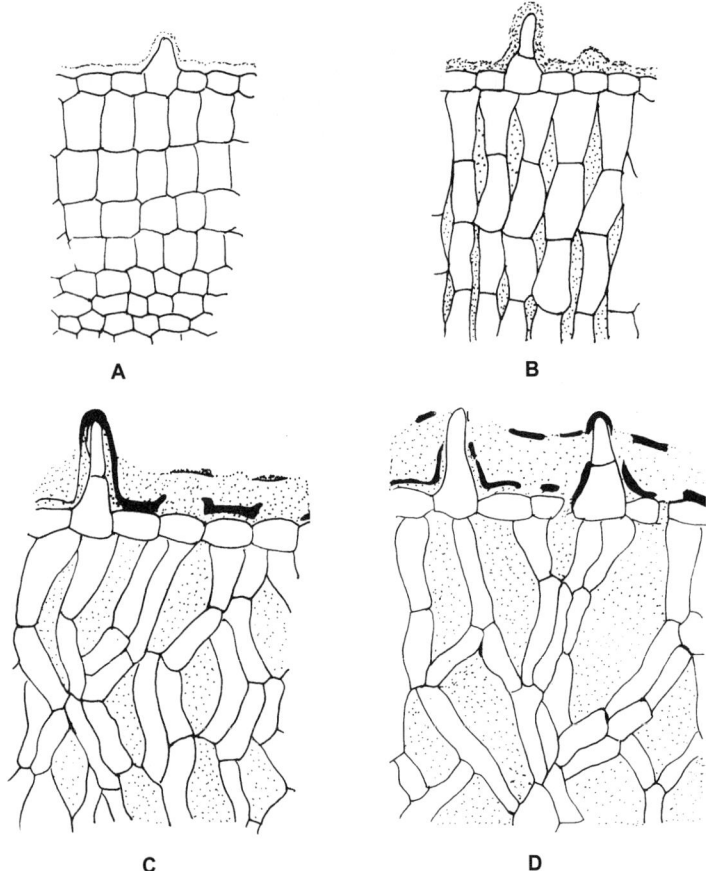

Fig. 8.3 A-D. Diagrammatic representation of the origin of stigmatic exudates in taxa characterized by wet stigma and solid style at different developmental stages. **A.** A very young stage. Only a thin cuticle-pellicle layer is visible. **B.** Secretion products have accumulated between the cuticle-pellicle layer and epidermal cells as well as in the intercellular spaces. **C.** Continued accumulation of secretion products has resulted in the enlargement of intercellular spaces and rupturing of cuticle-pellicle layer. **D.** Mature stigma (After Shivanna and Sastri, 1981).

Electron microscopic evidences indicate that exudate is secreted by the ER and is extruded by a process of exocytosis (Fig. 8.4). Golgi apparatus has no roles in the secretion of exudates. Besides exocytosis, the secretion of exudate appears to take place through the involvement of the vacuole also. During later stage, some of the ER vesicles are phagocytotically incorporated into the large cell vacuole. Following membrane dissolution, the contents of ER vesicle get mixed with the substance of the vacuole and result in the holocrine secretion following the eventual degeneration of the protoplast. This type of secretion is predominant towards the later stage of development, when the degeneration of the protoplast initiated (Kristen et al. 1979).

In hollow styled stigma, however, the detailed investigation is lacking. In these taxa stigma is non-secretory and the exudates which is present on the surface of stigma emanate

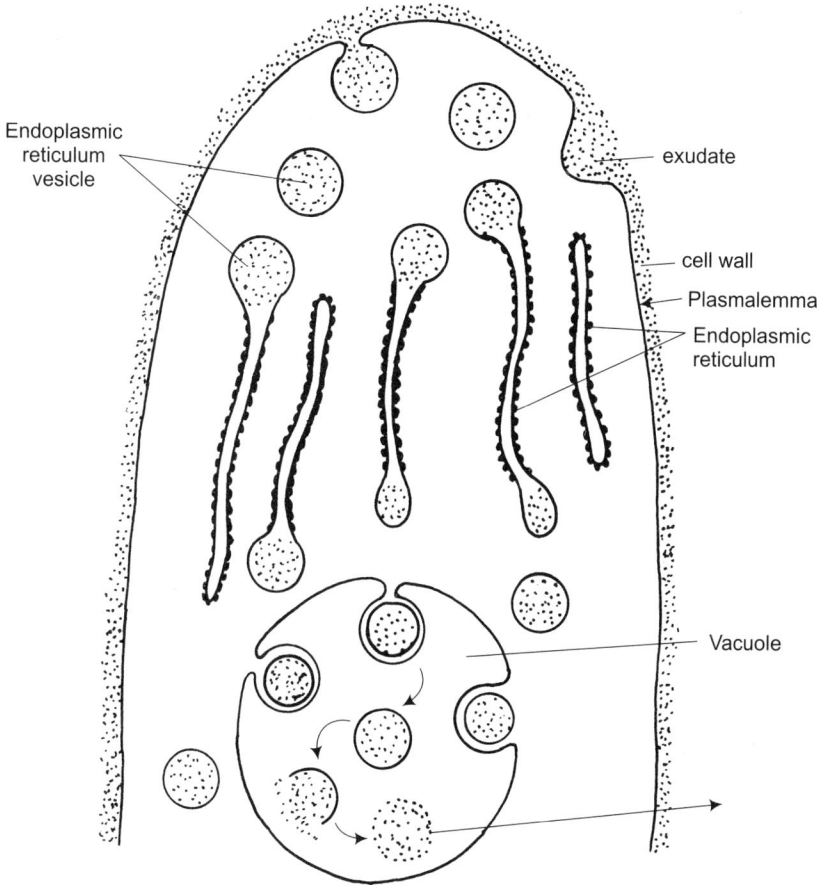

Fig. 8.4 *Aptenia cordifolia*, diagram of postulated secretion pathway in stigmatic papilla showing granulocrine secretion of exudates via exocytosis and holocrine excretion of mixed exudates via degeneration of protoplast. (After Kristen *et al.*, 1979).

from the glandular tissues lining the stylar canal. However, studies on in *Lilium longiflorum* (Dickinson et al. 1982) revealed that the viscous secretion present on the stigmatic papillae is secreted by the papillae themselves and is not derived from the stylar canal.

STRUCTURE OF STYLE

In the most primitive dicotyledons (e.g. in Ranalian families such as Winteraceae and Degeneriaceae) the carpels do not develop styles and the pollen tubes reach the ovules by growing through the hairs present on the unfused margins of the carpels. In the phylogenetically advanced forms the carpel margin fused, a style is developed and the stigmatic tissue is reduced to the upper portion of the style only. Three types of styles have been recognized: (i) solid, (ii) open and (iii) semi-solid half closed.

(i) Solid Style

The solid style which occurs mainly in dicots is characterized by the presence of central strand of specialized cells called transmitting tissue, termed by Arber (1937), through which growth of pollen tube takes place. The transmitting tissues of solid style have been studies in large number of taxa. It comprises elongated cells with their transverse wall traversed by plasmodesmata (Fig. 8.5 A-C). They exhibit normal electron microscopic profiles with numerous mitochondria, active dictyosomes, RER, plasmids and ribosomes. It is believed that ER and dityosomes are involved in the secretion. The intercellular substance in many solid-styled plants comprises of carbohydrates, proteins, glycoproteins and some enzymes such as acid phosphatase, peroxidases and esterases. Traces of Ca-salts may also occur. The proteins present in the intercellular substance besides being of nutritional value may also play some role in incompatible reaction. The cells of the transmitting tissues are surrounded but intercellular substance which is comparable to the secretion fluid of the stylar canal cells. In grasses, paramular bodies seem to be involved in the secretion of the components of the intercellular substance. The cells of the transmitting tissue of cotton, however, are not surrounded by any interecellular substance. The lateral walls of the transmitting cells are 7-10 μm thick and consists of four distinct layers (Jensen and Fisher, 1969): (a) an innermost layer composed primarily of pectic substances (b) outside it a thin darker appearing layer, richer in hemicellulose (c) still further to the outside a thick loose textured wall showing concentric rings of fibrous material, relatively poor in hemicellulose but rich in pectic substances and also containing cellulose and non-cellulosic polysaccharides outermost layer is a thick middle lamella consisting primarily of pectic substances.

In *Petunia hybrida* (Herrero and Dickinson, 1979) two distinct regions are distinguishable in the stylar transmitting tissue. The cells in the 'neck' region (just below the stigma) are large

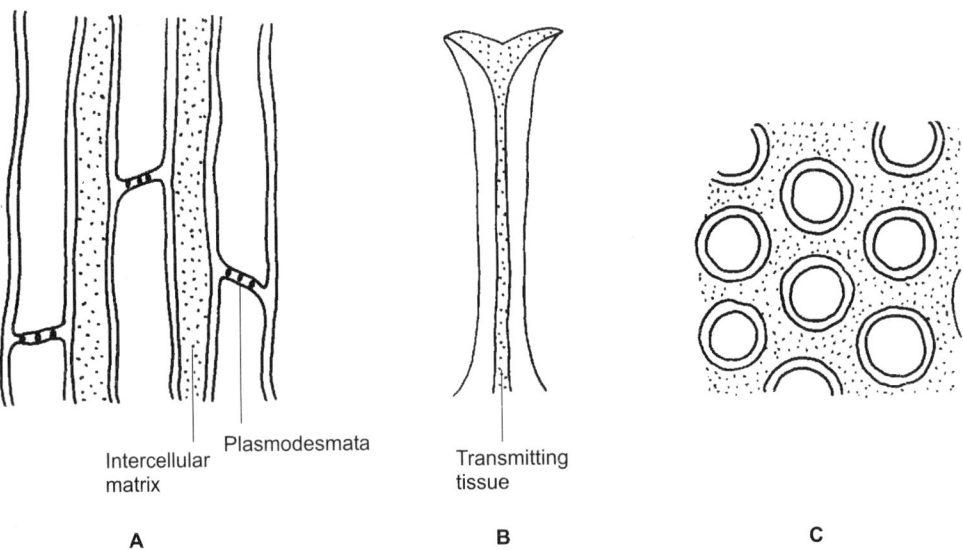

Fig. 8.5 Diagrammatic representation of solid style. **A.** Longitudinal section. **B.** Transmitting tissue in longitudinal section. **C.** Transverse section (After Heslop-Harrison, 1980)

and spherical with characteristic ridge on their walls which cause them to 'key' into the adjacent cells. The cells in the remaining part of the transmitting tissue are elongated, loosely packed and do not have superficial ridges. In both types the ER is frequently associated with vesicle which may merge with the plasmalemma. In this taxon the cells seem to pass through many phases of synthesis: the first takes place prior to pollination, the second is stimulated by pollination and occur prior to pollen tube passage and in the final phase the mobilized stylar reserves are transferred to pollen tubes.

(ii) Open Style

In open type there is wide stylar canal and the inner epidermis itself becomes secretory and assumes the functions of the nutrition and conduction of pollen tube as in Papaveaceae, Aristolochiaceae, Ericaceae and many monocotyledons. The canal is lined by single layer of glandular canal cells which may become multinucleate and polyploidy. The canal in mature style may contain secretion fluid as in *Amaryllis* and *Lilium* or may remain dry as in *Crocus* and *Gladiolus* (Fig. 8.6 A-C). The most striking structural feature of the canal cell is the presence of 8-14 µm thick secretory zone on the side facing the canal. The inner wall of the canal cells (facing the canal) is lined with a layer of cuticle. In *Lilium* cuticle is later disrupted and releases the secretion product into the canal (Ciampolini *et al.* 1981). The stylar canal contains a large number of vesicles and fibrils. The secretion from the canal cells contains proteins, carbohydrates and lipids and shows activity for esterase and acid phosphatases.

In some taxa, the hollow style is derived secondarily from the solid style by the dissolution of the transmitting tissues. A number of taxa exhibit a transitory condition between the solid and hollow style. The style of *Oenothera* and *Vitis* neither has an organized transmitting tissues nor a stylar canal. The style (in *Vitis*) contains a central canal into which numerous cells extend to form loose transmitting tissues. In the family Fabaceae, the style is hollow to different degree. *Vigna unguiculata* show a transition from solid to hollow condition (Ghosh and Shivanna, 1982). The stigma and 4-6mm of the upper part of the style are solid; gradually a canal develops by the dissolution of the transmitting tissues so that base of the style is hollow and lined with canal cells. In several other legumes, such a *Crotolaria* and *Cajanus*, the style is hollow throughout. At no stage do the canal cells possess a cuticle on the wall facing the canal. They do have a lining of proteinous material on this wall. After pollination the canal is filled with secretion, which serves as a nutrition for the growth of pollen tube.

The secretary zone consists of three regions: (i) an outer thick wall layer, made up of cellulose fibrils, (ii) the middle 7-12 mµ thick wall layer shows cellulose microfibrils. The granules in this region are pectic in nature probably complexed with protein and (iii) an innermost layer contains aggregates of tubules and vesicles called "paramular bodies".

(iii) Half-Closed Style

Half-closed or semi-solid type style occurs in the family Cacataceae and *Artabotrys*, which show intermediate features. The transmitting tissue is limited to only one side of the stylar canal.

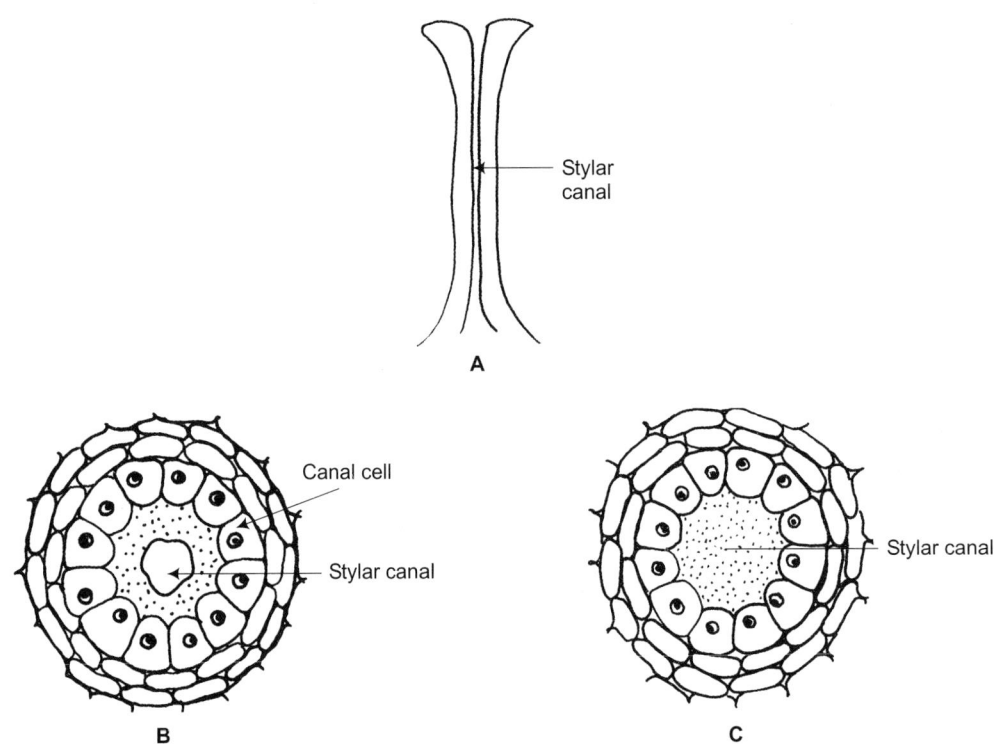

Fig. 8.6 Diagrammatic representation of the hollow style. **A.** Longitudinal section. **B.** Transverse section of style in which cuticle remains intact and secretion product accumulates between the cuticle and canal cell. **C.** Transverse section where stylar canal is filled with secretion produced following the disruption of cuticle bordering the canal (After Heslop-Harrison, 1980).

POST-POLLINATION EVENTS

Following pollination the pollen grains shows a series of events before it discharges the gamete near the egg (Fig. 8.7).

(i) Role of Exudates

Several important roles have been ascribed to the stigmatic and stylar exudates:
- (a) The lipoidal substances of the stigmatic surface help in trapping the pollens and in protecting the stigma from desiccation or wetting.
- (b) The exudates provides medium for pollen adhesion, hydration and germination.
- (c) The phenolic compounds presumably help in protecting the stigma from the microbes and pests.
- (d) These compounds have also been implicated in pollen nutrition and selecting promotion or inhibition of pollen grains on the stigma (Tara and Namboodiri, 1976).

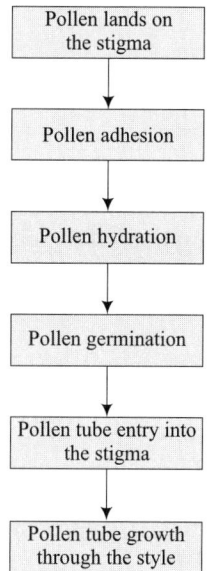

Fig. 8.7 Schematic representation of post-pollination events on stigam before the pollen tube enters into the embryo sac.

(ii) Stigma Receptivity

Receptivity of stigma is critical for successful completion of post-pollination events. Generally soon after the anthesis receptivity is maximum. The period of receptivity varies from species to species and is influenced by temperature and humidity. In wheat, rye and *Triticale* the stigma does not show much change in the receptivity up to 3-4 days after emergence. Later, there is a significant reduction in percent seeds set. In *Lilium*, stigma remains receptive for over 8 days after anthesis.

(iii) Pollen Adhesion

In addition to receptivity of stigma the adhesion of the pollens on the stigma is equally important for successful pollination. It is determined largely by the extent of stickiness of pollens and stigma and on the pollen wall structure. Wet stigma supports adhesion of both powdery and sticky pollens, which is a purely mechanical function. However, some are of the belief that electrostatic force is also implicated in pollen transfer and adhesion (Corbet *et al.*, 1980).

(iv) Pollen Hydration

The stigma provides the necessary moisture for pollen hydration. The rapidity of hydration depends upon the nature of stigma. In dry type hydration is gradual and controlled by the water potential of the stigma and pollens.

Following hydration pollen wall proteins are released on the stigmatic surface. In dry stigma the pellicle is the receptor site for pollen wall proteins, whereas in wet stigma the pollen wall proteins come in direct contact with exudates.

(v) Pollen Germination

The stigma provides the substances required for the germination that are absent or deficient in the pollen. The pollens of *Vitis vinifera* require boron for *in vitro* germination, thus suggesting the active role of boron in germination. Extracellular components present in the pellicle/exudates or the enzymes present in the pollen itself seem to play an important role in germination. The time taken for germination of microspore is variable; in *Saccharum officinarum* and *Sorghum vulgare* germination take place almost immediately; within 5 minute in *Zea mays* and *Hordeum distichon*; 2 hours in *Beta vulgaris*; 3 hours in *Reseda* sp. and 2 days in *Garrya elliptica*.

(vi) Pollen Tube Entry into Stigma

In species with solid style the pollen grains comes directly in contact either with the cuticle of stigma or papillae present on the stigmatic surface. The cuticle is eroded with the secretion of cutinase released by the pollens. The pollen tube finally grows through the interecellular substances of the stigma and transmitting tissues of style. In hollow-styled-system the pollen tubes grow down the surface of canal cells.

(vii) Pollen Tube Growth

Ultrastructural details associated with the pollen germination and tube growth are basically similar to those described for *in vitro* grown pollens (Dickinson, 1980). The activation of cytoplasm includes an increase in the number of dictyosome vesicles, RER and polysomes. The cell wall at the tip is fibrous and disorganized. The wall components are contributed by membrane–bound vesicles which presumably originate from dictyosomes (Fig. 8.8). The developing tube is continuous with the intine of the pollen wall. The pollen tube utilizes nutrients from the pistil for its growth. In taxa such as *Aegle marmelos*, *Fritillaria* and *Lilium*, the stylar tissue of the pistil contains abundant starch but following pollination and pollen tube growth starch is used up (Vasil, 1974).

Generally, a larger number of pollen tubes enter into the ovary than the number of ovules, however, only one pollen tube enters finally into each ovule. In *Persera americanum*, a uniovulate system, about 66 pollens germinate on the stigma, most of the tubes cease their growth after traversing various distances in the style and only one finally reaches the ovule. In *Prunus* several hundreds pollen grains germinate and their tubes grow into the style, although the first tube reaching the ovule affects the fertilization, while others stop further growth.

Most of the pollen grains are **monosiphonous**, producing single pollen tube as in Malvacaeae, Cucurbitaceae, while Campanulaceae shows **polysiphonous** condition where some species such as *Alathaca rosea* and *Malva neglecta* show emergence of 10 and 14 pollen tubes, respectively from the same pollen grains. However, only one of them makes further growth. Sometimes, the same pollen tube may be branched as in *Amentiferae*. In those plants, whose pollen grains are either united into tetrad or into pollinia, several pollen tubes are produced at the same time. Pollen grains may even germinate on the other parts of the flower besides stigma as in *Aeginetia indica*, where pollens germinate on the moist surface of corolla

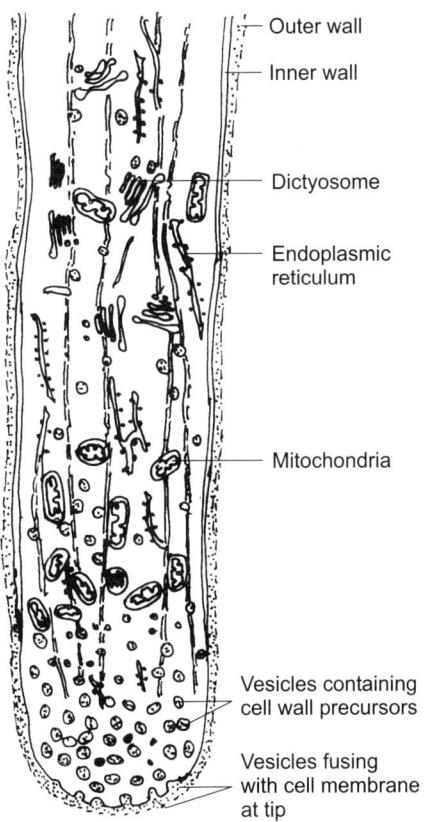

Fig. 8.8 A diagrammatic median longitudinal section through the tip region of the growing pollen tube shows wall structure and distribution of organelles (After Mascarenhas, 1993).

tube. An interesting case is reported in some members of Malpighiaceae (Anderson, 1980), which produce cleistogamous and chasmogamous flowers. In former, the pollen germination occurs inside the filament of the anther and pollen tube eventually reaches the ovule, thus they completely bypass the stigma and style. Such type of fertilization is called "**cryptic self-fertilization**".

(viii) Post-Pollination Change in the Stigma and Style

Pollination initiates many structural and physiological changes in the pistil. The growth of the pollen tube usually brings about the degeneration of the adjoining cells of the stigma and the transmitting tissue (J. and Y. Heslop-Harrision, 1980). The plasomodesmata interconnecting the protoplast of the papilla with neighbouring cells become occluded. Post-pollination secretion of exudates appears to be widespread, as even dry stigmas of many taxa become wet following pollination.

Ultrastructural studies on the transmitting tissue following pollination have been made in *Petunia hybrida*. Pollination initiates many changes in the 'neck cells' even before the arrival of pollen tubes. The number of polyribosomes increases accompanied by the appearance of a large number of single membrane cytoplasmic inclusions. These inclusions become associated with the plasma membrane to form characteristic embayment. This configuration of the plasma membrane is transitory and it reverts to the original configuration within 4 hours. As the pollen tubes reach the region of '**neck**', many cells become necrotic with dark granular cytoplasm, a large number of vesicles and disorganized plastids.

Pollination increases the respiratory activity in the pistil, changes pattern of RNA and protein synthesis and initiates marked increase in the activity of several enzyme and growth substances. Some of the physiological changes are initiated even in those regions of the pistil into which pollen tubes have not yet entered. Therefore, in the pistil the stimulus precedes the growing pollen tube and initiates response for the normal growth of pollen tube. The nature of the stimulus is not clear. It seems to be in the form of an electrophysiological signal as voltage variations following pollination have been recorded in the pistils of *Lilium longiflorum* (Spangers, 1978).

In cotton and several other taxa, pollination induces degeneration of one of the synergids, before the tube reaches the ovule. Synergid degeneration also occurs when ovules from unpollinated flowers are cultured on a GA_3 containing medium but not on IAA or cytokinin medium (Jensen and Ashton, 1981). Pollen grains seem to increase GA_3 content in the pistil triggering the breakdown of synergid.

Since long it has been thought that the pollen tube is guided from the stigma to the ovule by the gradient of chemotropic substance present in the pistil. Many investigators have demonstrated, using *in vitro* experiments, the positive chemotropic effect of various parts of the pistil to pollen tube. However, Mascarenhas and Machils (1962, 1964) suggested that Ca-ions to be chemotropic factor. They noted gradation in the distribution of total Ca^{2+} from the stigma to the ovules in *Antirrhinum majus*. However, subsequent studies failed to establish the Ca-ion gradients from stigma to ovule in different taxa or found to be inactive in *Oenothera* and *Lilium*. Mascarenhas (1966), therefore, suggested that besides Ca-ions some other factors might be involved in directing the pollen tube. The present consensus is that the cells of the transmitting tissue, with their file-like arrangement, provide a path of least mechanical resistance for the growth of pollen tubes, thus there is no need for a chemotropic gradient in the style (Jensen and Fisher, 1969). Regarding the change of direction of pollen tube in ovary Mascarenhas (1975), further proposed that a chemotropic factor is necessary in the ovary for changing the direction.

Many suggestions have been put forward to explain the nature of the chemotropic factor which originates in the ovule itself. Some investigators consider synergids, particularly the filiform appararus to be the source of chemotropic factor (Coe, 1971, 1977). According to Chao (1971, 1977), the chemotropic substance in *Paspalum* is produced by the dissolution of the integumentary and nucellus cells at the region of micropyle. According to Jenson and Ashton (1981), the calcium is localized in the vacuole of the synergid. Following the degeneration of the synergid the calcium is released and chemotropically attracts the pollen tube.

(ix) Entry of the Pollen Tube into the Ovule

After arriving at the top of the ovary, the tube may enter the ovule through the following route and accordingly the entry is of the following three types:

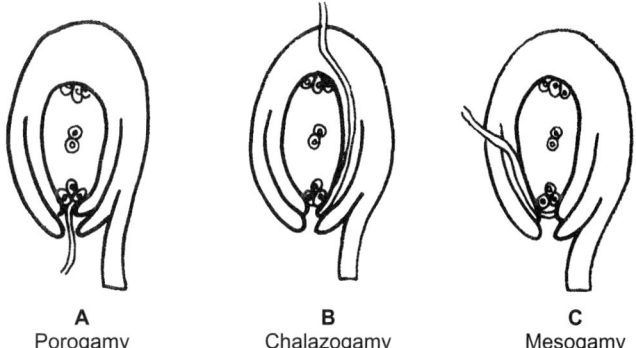

Fig. 8.9 Diagrams showing mode of pollen tube entry into embryo sac.

(i) **Porogamy:** This refers to a situation, where pollen tube enters through micropyle. It is the most common condition (Fig. 8.9 A).

(ii) **Chalazogamy:** In this type, the pollen tube enters through the chalaza (Fig. 8.9 B). This case has been for the first time reported by Treub (1891) in *Casuarina*.

(iii) **Mesogamy:** In this type, the pollen enters either through the funiculus (*Pistacia*) or through the integuments (*Cucurbita*; Fig. 8.9 C).

Sometimes more than one type may be seen even in the same species. Obturator is considered to be an organ of special interest, which facilitates the entry of pollen tube into the ovule.

In several members of the family Loranthaceae, there is no integument, therefore, the embryo sac undergoes remarkable elongation and meets the pollen tube at some point in the stylar region. The pollen tube enters the embryo sac through the nucellus either making its way between the cells (intercellularly) or destroying tissues, which lie in their way.

(x) Entry of the Pollen Tube into Embryo Sac

After penetrating the ovule, the pollen tube enters into the embryo sac by any one of the following passage: (i) between egg and one of the synergids as in *Fagopyrum* (ii) between the embryo sac wall and one of the synergids as in *Cardisopermum* (iii) directly into the synergid as in *Oxalis*.

Detailed studies of the third type show that the pollen tube enters into the synergid through the filiform apparatus. Mostly the penetrable synergid starts degenerating before the arrival of the pollen tube or after arrival of pollen tube as in *Petunia*. (Fig. 8.10 A-B). The content of the pollen tube is discharged into the synergid through the pore situated either sub-terminal or terminal (Fig. 8.10 C). Fagerland (1939) noted some embryo sacs of *Peperomia*, in which the tube had divided into two short branches. Later, Cooper (1940) refers to a similar bifurcation of the pollen tubes in other taxa (Fig. 8.10 D). That one of the branches become closely appressed to the egg and the other extending in the direction of polar nuclei and suggested that the two male gametes reach their destination by way of these separate branches.

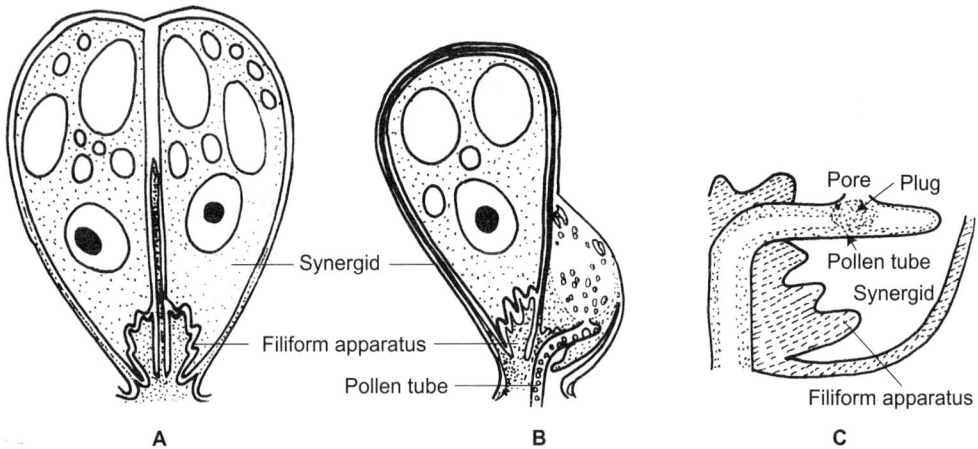

Fig. 8.10 Diagrammatic summary of the changes in the synergids after pollen tube discharge. **A.** Synergid in a pollinated flower. **B.** Synergid after pollen tube discharge **C.** Pollen tube discharge from sub-terminal pore (After Jensen and Fisher, 1968).

Increase in the GA_3 concentration in the ovule as a consequence of the pollen tube-pistil interaction has been suggested as a possible cause of the breakdown of the synergid in cotton. This has been further supported by the culture of ovules from unpollinated pistil in a medium containing GA_3, where synergids degenerate similar to pollinated flowers (Jensen and Ashton, 1981). The mode of transfer of sperms has been described in detail by Jensen (1973). The pollen tube enters one of the synergids through the filiform apparatus. The differences in the chemical composition of two synergies may be responsible for attracting the pollen tube to degenerating/degenerated synergids. However in several species both the synergids remain healthy until the entry of pollen tube (Fig. 8.11 A-C). One of sperm cells is first attracted by egg and the second sperm cell moves to the central cell later. However, the fusion process of egg and sperm take longer time than the fusion secondary nucleus and sperm cell.

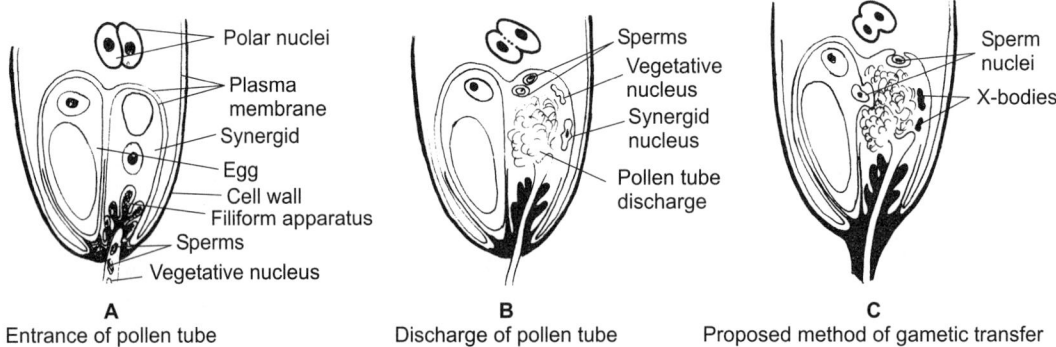

Fig. 8.11 Diagrammatic representation of fertilization and X-bodies in pollen discharge (After Jensen, 1973).

(xi) Rate of Growth of Pollen Tube

The time taken between pollination and the entry of the pollen tube into the embryo sac is variable and is also affected to an appreciable degree by the prevailing environmental conditions. The time period may be several months in Fagaceae and Betulaceae, while in some species of *Quercus*, this period may be as long as 12-14 months. However, there are a few examples where this period is relatively very short as in *Paphiopedilum maudae* about 19-20 weeks; *Cypripedium parviflorum*, about 3-4 weeks; *Orchis machulatus*, about 2 weeks; *Carica papaya*, about 10 days; *Carya illinoensis*, 4-7 days. The shortest period is recorded in *Taraxacum koksaghys* where fertilization occurs within 15-45 minutes after pollination.

DOUBLE FERTILIZATION

Following the discharge of the pollen tube, one of the sperms enters the egg and the other into the central cell. The forces involved in bringing about movement of the male gametes are not clear. They move passively along with the streaming cytoplasm.

In majority of the taxa studied, there are regions in the synengids where the plasma membrane is in direct contact with the plasma membrane of the egg and of the central cell without the intervening cell wall. Usually the sperms are discharged within the synergid. In *Plumbago zeylanica* (Russell, 1982) the sperms are discharged into the intercellular space between the egg and the central cell. Russell (1982, 1983) has studied the details of fertilization in *P. zeylanica*. The two sperms of pollen grains exhibit structural differences (Russell and Cass, 1983). One is associated with the vegetative nucleus has 30 µm long projection wrapped round the vegetative nucleus containing only a few plastids (0-2) and a large number of mitochondria (average 256). The outer sperm is connected to the first sperm through plasmodermata and is not closely associated with the vegetative nucleus, lacks a prominent cellular projection and contains 8-48 plastids and an average of 40 mitochondria. The plastid-rich sperm seems to be preferentially fused with the egg and the plastid poor sperm with the secondary nucleus.

Thus, the fusion of the gametes does not seem to be random; a recognition mechanism seems to operate in the fusion of two male gametes.

Different scientists have studied the mode of fertilization in different taxa. According to them soon after the male gamete makes contact with the plasma membrane of the egg (or the central cell), the plasma membrane of both the cells fuse forming a bridge through which the male nucleus enters. According to Jensen (1973), the fusion of the plasma membrane of the egg (or central cell) with that of the male gamete triggers off a reaction that makes the plasma membrane of that particular cell non-receptive to the fusion of a second gamete. The triple fusion (1-male nucleus and 2-polar nuclei) is comparatively quicker than syngamy because of the active state of the central cell.

Syngamy: However, the details of fertilization on the basis of electron microscopic studies involve the fusion of the whole of gamete (Jensen, 1973) and not just the nucleus. It is quite likely that the male cytoplasm including its organelles would be soon degraded after the entry into the egg or central cell. Such a phenomenon has also been documented in lower plants (Kuroiwa et al., 1981). However, there are many examples in which biparental inheritance of plastids has been genetically documented. Apparently, in such taxa, the functional organelles of the male cytoplasm are transferred to the egg and central cell and are not subjected to degradation.

Since the classical studies of Gerassimova (1933) on nuclear fusion in *Crepis capillaris* has been extended to many other systems largely by Russian scientists. Most of the investigators reported the disappearance of entirely or partially of the nuclear membrane preceding nuclear fusion. Based on the stage of male gamete at the time of nuclear fusion Gerassimova-Nawaschin (1960) recognized three types of fusion: (i) pre-mitotic (ii) post-mitotic and (iii) intermediate. In pre-mitotic type the sperm nucleus fuses immediately on coming in contact with the egg nucleus and the zygote nucleus divides subsequently e.g., Poaceae and Asteraceae. In post-mitotic type the sperm nucleus and the egg nucleus remains in contact for a while and fuse only after both the nuclei have entered into division (zygotic mitosis) e.g., *Lilium, Fritillaria*. In intermediate type the sperm nucleus fuse with the egg nucleus after completing its previous mitosis and the male nucleus is at interphase. Even after the fusion, the nuclear membranes, the contents of the two nuclei showed incomplete mixing. At the prophase of zygotic mitosis often the two sets of chromosomes can be seen separate, e.g., *Impatiens*.

X-bodies: Nawaschin has observed certain densely staining structures in the tip of pollen tube or the pollen discharge in the synergid. Since their exact nature could not be determined, he called them as X-bodies. From time to time these bodies have been variously interpreted, as remains of vegetative nucleus, remains of synergid nucleus, adjacent nucellar cells, remains of the degenerated megaspores etc. Based on their shapes and distribution and the fact they contain DNA, Jensen (1972) interpreted one of them as the remains synergid nucleus and the other as the remains of vegetative nucleus (Fig. 8.10 C). The finding has been supported by Russell (1982), who observed only one X-body in *Plumbago*, which represents the remains of vegetative nucleus, as there being no synergids.

UNUSUAL FEATURES

Polyspermy

The polyspermy is unusual situation, where more than two sperms are discharged within an embryo sac. This may result because of the formation of more than two sperms in a pollen tube or due to penetration of embryo sac by more than one pollen tube. The former abnormality originates either in the pollen grains or in the pollen tube. However, in the latter situation, two pollen tubes have been recorded in *Elodea, Ulmus, Oenothera, Sagittaria* etc; three in *Statica, Gossipium* etc. and as many as five in *Juglans*.

The presence of extra sperms inside the embryo sac may either result in a polyploidy offspring or other components of the embryo sac, such as synergids or antipodals may be fertilized resulting polyembryony.

Persistence and Branched Pollen Tubes

In the majority of angiosperms, the pollen tube after discharging its contents starts degenerating immediately after the fertilization is over. However, in some cases, the pollen tube may persist for as long as three weeks till the time of embryo development. In *Cucurbita* and some members of Onagraceae, the pollen tube after entering into the embryo sac initiates branching at the tip. These branches grow between and around the inner and outer integuments and penetrate the nucellus and other ovular tissues, behaving in haustorial fashion, called **haustorial pollen tube** However, some of the workers are of the opinion that the branches simply ramify in ovular tissue and do not have any haustorial function.

Heterofertilization

A different type of abnormality is the heterofertilization, where two sperms fusing with egg and secondary nucleus respectively are derived from two, different pollen tubes. The heterofertilization has not yet been cytologically demonstrated but Sprague (1932) has inferred in Zea *mays* on the genetical ground.

9
Sexual Incompatibility

In nature seed setting is an outcome of some sequential events, such as pollination, growth of the pollen tube, penetration of the pollen tube into embryo sac, release of sperms and finally fertilization are the essential steps. Of the large number of viable pollens landing on the stigma, only the right mating type is recognized by the pistil and allowed to function normally to establish seed setting, however, other undesirable pollens are discarded. Here, the incompatibility between the pollen and pistil appears to be because of some biochemical reaction and the phenomenon is called sexual incompatibility. This is an excellent mechanism to reject undesirable pollens and accept those pollens which are of right mating types.

Sexual incompatibility may occur either between individuals of two species, known as **interspecific incompatibility** or between the two individuals of same species, called **intraspecific incompatibility (Self-incompatibility)**. Sexual incompatibility has a great biological significance as it is an adaptation to prevent self-pollination and could be fertilized by the pollens of other plants only.

SELF-INCOMPATIBILITY

Self-incompatibility (SI) was first reported by Koelreuter in the middle of eighteenth century. More than 300 species belonging to 70 families show self-incompatibility. In order to prevent self-pollination, plants in nature have developed different types of adaptation such as unisexuality, self-sterility, heterostyly and herkogamy etc. but the most effective device is self-incompatibility. It is of the following two main types:

(i) **Heteromorphic Incompatibility:** In this category a species show either dimorphic flowers or trimorphic flowers depending upon the length of style and accordingly called distyly (Fig. 9.1) or tristyly (Fig. 9.2), respectively. The former type is found in e.g. *Primula* which shows two types of flowers: **Pin flowers** have long style and short stamen; **Thrum flowers** have short style and long stamen. The compatible mating is between the flowers having styles and stamens of the same lengths i.e. between pin and thrum types only.

Here, the genetical basis of incompatibility is single gene with two alleles **S** and **s**. The allele for short-style or thrum forms (S) is dominant over the allele for long- style or pin

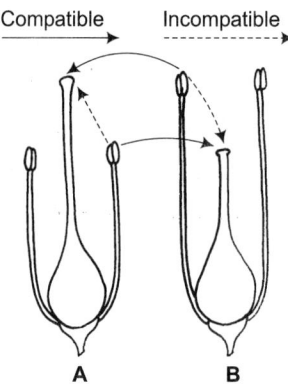

Fig. 9.1 Diagrammatic representation of heteromorphic incompatibility (distyly). (After Shivanna, 1982).

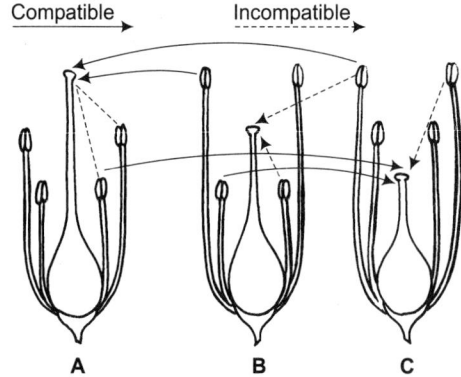

Fig. 9.2 Diagrammatic representation of heteromorphic incompatibility (tristyly). (After Shivanna, 1982).

forms (s). Long-styled pin plants are homozygous recessive (ss) and the short styled thrum are heterozygous (Ss). Incompatibility in pollen is sporophytically determined i.e. the incompatibility reaction of pollen is determined by the plant producing them. Since the allele **S** is dominant over **s**, therefore, the incompatibility system is **heteromorphic-sporophytic**. The pollen grains produced by pin flowers would all be **s** in genotype as well as incompatible in reaction. The pollens produced by the thrum flowers would be of two types genotypically; **S** and **s** but all of them would be **S** phenotypically. The mating between pin and thrum plant would produce Ss and ss in equal frequencies (Fig. 9.3).

Similarly, *Lythrum salicaria* offers an excellent example of tristyly where the styles are of three different lengths: long, medium and short. This feature is controlled by two dominant genes (**S** and **M**) with two alleles each. Gene **S** shows masking effect on the expression of gene M. Alleles S specifies short style, while allele **M** determines medium style. Thus flowers having short styles may have the genotypes **SsMm**, **Ssmm**, or **SsMM**, those with medium style may be **ssMM** or **ssMm**, while those with long styles will be **ssmm**. In this case

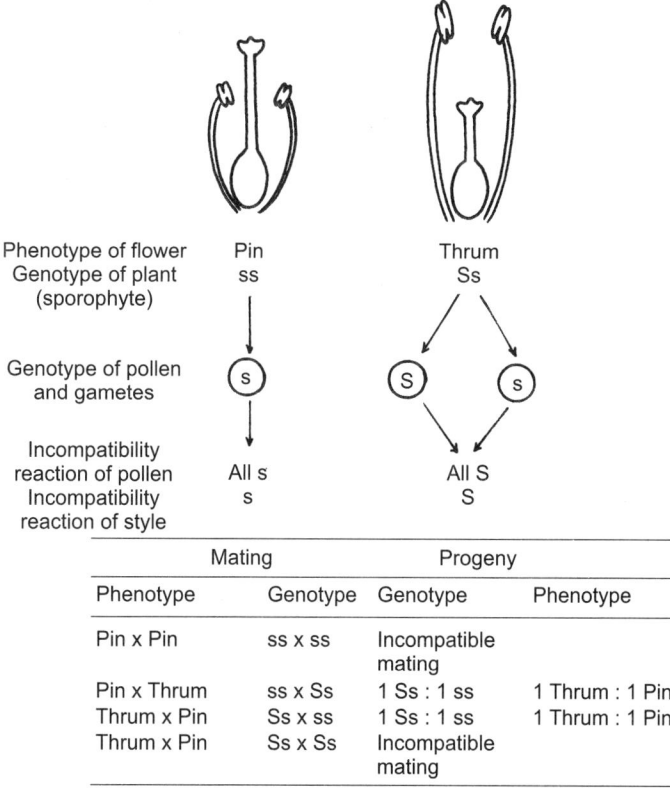

Fig. 9.3 Heteromorphic-sporophytic system of incompatibility.

compatible mating takes place between flowers showing the style and stamens of same lengths i.e. long x medium, long x short and medium x short.

(a) Zone of Inhibition

Although extensive studies have been conducted on the seed-set following legitimate and illegitimate pollination, the details of pollen germination and tube growth are confined to a limited number of taxa. The inhibition may either takes place on the stigma or in the style. There is general correlation between pollen cytology and zone of inhibition: 3-celled pollen species such as *Linum* and *Limonium* show stigmatic inhibition and 2-celled pollen species such as *Primula* show stylar inhibition.

The zone of inhibition may differ in different floral morphs of the same species. In *Fagopyrum*, for example, the inhibition in thrum x thrum pollination is in the stigma, whereas in pin x pin pollination the inhibition is in the style although pollens of both the morphs are 3-celled.

In *Limonium* pollen grains following incompatible pollination either fails to germinate or pollen tubes fail to enter the stigma. In species of *Linum* thrum stigma (wet) supports adhesion and germination of both thrum and pin pollens. Thrum pollen tubes, however, are effectively inhibited on the stigma and only pollen tubes from pin morphs grow through the style. Pin pollens fail to adhere to pin stigma. This is because of the lack of exudates on the stigma and insufficient pollen-coat substance on the pollens of pin form. Pin stigma, however, supports adhesion and subsequent germination of thrum pollens because of the presence of sufficient coat substance on the thrum pollens. Shivanna et al., (1981), working on *Primula vulgaris*, have shown that incompatible pollens may be inhibited at several sites: on the surface of the stigma due to failure of germination or pollen tube penetration in the stigma head or in the transmitting tract of the style through inhibition of the growth of tube. None of these barriers is complete and incompatibility is, therefore, the result of cumulative screening effect at different levels.

(b) Mechanism of Inhibition

Earlier experiments concerned with the mechanism of inhibition were conducted by Lewis (1943) on *Linum grandiflorum*. He reported that pin x pin pollination does not support pollen adhesion; even those pollens that remain on the stigma fail to be hydrated, whereas thrum x thrum pollination supports pollen hydration and germination. However, pollen tubes swell and burst soon after they enter the stigma. He also determined the osmotic pressure of the two morphs. By fixing a reference value of 1 to thrum style, the osmotic pressure value were 1.75 for pin styles, 7 for thrum pollens and 4 for pin pollens. Thus, in compatible pollination the compatible osmotic pressure ratio of pollen: style is 4:1. In incompatible pin x pin pollination (in which pollen does not absorb water from the stigma) it is 5:2 and in thrum x thrum pollination (in which pollen tubes burst soon after they enter the stigma) it is 7:1. The osmotic pressure ratio of 4:1 between pollen and style is optimal for normal pollen hydration and growth of pollen tube. If the ratio is too high (thrum x thrum) the pollen tube bursts, pollen hydration is prevented (pin x pin) even if it is too low.

In addition to the differences in the osmotic pressure and surface feature of the pollen and stigma of the two morhs, physiological mechanism is similar to those occurring in homomorphic system.

(ii) **Homomorphic Incompatibility:** In this category species produce flowers, which are morphologically indistinguishable although it possesses numerous mating types. Homorphic incompatibility has been reported in 250 genera belonging to 71 families (Crowe, 1964). However, cytology and physiological details have been worked out only in a limited number of taxa belonging to Solanaceae, Liliaceae, Brassicaceae, Asteraceae and Poaceae: *Nicotiana, Petunia, Lilium, Brassica* and *Raphanus* are well known examples. This type of incompatibility is controlled by multiple alleles.

Self-incompatibility may be categorized into two types depending upon the origin of factors which determine the mating type on the pollen side.

(a) **Gametophytic Self-incompatibility (GSI):** The incompatibility reaction is determined by the genotype of male gametophyte (pollen) itself and not by the genotype of the donor plant, e.g., Poaceae, Liliaceae and Solanaceae.

(b) **Sporophytic Self-incompatibility (SSI):** The incompatibility reaction is controlled by the genotype of the sporophytic tissues of donor plant from which the pollen are derived e.g., Asteraceae, Brassicaceae and Convolvulaceae.

(i) Genetics of Gametophytic Self-Incompatibility

The most suitable hypothesis about the genetical basis of self-incompatibility is one which has been proposed by East and Mangelsdorf (1925) in *Nicotiana sanderae*, referred as "opposition **S-allele**". According to this hypothesis it is the S-allele of the pollen or the male gametophyte which controls the incompatibility reaction.

GSI can best be explained by visualizing a situation where the pistil has two alleles S_1 and S_2. This may be pollinated by plants having genotypes, S_1S_2, S_2S_3 or S_3S_4. As per the opposition S-allele concept, pollens carrying S_1 or S_2 allele will not be able to penetrate the style, on the other hand, pollen with allele S_2 or S_3 will show functional activity of S_3 pollens only. However, all pollens carrying either S_3 or S_4 allele will function normally and bring about fertilization of S_1S_2 pistil (Fig. 9.4).

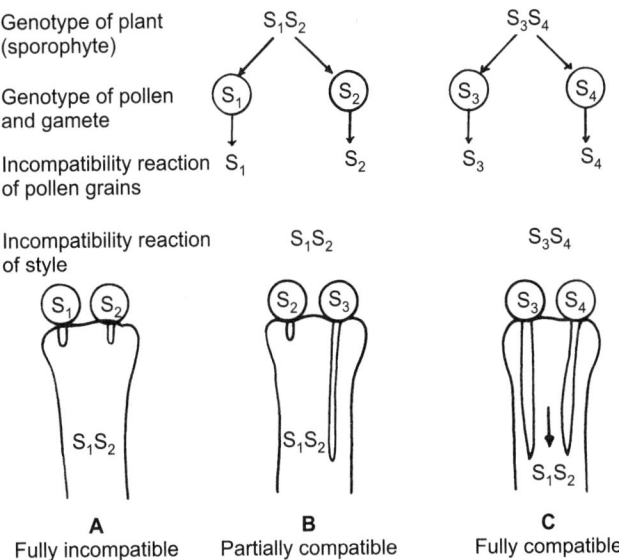

Fig. 9.4 Responses of pollen on a pistil of $S_1 S_2$ plant showing gametophytic self-incompatibility (GSI). **A.** None of the pollen from S_1S_2 plant is able to affect fertilization. **B.** From S_2S_3 plant only S_3 pollen succeed in fertilizing the ovule. **C.** All the pollens from S_3S_4 plants bring about fertilization.

(ii) Genetics of Sporophytic Self-Incompatibility

In SSI system all pollens of a plant will behave similarly irrespective of the S-allele they carry. It was first reported by Hughes and Babcock (1950) in *Crepis foetida* and by Gerlstel in *Parthenium argentatum* (1950). For instance, plant-carrying genotype S_1S_2 will produce pollen carrying S_1 or S_2 would behave as S_1 if S_1 is dominant or S_2 if S_2 is dominant; if there is no dominance both will behave as $S_1 + S_2$. In other words, presence of even one of the alleles in the stylar tissue similar to the sporophytic tissues of male parent would render all the pollens non-functional with respect to particular style. A S_1S_2 plant would, therefore, be completely incompatible to plants to pollens S_1S_2, S_1S_3, S_1S_4 or S_2S_3, S_2S_4, S_2S_4 and so on but would show 100% compatibility with a plant showing pollens S_3S_4, S_3S_5 and so on (Fig. 9.5).

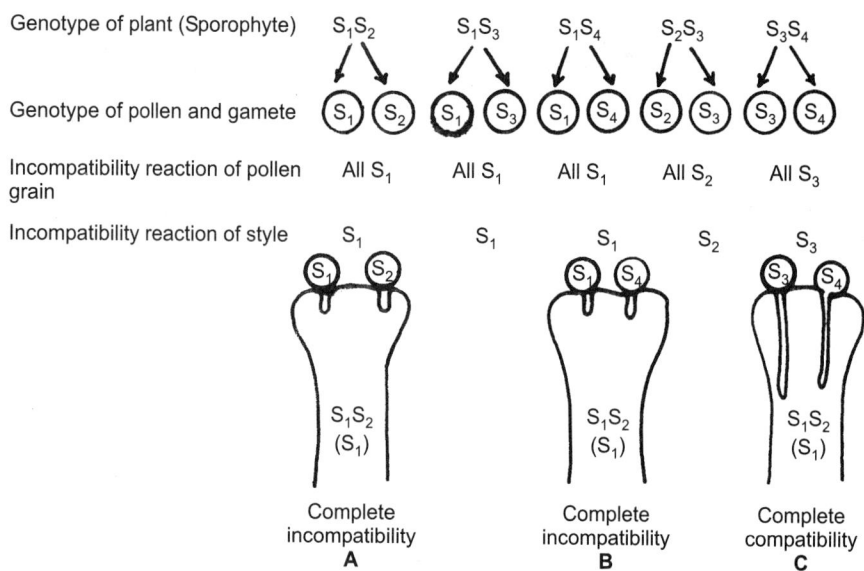

Fig. 9.5 Responses of pollen on a pistil of S_1S_2 plant showing sporophytic self-incompatibility (SSI). None of the pollen from S_1S_2 **(A)** or S_1S_4 **(B)** plant can bring about fertilization but every pollen of a S_3S_4 **(C)** plant are capable of fertilization.

In self-incompatibility system, the pollen or male gametophyte is inhibited anywhere, starting from germination on stigma, penetration of pollen tube into the stigma, growth of the pollen tube through the full length of style, entry into ovary, ovule till the penetration of the embryo sac. Both the SSI and GSI systems show remarkable differences corresponding to nature of pollens and stigma. In the SSI type the pollens are shed at 3-celled stage, *in vitro* germination is difficult, short-lived, respiration rate is high, stigma is dry and rejection reaction is attained on the stigma itself. However, in GSI system pollens are shed at 2-celled stages, *in vitro* germination is easy, long-lived, respiration rate is low, stigma is wet and the rejection reaction takes place in the style (Fig. 9.6).

Zone of Inhibition: The zone of inhibition of incompatible pollen/pollen tube is generally the stigma or the style. There is close correlation between the cytology of the pollen, genetics of incompatibility and the zone of inhibition (Brewbaker, 1957). The species which are shed at the 2-celled stage show gametophytic type of incompatibility and the zone of inhibition is in the style. The species which are shed at the 3-celled stage show sporophytic type of incompatibility and the zone of inhibition is in the stigma.

J. Heslop-Harrision (1978), on the basis of protein synthesis during meiotic division, suggested that in gametophytic species synthesis of S-allele specific proteins and those required for the cellular metabolism are initiated together following meiosis and consequently it results in competition for ribosome sites. This is responsible for the delaying the second mitotic division resulting in 2-celled pollen. In the sporophytic species, on the other hand, the function of S-protein synthesis is taken over by the tapetum and this permits the microspore to complete second mitotic division resulting 3-celled stage before pollen maturation/dehiscence.

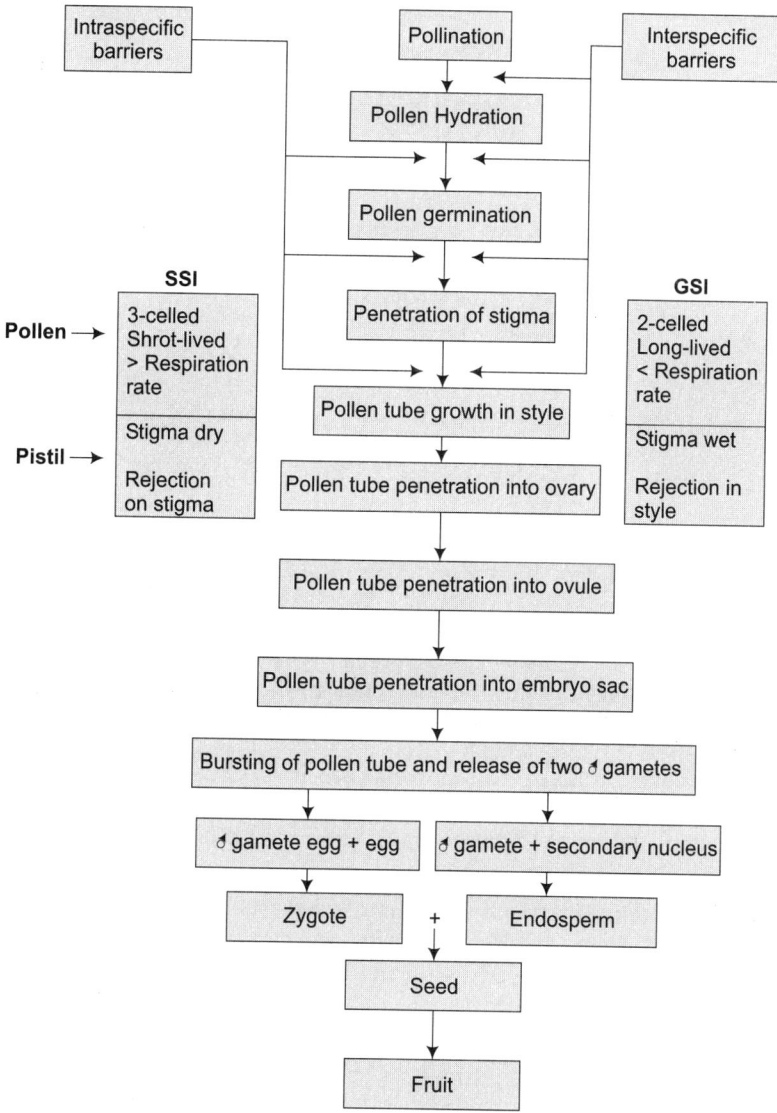

Fig. 9.6 Schematic representation of GSI and SSI systems in Plant.

According to Pandey (1979), in 3-celled pollens incompatible substance in the sporophytic species is present in pollen (the **S-genes** is activated in meiocytes before the completion of meiosis), therefore, the incompatible reaction and inhibition of pollen tube takes place in the stigma itself. However, in the gametophytic species with 2-celled pollens, incompatible substance synthesis is delayed (S-gene activation is delayed until the completion of meiosis in pollen tubes after pollen germination); therefore, the pollen tube inhibition takes place in style only (Fig. 9.7).

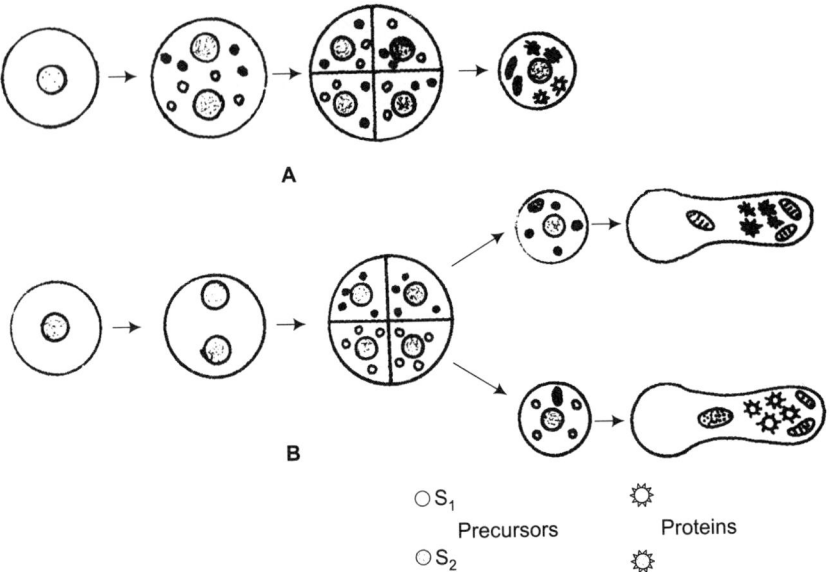

Fig. 9.7 The time of S-gene action in sporophytic and gametophytic systems. **A.** Sporophytic systems S-allele-specific precursors are synthesized before completion of meiosis, and S-allele-specific proteins before pollen dispersal. **B.** Gametophytic systems. S-precursors are synthesized after the completion of meiosis and S-proteins in the pollen tube after pollen germination. (After Pandey, 1979).

There are many exceptions to this correlation. In Poaceae taxa, the incompatibility is gametophytic but the pollen grains are invariably shed at the 3-celled stage. *Oenothera organensis* has 2-celled pollen and gametophytic incompatibility but the zone of inhibition is in the stigma.

Investigations of Heslop-Harrison (1978) and scientists have shown that, at least in some sporophytic species, S-allele specific products are synthesized in the tapetum and are incorporated into the pollen exine following the breakdown of the tapetum (Fig. 9.8).

In some species *Hemerocallis, Lilium, Annona, Gasteria* and *Ribes*, incompatible tubes are inhibited in the ovary. All these taxa are characterized by the presence of hollow style. According to Brewbaker (1957), the intimate contact between pollen tubes and stylar tissue is necessary for inhibition to occur. In hollow- styled systems such a contact is established only in the ovary.

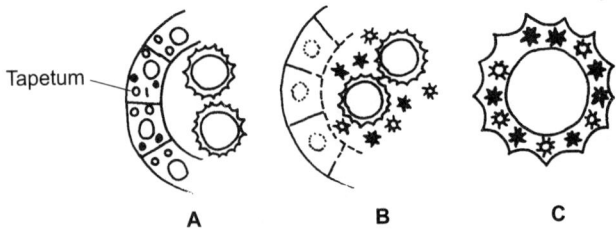

Fig. 9.8 Tapetal origin of S-allele component and their incorporation into pollen exine.

Cytology: Cytological details of pollens have been worked out in a number of systems. In sporophytic systems incompatible pollen grains either fail to germinate or the small tubes that emerge, at the most, penetrate the cuticle (Dickinson and Lewis, 1973). Inhibition of the pollen tube is generally associated with the deposition of excessive amount of callose in the pollen tube, particularly at the tip.

The stigmatic papillae which are in contact with the incompatible pollen also develop a lenticular plug of callose at the tip between the cell wall and the plasma membrane (Fig. 9.9). No such plug develops following compatible pollination. The deposition of the callose plug in the papilla is very rapid and often visible within 10 min after pollination. It is a clear manifestation of incompatible pollination.

In gametophytic system although there are no critical studies, it is generally presumed that germination of incompatible pollen is not affected at least. Pollen tubes are generally inhibited before growing half-way down in the style. In gametophytic system also pollen tube inhibition is associated with extensive deposition of callose in the tube.

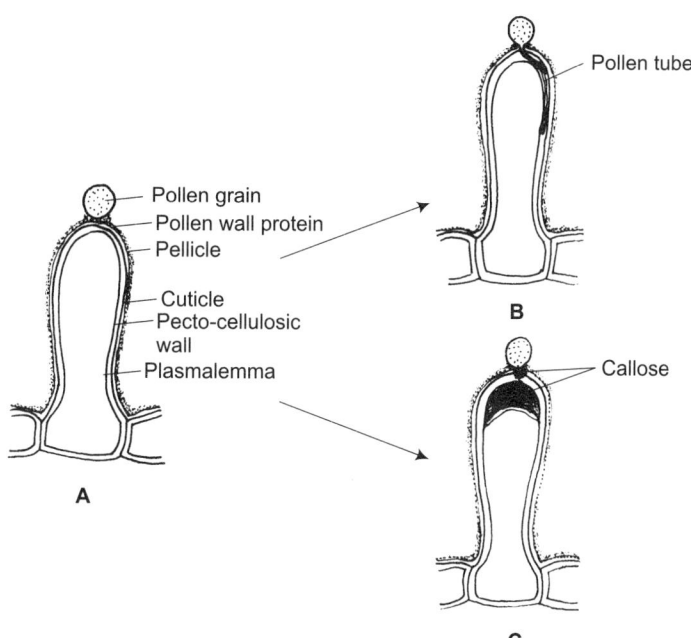

Fig. 9.9 Diagrammatic representation of pollen-pistil interaction in sporophytic systems. **A.** Compatible pollination. **B.** Incompatible pollination. Note the deposition of callose plug between the plasmalemma and the cell wall in the stigmatic papilla below the incompatible pollen (After Shivanna, 1982).

MECHANISM OF SELF-INCOMPATIBILITY

Physiology and Biochemical Studies

Extensive studies have been carried out on the physiology and biochemical aspects of self-incompatibility. There are many important differences in the metabolism between self-pollinated and cross-pollinated pistils.

Stigmatic pollination triggers many physiological changes in the lower part of the style and ovary even before the pollen tubes reach the base of ovary. Protein contents of the style undergo characteristic changes (Linskens and Tupy 1966). Following cross-pollination, a single new glycoprotein fraction appears in the style. After selfing, however, two glycoprotein fractions appear with different electrophoretic mobilities.

Attempts have been made to isolate, purify and characterize the S-allele specific glycoproteins (Nishio and Hinata, 1979). Starting with 10,000 stigmas, 230μg proteins was obtained as purified S- allele specific glycoprotein. It gave a single band following SDS polyacrylmide electrophoresis. Its molecular weight was estimated to be 57,000 from its mobility. The carbohydrate to protein ratio was 1:2. In *Brassica oleracea* (Ferrari *et al.*, 1981), carbohydrate to protein ratio was 1:3 and molecular weight was 54,000. The S-allele specific glycoproteins purified from both the species (*B. campestris* and *B. oleracea*) appear to be similar. The predominant amino acids for both glycoproteins were serine, glutamine, glycine and asparagines. The less frequent were cystine, methionine, phenylalanine, tyrosine and isoleucine.

Van der Donk (1974) studied the synthesis of RNA, DNA and the size of free nucleotide pool following self and cross- pollination. He concluded these differences as a result of pollen recognition and takes place at a very early stage during pollen stigma interaction. A difference in the protein metabolism in ovary has also been observed by some other scientists.

Injection of 6-methylpurine or actinomycin D, inhibitors of RNA synthesis, into the stylar canal, permits incompatible tubes to grow to the same length as compatible tubes. Thus RNA synthesis in the style is necessary for the incompatible reaction (Ascher, 1974).

POLLEN-PISTIL INTERACTION

(i) Recognition and Rejection Reactions

The recognition is the first step of sexual compatibility. The wall layers, exine and intine contains large amount of mobile proteins which are released on the stigma surface and comes in contact with stigmatic proteins. If compatible, the pollen germinates and shows normal growth through the style. If incompatible the pistil will initiate rejection reaction.

In species with wet stigma and solid style the recognition occurs on the stigma itself, but in wet stigma with hollow style the rejection occurs in the style (Shivanna, 1980). In this, the pistil rejects incompatible pollens either right on the stigma (SSI-system) by preventing germination or in the style (GSI-system) by inhibiting the growth of pollen tube.

Most of the studies on the intra- or interspecific incompatibility have focused the importance of the pollen wall and its proteins in pollen- pistil interaction (Petrovskaya-Baranova, 1961). The studies revealed that in *Oenothera organensis* GSI-system works but the inhibition occurs on the stigma. They further suggested that the recognition proteins of incompatibility are held extra- cellularly in the pollen wall. It has been suggested that in GSI-plants these proteins are present in the intine and in SSI-plants in the exine (Heslop-Harrison, 1968, 1975).

Notwithstanding the progress made in isolation and characterization S-allele product, very little information is available on the biochemical basis of self-incompatibility. Many models have been proposed by the scientists. Basically, these models envisage the production of an S-allele specific polypeptide identical in pollen and style. Of the various models proposed, that of Pandey (1975) is so far the best. According to Pandey (1975), S-allele specific proteins coded by the specific part (which is identical in pollen and style) unites with the tissue specific complementary protein in the style and pollen. Thus, the S-allele products of the pollen and pistil have complementary configuration. This model explains the failure of the stylar or the pollen proteins to react amongst them but allows for the interaction between them. The model is based largely on genetical analysis but the experimental evidences are not available.

In *Petunia hybrida* (Sharma and Shivanna 1982, 1983), crude pistil extract from unpollinated pistils selectively inhibited *in vitro* germination and tube growth of self-pollinated and not of cross-pollens. On the other hand, pistil extract from the bud (which allows incompatible pollens to grow through) did not inhibit *in vitro* germination and tube growth of self-pollens. This finding has been supported by investigation on other plants also. Thus, the synthesis of S-allele-specific molecule is formed after anthesis and is not a pollination induced response.

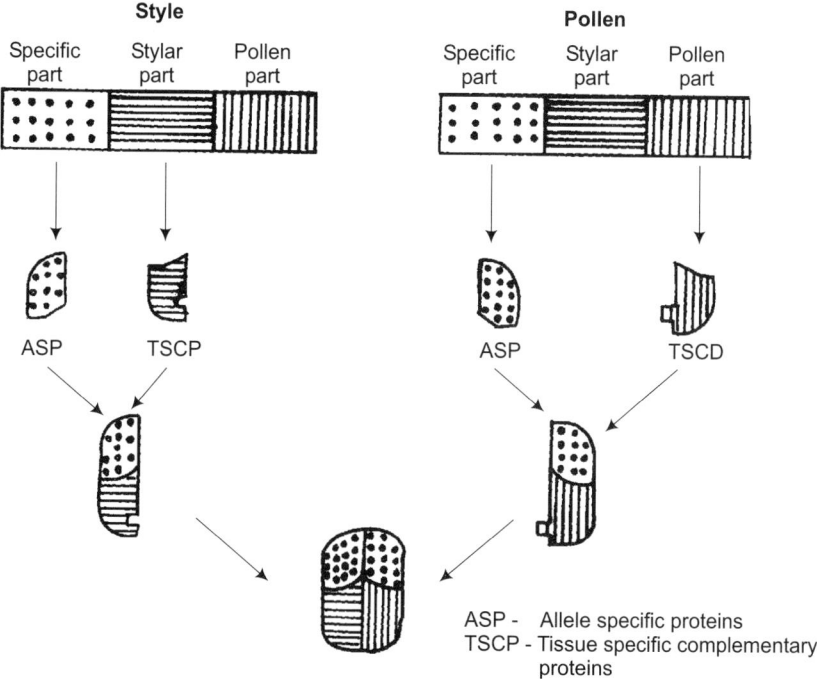

Fig. 9.10 Model depicting the nature of S-gene products based on the tripartite nature of the S-locus (specificity part common to both pollen and pistil; stylar part active in pistil and pollen part active in pollen). Identical specificity proteins, in the pollen and style, form a complex with tissue -specific complementary proteins. This model explains the formation of mutually reactive S- allele- specific proteins in pollen and style (After Pandey, 1975).

9.12 Plant Embryology: Classical and Experimental

Using an *in vitro* assay, Sharma and Shivanna (1983) have investigated the biochemical basis of self-incompatibility recognition in *Petunia hybrida*. Pollen grains were treated either with the sugar or with lectins before culturing in a medium containing pistil extract. Inhibition of self-pollen was overcome when pollen was treated with glucose or lactose but not with lectins, indicating that lectin-like components of pollen is involved in self-incompatibility recognition. Blocking these molecules with complementary sugars make them ineffective in establishing recognition and thus, overcome inhibition. Similarly, incorporation of specific lectins into medium containing pistil extract was also effective in overcoming inhibition of pollens. Apparently, recognition factor in the pistil are saccharides-containing molecules; binding of the saccharides with specific lectins make them ineffective in recognizing self- pollen (Fig. 9.11). This is the first experimental evidence to implicate the involvement of lectins and the saccharides in self-incompatibility recognition. This has paved the way for further experimentation to isolate and characterize the recognition factor and to understand the mechanism of inhibition.

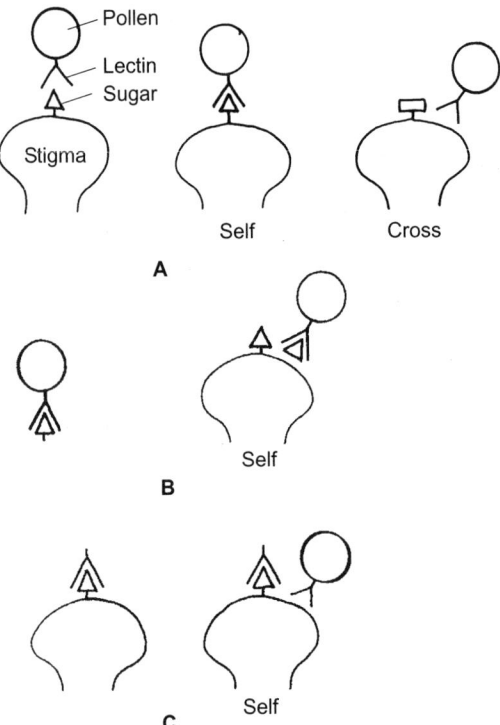

Fig. 9.11 Diagrammatic representation of the self-incompatility recognition through complementation between pollen-lectin and specific sugar moiety of the pistil. **A.** There is no complementation following cross- pollination. **B.** Pollen-lectins blocked by treating them with specific sugars before pollination. **C.** Pistil recognition molecules blocked by treating them with specific lectins. There is no self self-incompatibility recognition in **B** and **C**.

(ii) Stigma Surface Inhibition

Stigma functions as a platform for the landing of pollens. Anatomical studies of a young stigma reveal that it is composed of columnar tissues with a slight depression in the centre and has two distinct zones; an upper region with epidermis forming the secretory zone and lower region is storage zone. Epidermal cells give rise to stigmatic papillae. The main function of the stigma is to provide water and necessary exudates for germination of pollens. The chief components of the exudates secreted by solid style are lipids, amino acids and polysaccharides, whereas those with hollow style possess only polysaccharides (Ciampolini *et al.*, 1981). The stigma may further be classified as wet (*Petunia*) or dry (cotton) depending upon the presence or absence of exudates. In the former type there is intact cuticle-pellicle, whereas in the latter type the cuticle-pellicle is disrupted (also chapter-8). The pellicle has been suggested as a receptor site for the exine proteins on the stigma and the interaction between the two proteins determine the incompatible reaction. In SSI-system (except in family Poaceae and *Oenothera*) the recognition and rejection occur on the stigma.

Despite experiencing the rejection reaction and inhibition of pollen germination on the stigma surface, the incompatible pollens sometimes show some abnormal behaviour, such as:

(a) The pollen tube is very short.
(b) Development of callose plug between plasma membrane and pectocellulosic layer.
(c) Development of callose plug at the tip of pollen tube.

(iii) Stylar Inhibition

This type of inhibition is evident in GSI-plants where the recognition factor develops by the male gametophyte itself in contrast to SSI-plant where the recognition factor is contributed by the sporophytic tissues or the tapetum. In GSI plants the rejection site is the style, which suggests that the pollen factor to be recognized by the pistil is released by the pollen tube in the style.

In *Petunia hybrida*, the extracts from the stigma and 2-4 mm of the style, just below the stigma, selectively inhibit *in vitro* germination and tube growth of self-pollen. The extract from the lower part of the style, on the other hand, has no inhibitory effect. It appears that incompatibility factors in *Petunia* are present on the stigmatic surface, the transmitting tract of the stigma and a short length of the style just below the stigma. This may be true in other solid-styled system also. Variation in the zone of inhibition between individuals/species possibility depends on the intensity of secretion of the pistillate factors. It may also to some extent, be controlled by environmental factors as well as the genetic background of the plant.

In *Lilium*, with a hollow style, the stigma does not seem to play any role in pollen recognition and rejection. By a series of grafting experiments on *Prunus avium* and *Lilium henryi*, Lawson and Dickinson (1975), however, demonstrated that in *L. henryi* the incompatibility message is not perceived in the stigma but is registered only after the pollen tubes have grown through a quarter of length or grafting of incompatible stigma on compatible style was not effective in inhibiting incompatible pollen; but grafting of incompatible stigma together with upper part of the style was effective (Fig. 9.12). Similar findings have been recorded in *L. longifolium*. Additional evidence is provided from experiments involving hot water treatment of the pistil (Fett *et al.*, 1976). In *L. longifolium*

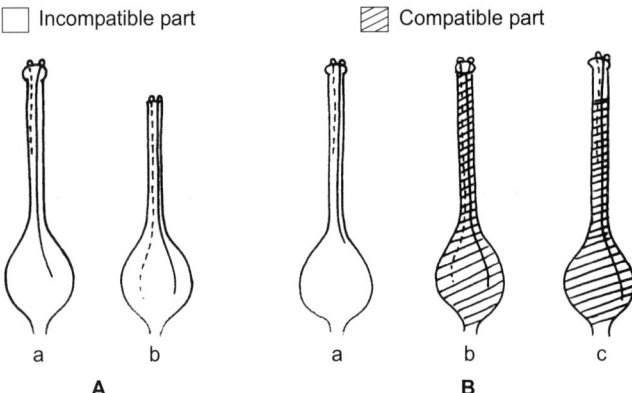

Fig. 9.12 *Prunus avium* **A.** and *Lilium henryi* **B.** Diagrammatic representation of the response of self (broken line) and cross (solid line) pollen tubes to various treatments. In both taxa incompatible pollen tubes are inhibited in the style. In *P. avium* removal of stigma was effective in overcoming pollen tube inhibition indicating that incompatible message is perceived in the stigma. In *L. henryi* grafting of incompatible stigma on compatible style was not effective in inhibiting incompatible pollen; but grafting of incompatible stigma together with upper part of the style was effective, indicating that stigma has no role in pollen recognition in this taxon. (A. modified from Raff and Knox, 1977; B. from Lawson and Dickinson, 1975)

treating the pistil with hot water at 50°C for 6 min, prior to pollination, is effective in overcoming self-incompatibility.

The incompatibility reaction occurs due to the production of polarized glycoprotein which is under the control of S allele and secreted by the stylar cells (Dickinson, 1982).

Incompatible reaction in the style is an active process as revealed by the studies of pollen tube growth in *Petunia hybrida* and *Lycopersicon peruvianum*. Following recognition, pollen tube growth in the style may be arrested in the following manners:

(a) Retardation of pollen tube growth.
(b) Inner wall of the pollen tube disappears.
(c) Disappearance of the inner wall is associated with appearance of small particles.
(d) Outer wall becomes thickened and the tip swells up.
(e) Pollen tube bursts releasing its contents in the style.

Shivanna and Johri (1989), summarized on the basis of existing data, the location of incompatibility factors in the sporophytic and gametophytic systems (Table 9.1).

METHODS OF OVERCOMING INCOMPATIBILITY

(i) Mixed and Mentor Pollens

A mixture of compatible (termed **Mentor** or **Recognition pollens**) and incompatible pollens are pollinated on the stigma in order to prepare the stigma, through the secretions of mentor pollens, to accept the incompatible pollens and enable the latter to behave like compatible

Table 9.1 Location of incompatibility factors and the zone of inhibition

	Pollen factors	Pistil factor	Zone of inhibition	Some examples
Sporophytic system	Exine (and pollen Cytoplasm?)	stigma surface	stigma surface	members of Cruciferae and Asteraceae
Gametophytic System	Intine/or Pollen cytoplasm	stigma surface and/or transmitting tract of the stigma and upper 2-4 mm of style	stigma surface transmitting tract of the stigma upper part of the style	some members of Poaceae Poaceae and Commelinaceae Oenothera Prunus, Petunia, Nicotiana
	Pollen cytoplasm	secretion of the stylar canal	stylar canal	Lilium

pollens. To prevent fertilization by the mentor pollens they are either inactivated by irradiation or killed by certain chemicals or subjecting them to repeated freezing and thawing.

Pandey (1977, 1978) was successful in overcoming both intra- and interspecific incompatibility by using mentor pollens. The mechanism in overcoming incompatibility by mentor pollen methods is summarized as follow:

(i) Mentor pollen provides recognition proteins which permit incompatible pollen to germinate.
(ii) Mentor pollen provides P-factor which interacts, with S-factor from the stigma to render it accessible to incompatible pollens.
(iii) Mentor pollens provide a pollen growth promoting substance which permits incompatible pollens to sustain tube growth.
(iv) After tube penetration by incompatible pollens, mentor pollens provide substances critical for substantial growth of ovule, ovary and other fruit tissues.

Often, application of mentor pollens before incompatible pollens, termed pioneer pollens, is more effective than the mixture of mentor pollen and incompatible pollens.

(ii) Bud-Pollination

Bud pollination is, by far, the most successful technique for overcoming self-incompatibility. It has been successful in both sporophytic and gametophytic systems, such as *Brassica, Raphanus, Nicotiana* and *Petunia*. The degree of success varies from species to species and the optimal stage of bud for overcoming self- incompatibility ranges from 2-7 days before anthesis. In *Nicotiana* and *Petunia* the application of the stigmatic exudates, collected from a mature flower, on the stigma of the bud before pollination promoted seed set after self- or cross-pollination, irrespective of the fact that the exudates was collected from the same plant or a different plant. This suggests an explanation that incompatible substance is either absent or present in insufficient amount in the bud or pollen tube grows normally before the appearance of inhibitory factor in the style either at the time of anthesis or just before it.

(iii) Stub Pollination

In *Ipomoea tricocarpa* the inhibitory factor is found in the stigma, therefore, removal of stigma and pollination directly to the cut end of style showed uninhibited growth of pollen tube into ovary (Charles *et al.*, 1974).

In *Nicotiana tabacum*, the length of the style is too long to be traveled by pollen tube i.e. pollen tube fails to reach the ovary. In order to overcome this barrier the style is shortened by trimming and its open surface is smeared with agar-sucrose medium to create artificial stigma, which provide a substrate for pollen germination and subsequently fertilization could be achieved (Swaminathan and Murty, 1957).

(iv) Intra-Ovarian Pollination

This method provides an effective alternative to overcome the barrier imposed by stigma. Kanta *et al.* (1962) for the first time established the intra-ovarian pollination. In this technique, the ovary is surface sterilized with ethanol and two punctures are made in the wall, one for introducing the pollen suspension (prepared in sterile distilled water) and the other to permit the escape of air present in the ovarian cavity. Subsequently, both the holes are plugged with petroleum jelly. Pollens germinate within the ovary and bring about the fertilization.

Viable seeds following intra-ovarian pollination have been obtained in *Papaver somniferum, P. rhoeas, Argemone mexicana, A. ochroleuca* and *Eschscholtzia californica*. Using this technique, Kanta and Maheshwari (1963) developed an interspecific hybrid between *Argemone mexicana* and *A. ochroleuca*.

(v) *In vitro* Fertilization

In addition to intra-ovarian pollination, test tube fertilization has been successfully employed to overcome self-incompatibility. In this technique, the unfertilized ovule is dusted with pollens and are directly cultured *in vitro* till seed maturity. Test tube fertilization has been first reported in *Papaver somniferum* (Kanta *et al.*, 1962), where a well-differentiated dicotyledonous embryo develops within 22 days. The technique is more effective if the ovules are cultured with the placental tissues intact. *In vitro* fertilization has been successfully employed to many taxa, such as *Argemone mexicana, Eschscholtzia californica, Nicotiana rustica* and *N. tabacum* (Kanta and Maheshwari, 1963). Utilizing this technique, Zenkteler and associates (1967, 1975, 1978, and 1980) made successful interspecific, intergeneric and interfamily crosses through placental pollination and developed viable seeds. However, in the present context, *in vitro* tertitization refers fusion of sperm with egg only (Chapter 19).

(vi) Irradiation

Induction of mutation through X-ray irradiation of flower buds at pollen mother cell stage was one of the earliest successful methods for breaking the self-incompatibility. Lewis (1949) and Crowe (1953) worked extensively on spontaneous and induced mutations at the S-locus in *Oenothera organensis* and *Prunus avium* affecting the activity of S-allele, in the pollen and style. Sexual incompatibility can be overcome by irradiating the pistil to different doses of gamma- or x-rays. In *Lilium* style high dose of irradiation ranging from 6,000 to 24,000 rads was required to overcome the barrier of incompatibility (Hoper and Peloquin, 1968). However, in *Lycopersicon peruvianum* exposure to low dose of gamma rays during entire flowering season is very effective for seed and fruit set. This has been successfully achieved in many taxa *viz.*, *Rubus, Lycopersicon, Petunia* and *Nicotiana* etc. In orchids UV-rays was effective in inducing selfed seed-set (Thimmappaiah, 1982).

Irradiation of pollen grains rather than style is more convenient method for overcoming incompatibility.

(vii) Hot Water and High Temperature Treatment

High temperature treatment has been found to be beneficial in overcoming the incompatibility as reported in plants like *Malus, Pyrus, Prunus, Oenothera* and *Lilium* etc. In *Lilium longiflorum*, self-incompatibility can be overcome by pre-pollination treatment of the style with hot water at 50°C for 6 minutes (Hopper *et al.*, 1967). According to Pandey (1973), the underlying mechanism is the inactivation of isoenzymes which are associated with the self-incompatibility.

Townsend and his associates (1971) have investigated extensively the effect of high temperature on self-incompatibility in *Trifolium*. He reported that different species showed variation in their response to high temperature, however, there are species which are not sensitive to high temperature. Genetically it has been shown that sensitivity is due to a dominant gene designate T-gene.

(viii) Electrical-aided Pollination

This is novel method devised to induce selfed seed- set in *Brassica oleracea* (Roggen *et al.*, 1972). The application of direct electric potential of 100 V between the pollen and stigma, during pollination, for 2-3 seconds was very effective in overcoming self-incompatibility.

(ix) Parasexual Hybridization

The production of naked protoplast with the application of enzymes (cellulase, pectinase, macerozyme etc.) opened a new avenue for genetic manipulation through fusion of protoplasts. This phenomenon has been called as 'Parasexual hybridization' or 'Somatic hybridization'. This technique is successfully applied to overcome the barrier of self-incompatibility. The first parasexual hybrid between *Nicotiana glauca* and *N. langsdorfii* was produced by Carlson *et al.* (1972). Melchers and Labib (1974) are credited for the production of first intraspecific hybrid of *Nicotiana tabacum*.

(x) Introduction of Autotetraploidy

When tetraploidy is induced in the diploid self-incompatible species, the two alleles present in the diploid pollens often show competitive interaction result in the weakening of self-incompatibility and, therefore, diploid pollens can grow through any tetraploid or diploid style. This mutual weakening in diploid pollen occur only in the gametophytic system such as *Petunia, Pyrus, Antirrhinum, Lolium, Gibasis* and *Arrhostoxylus* but not in sporophytic systems.

(xi) Delayed Pollination

Delay in the pollination for a few days (3-9 days) often allows incompatible pollen tube to pass through style. Delayed pollination has been successfully used to overcome self-incompatibility in *Brassica* and *Lilium*, however, in *Petunia* delaying pollination up to 7 days or pollinating fresh pistil with pollens stored up to 6 weeks, failed to overcome self-incompatibility.

(xii) Use of Specific Sugars and Lectins

Studies of Sharma and Shivanna (1983) implicating lectin—like components of pollen and specific saccharide moiety of the pistil in self-incompatibility recognition has opened a new approach to overcome self-incompatibility. Self-incompatibility can easily be masked by treating the pollens with sugars or the pistils with lectins, before pollination. Selfed seeds have been successfully obtained following such treatments in *Petunia hybrida*. This approach has considerable potential in overcoming self-incompatibility, particularly in systems in which the zone of inhibition is confined to the stigma.

(xiii) Application of Growth Substances and Other Chemicals

There are many reports of the application of growth substances such as NAA, IBA, and GA_3 to overcome self-incompatibility in *Petunia, Tagetes, Trifolium, Brassica, Lilium* and *Lycopersicon*. Generally, it is presumed that the effect of growth substances is brought about by inhibiting floral abscission and thus allowing the growing pollen tubes to reach ovary. However, critical experiments have not been conducted to understand the mode of action of growth substances in overcoming self-incompatibility. Matsubara (1973) studied the effect of different cytokinins on self-incompatibility in *Lilium* and observed that benzylaminopurine was most effective in inducing seed-set.

Many other chemicals have also been used to overcome self-incompatibility. In *Brasssica oleracea* the treatment of the stigma with hexane, before pollination, was effective (Ockendown, 1978). Hexane has been suggested to remove or inactivate self-incompatibility. In *Petunia hybrida* injection of RNA and protein synthesis inhibitors, such as olivomycin and cycloheximide, into the flower buds 2-3 days before anthesis could overcome self-incompatibility (Kovaleva *et al.*, 1978). Amongst different inhibitors tested, olivomycin gave the best response and treatment of only pollen or pistil or both was effective. There is also report that spreading of paraffin oil on the stigma of *Brassica*, before pollination, self-incompatibility can be overcome.

(xiv) Increases Atmospheric Humidity/ CO_2 Level

In *Brassica* increasing the humidity of the surrounding atmosphere, after pollination, was effective in overcoming self-incompatibility. Increase in the level of CO_2 in the surrounding atmosphere was also found effective in inducing selfed seeds in *Brassica*.

BIOLOGIOCAL SIGNIFICANCE OF INCOMPATIBILITY

Self-incompatibility is one of the most effective outbreeding mechanisms and is considered to be one of the main causes for the rapid evolution of angiosperms. A continuous self-pollination may lead to homozygous individuals which are weaker progenies, therefore, their growth is poor and survival is at risk. The nature has, therefore, evolved sexual incompatibility as an effective device to prevent self-pollination so that heterozygosity is maintained within a species.

10 The Endosperm

Endosperm is the most common nutritive tissue for the developing embryo in angiosperms. Endosperm is a post-fertilized product developing as a result of fusion of secondary nucleus and one of the male gametes. The endosperm is predominantly a triploid tissue (**3n**). However, it shows variation in its ploidy level i.e. diploid (**2n**) in the members of the family Onagraceae; pentaploid (**5n**) in *Fritillaria* and nonaploid (**9n**) in *Peperomia*, respectively. The endosperm is comparable to female gametophyte of Gymnosperm which is haploid and pre-fertilized product and supports the development of embryo.

In many plants the endosperm is consumed by the developing embryos, as seen in the members of Papilionaceae. The seeds of such plants are called non-endospermic or **exalbuminous**, where the cotyledons become much thickened due to storage of food materials and give nutrition to growing points during germination. In other category of plants, e.g. *Ricinus, Triticum, Zea* etc., the seeds are called **endospermic or albuminous**, where endosperm provides nutrition to growing points. However, there are some angiospermic families which do not form endosperm at all for e.g., Orchidaceae, Podostemaceae and Trapaceae.

Based on the mode of development, three types of endosperms have been recognized:
(1) Nuclear endosperm
(2) Cellular endosperm
(3) Helobial endosperm.

NUCLEAR ENDOSPERM

The primary endosperm nucleus divides and redivides forming many nuclei, which is usually not followed by any wall formation. The first few divisions are synchronous, therefore, the nuclei are seen in multiples of two i.e. 4, 8, 16, 32 and so on. Later, the nuclear divisions are non-synchronous i.e. nuclei may be seen in different stages of divisions and the numbers of nuclei are not in multiples of two. The free nuclei, thus formed remain suspended in the cytoplasm of the embryo sac. After sometime, the nuclei are gradually

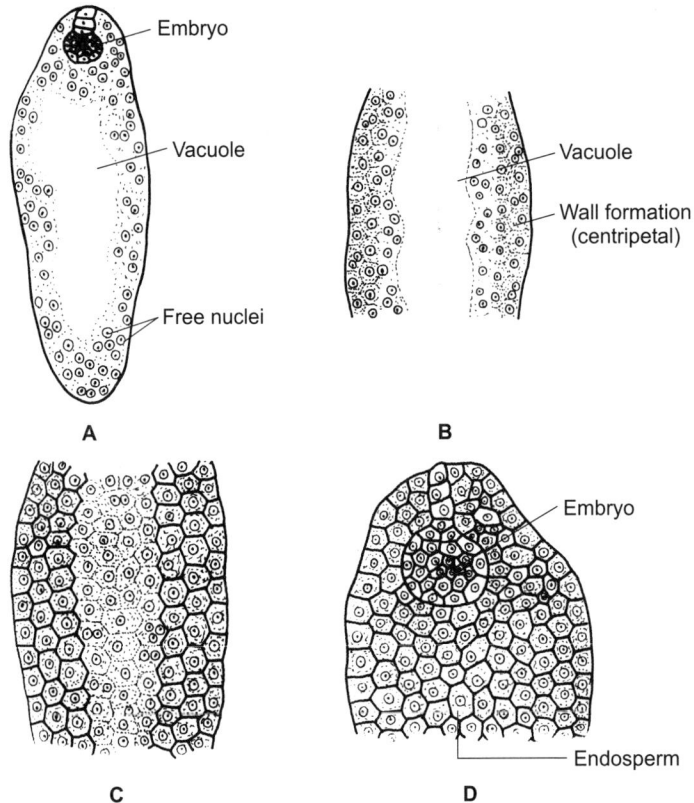

Fig. 10.1 Development of free nuclear endosperm. **A.** A central vacuole pushes large number of nuclei together with the cytoplasm towards periopher. **B.** Centripetal mode of wall formation. **C.** Development of multicellular endosperm. **D.** Mature embryo sac filled with cells and growing embryo.

pushed towards the periphery by an enlarging vacuole. A large number of nuclei accumulate towards the micropylar and chalazal end and forms a thin layer at the sides (Fig. 10.1 A).

Frequently, the endospermic nuclei towards the chalazal end have been observed to be larger than those towards the micropylar end either due to actual growth in their sizes e.g., *Colutea, Ranunculus* or due to fusion of the adjacent nuclei as in *Primula, Tilia, Malus* etc.

The number of free nuclei before wall formation varies in different species : (i) **in the first type** several hundred nuclei may be seen lining the embryo sac, (ii) **in the second type**, wall formation does not take place at all, (iii) **in the third type**, wall formation is very early, when only 8-16 nuclei are formed. The wall formation is either centripetal i.e. from periphery to the centre (Fig. 10.1 B). Or from its apex to the base. Less commonly, wall formation may take place simultaneously e.g., *Tacca*.

Irrespective of the mode of wall formation, eventually either the entire embryo sac is filled with cells (Fig. 10.1 C-D) or there are one or two layers of cells and the rest of the endosperm remains in free nuclear state or cell formation is restricted to the micropylar part of the embryo sac only. Sometimes, all three types occur in one and the same family e.g., Caryophyllaceae. The nuclear conditions vary from uninucleate to multinucleate.

An interesting condition has been reported in *Musa errans* (Juliano and Alcala, 1933), where some of the endosperm nuclei divide more actively than others forming isolated groups or **"nodules"**. They become invested with distinct cytoplasmic wall and move to the centre where they develop a separate endosperm mass (Fig. 10.2).

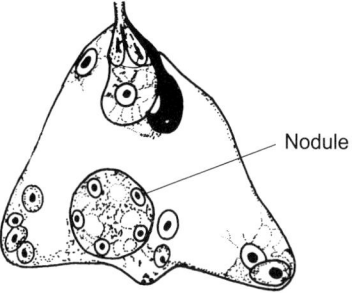

Fig. 10.2 Embryo sac of *Musa errans* showing free nuclear endosperm and a vesicle containing five endosperm nuclei.

CELLULAR ENDOSPERM

In the cellular type, the division of the primary endosperm nuclei is immediately followed by wall formation. The first wall is usually transverse but sometimes vertical or oblique and in a few cases the plane of division is not constant. On the basis of orientation of the walls following first two or three divisions, cellular endosperm has been classified mainly into two sub-types, namely (i) Vertical and (ii) Transverse.

(i) Vertical Type

Largerberg (1909) has described in *Adoxa*, where that first as well as second division of the endosperm mother cell is vertical resulting in the formation of four large cylindrical cells. The third division is transverse and results in eight cells arranged in two tiers (Fig. 10.3 A-C). The fourth division is also transverse but subsequent divisions are irregular. Fagerlind (1938) has observed that the first wall is longitudinal in *Helosis* but sometimes it may be oblique.

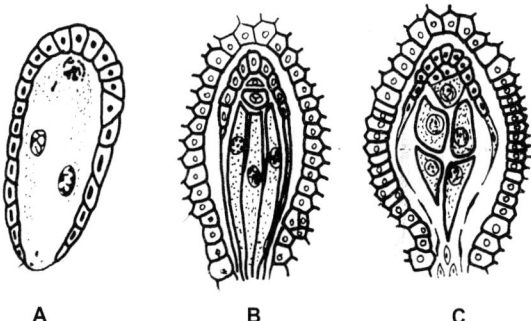

Fig. 10.3 A-C. Development of cellular endosperm by laying down of vertical wall.

(ii) Transverse Type

In Annonanceae, Aristolochiaceae, Gentianaceae, Boraginaceae etc., the first division is transverse followed by second and third division in the same plane, thus a row of four or more cells are formed (Fig. 10.4 A-F). More commonly, second division is vertical and subsequent walls are laid down in variable planes.

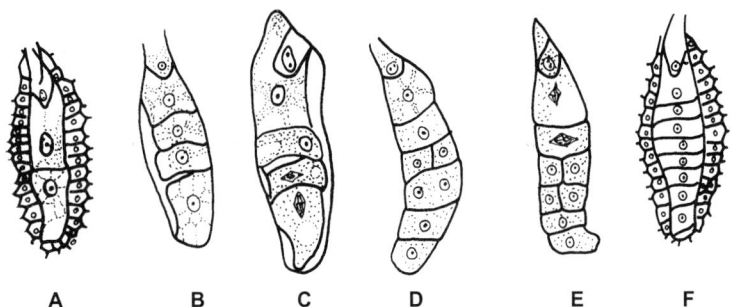

Fig. 10.4 A-F. Development of cellular endosperm by the laying down of transverse wall.

HELOBIAL ENDOSPERM

The Helobial type of endosperm is intermediate between the nuclear and cellular types. The first division of the primary endosperm nucleus is followed by the formation of transverse wall that divides the embryo sac into two chambers; the micropylar and the chalazal. The micropylar chamber is larger, where the nucleus undergoes divisions to form several free nuclei. The chalazal chamber is smaller and its nucleus may not divide or may undergo only a few divisions (Fig. 10.5 A-D). Later on, the wall formation takes place only in micropylar chamber which forms the main endosperm. The chalazal chamber eventually becomes crushed and its nuclei disintegrate. Generally, formation of the Helobial endosperm is confined to monocotyledonous families. If it occurs in dicotyledonous families, it is the only modification of the cellular or nuclear type of endosperm. However, typical Helobial type of endosperm has been observed in dicotyledonous families, such as Santalaceae and Saxifragraceae.

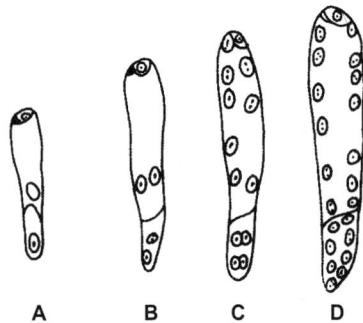

Fig. 10.5 A-D. Development of Helobial type of endosperm.

ENDOSPERM HAUSTORIUM

All the three types of endosperms described above may develop special structure called haustoria, which elongate or branched considerably and invade the tissues in the seed and placentae to absorb nutrients for proper growth of endosperm. There are variations in origin and shapes of haustoria. A few important types found in different types of endosperms are described here:

(i) Free Nuclear Type

Endosperm with chalazal haustorium: Kausik (1938, 1941) has reported some of the haustorial structures met within some members of the family Proteaceae. Here most of endosperm nuclei are distributed in the upper portion of the embryo sac. Cell formation is restricted to this region, while the lower portion of the sac remains free nuclear. In *Macadamia ternifolia* this part forms several prominent lobes or **diverticulae** which invades the nutritive tissue at the chalazal end (Fig. 10.6A). In another member of the Proteaceae, *Gravillea robusta* the lower coenocytic part of the endosperm grows in the form of a coiled and tubular worm-like structure, called the **"Vermiform appendage"** (Fig. 10.6B). The occurrence of chalazal haustoria is also reported in several other members. The longest haustorium is found in *Echynocystis lobata* of the family Cucurbitaceae.

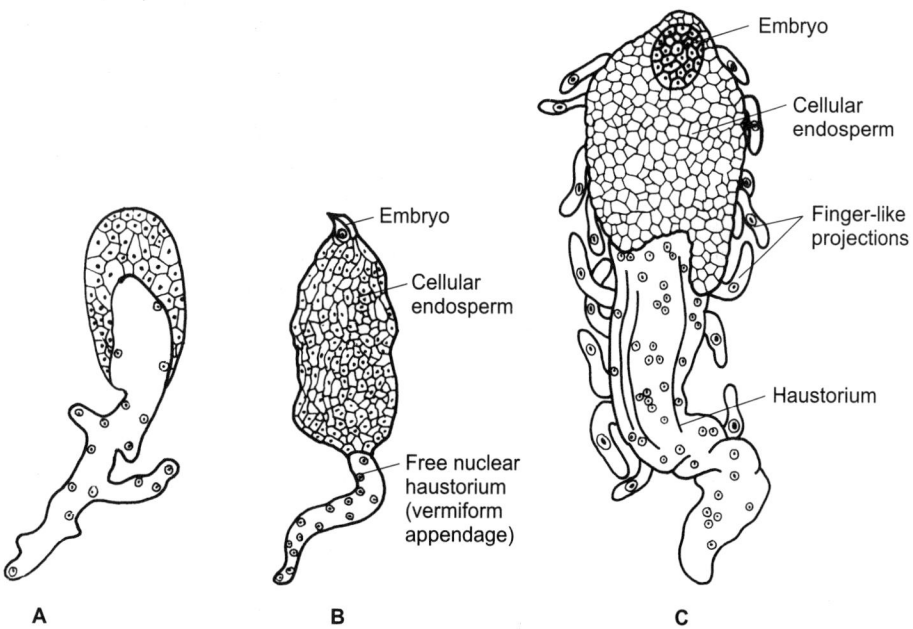

Fig. 10.6 Development of haustoria in free nuclear endosperm. **A.** *Macadamia ternifolia*, with chalazal haustorium showing lobes or diverticulae. **B.** *Gravillea robusta*. Endosperm with chalazal haustorium. **C.** *Lomatia polymorpha*. Endosperm with chalazal hautorium showing many finger-like projections.

In *Lomantia*, besides the main chalazal haustorium, numerous single-celled, finger-like projections are present all over the cellular endosperm. These increase the absorbing surface of the endosperm (Fig. 10.6 C).

(ii) Cellular Type

The occurrence of haustoria is a common feature in this type of endosperm also. The haustoria may be micropylar or chalazal. Occasionally, both types of haustoria are present on the same plant.

(a) **Endosperm with micropylar haustorium:** A very prominent and aggressive micropylar haustorium is seen in *Impatiens* (Fig. 10.7A). Here, the division of the primary endosperm nucleus is followed by transverse wall to form an upper smaller micropylar chamber and lower larger chalazal chamber. The micropylar chamber develops into an extensive, much branched haustorium. Its branches extend deep into the funiculus to derive nutition. In *Thunbergia alata* (Fig. 10.7B) the micropylar haustorium show profuse branching and secondary haustoria are also seen. (Wadhi, 1962).

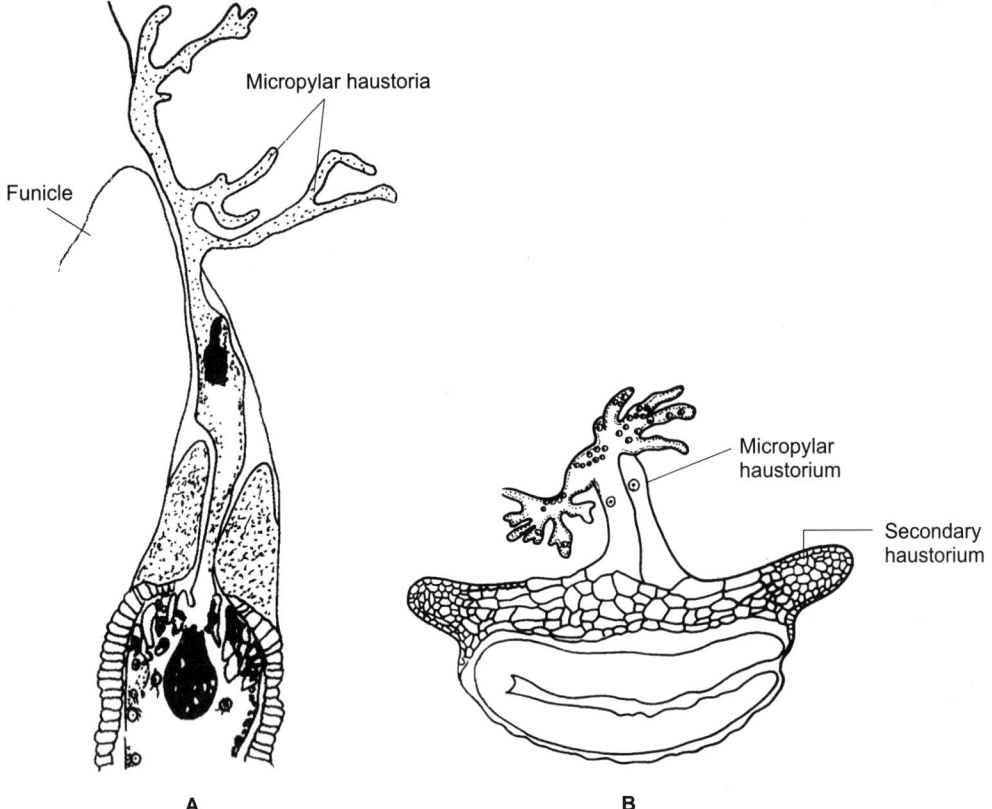

Fig. 10.7 Development of haustoria in cellular endosperm. **A.** *Impatiens roylei*. Branched micropylar haustorium penetrating into funiculus. **B.** *Thunbergia alata*. Micropylar and secondary haustoria.

(b) **Endosperm with chalazal haustorium:** In *Exocarpus sparteus* (Ram, 1959), the chalazal haustorium is prominent. Here, the division of the endosperm nucleus is followed by transverse wall. Of the two cells the lower divides vertically, while the upper divides by a transverse partition. The two cells of the lower tier do not divide any further but elongate considerably and eventually reach as far as the base of the ovary. They develop numerous tubular processes at their ends which serve as efficient haustoria (Fig. 10.8A).

A highly aggressive chalazal haustorium is found in *Iodina rhombifolia* (Bhatnagar and Sabharwal, 1969). The profuse branching at the free end of **caecum** gives the haustoium a corolloid appearance (Fig. 10.8B).

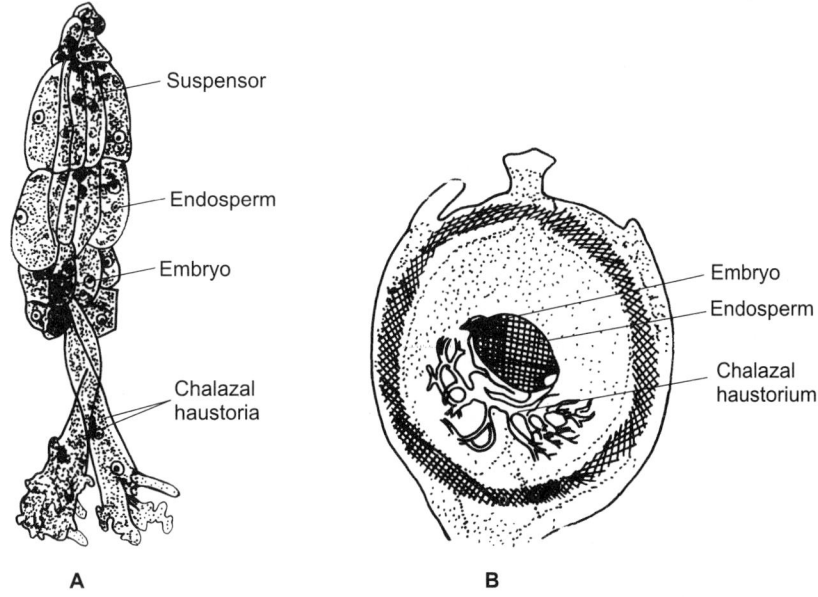

Fig. 10.8 A. *Exocarpus sprarteus*. Two large chalazal haustoria having finger-like projections at their tips. B. *Iodina rhombifolia*. Highly branched chalazal haustorium.

(c) **Endosperm with micropylar and chalazal haustoria:** Some plants develop haustoria from both the micropylar and the chalazal ends of endosperm, such as *Nemophila* and *Lobelia amoena* (Fig. 10.9A). The chalazal haustotium sometimes gives out a prominent lateral branch. In *Blumenbachia hieronymi* (Garcia, 1962), a member of the family Losaceae, the embryo sac has two haustorial processes, one at each pole (Fig. 10.9B). In another member, namely *Barleria cristata* (Mohan Ram, 1962) of the family Acanthaceae, endosperm shows both micropylar and chalazal haustoria (Fig. 10.9C).

(d) **Endosperm with secondary haustoria:** In *Centranthera hispida* of the Cucurbitaceae family, the micropylar and the chalazal haustoria are not very active rather secondary haustoria are formed close to micropylar haustorium (Fig. 10.10). Similarly, secondary haustoria are seen in *Veronica*, but here they arise from the chazalal region of the haustorium.

10.8 Plant Embryology: Classical and Experimental

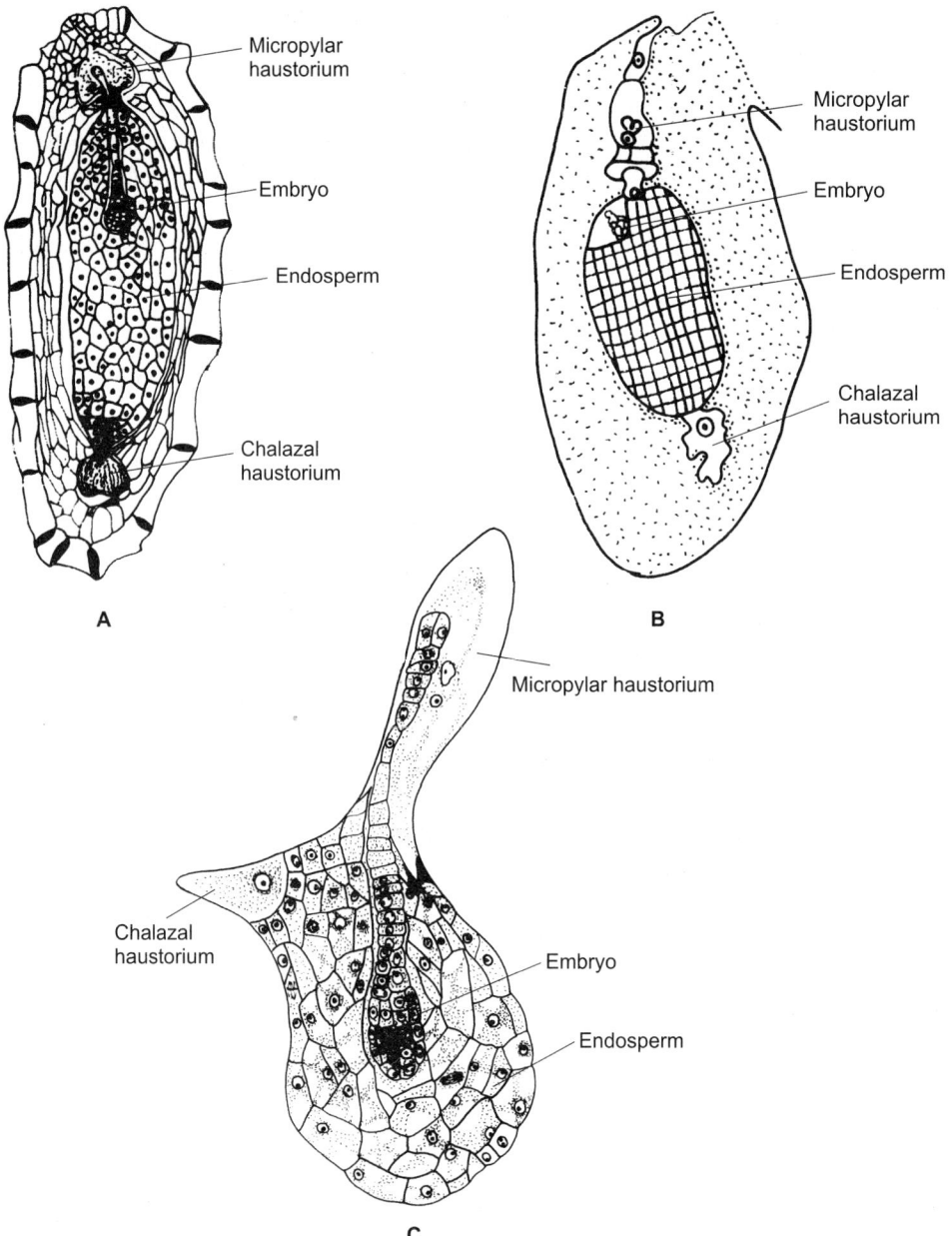

Fig. 10.9 **A.** *Lobelia amoena*. Micropylar and chalazal haustoria. **B.** *Bluemenbachia hieronymi*. L.S. of ovule showing micropylar and chalazal haustoria. **C.** *Barleria cristata*. Endosperm with micropylar and chalazal haustoria.

(e) **Endosperm with micropylar, chalazal and secondary haustoria:** In a few plants well developed and aggressive haustoria of all three kinds—micropylar, chalazal and secondary occur together and form a very efficient absorptive system. An interesting

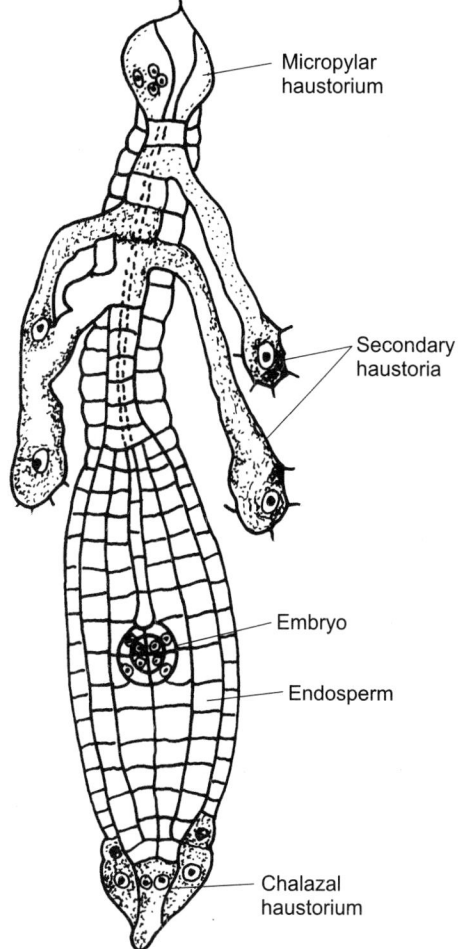

Fig. 10.10 *Centranthera hispida*. Development of secondary haustoria from micropylar chamber.

case of this kind has been described by Rosen (1940) in *Globularia vulgaris*. The micropylar haustorium consists of two to four separate cells but they soon unite to form a single entity. This composite structure often grows out of the micropyle and its hyphae-like ramification extends along the outer surface of the furniculus and placentae, later penetrating even the wall of the ovary. The chalazal haustorium also branches profusely and sucks the content of the integumentary cells, destroying their walls and causing the formation of large lacunae. In addition, the endosperm cells lying nearest to the chalazal haustorium also increase in size and send out secondary haustoria, so that the entire tissues of the chalazal and the integument becomes riddled by a number of haustorial processes.

(iii) Helobial Type

Endosperm with lateral haustoria: In *Monochoria karasakowii*, where the endosperm is of helobial type, the haustorium is neither micropylar nor chalazal but laterals (Fig. 10.11). The chalazal chamber ceases to grow further but the micropylar chamber shows the active nuclear divisions and develops into two lateral outgrowths, one on either sides of the chalazal chamber. These structures grow downwards and function as active haustoria invading the tissues of the chalaza. Subsequently, the main body of the endosperm enlarges considerably and fuses with the haustoria to form a compact mass of endosperm.

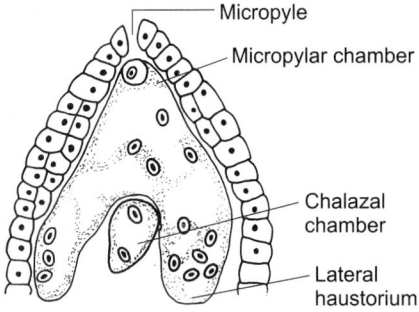

Fig. 10.11 *Monochoria korasakowii*. Development of lateral haustoria from micropylar chamber.

VARIANTS OF ENDOSPERM

Apart from the normal three types, variations may be found. A few interesting types are as follow:

(i) Composite Endosperm

In the family Loranthaceae, the sporogenous tissues located at the base of ovary develop several embryo sacs, which elongate considerably; some of them even enter the style. After fertilization, the primary endosperm nucleus of each embryo sac moves to the basal part, where it divides to form cellular endosperm. During further development, the endosperm of all the embryo sac in an ovary enlarges and become fused to produce composite endosperm mass (Figs. 10.12; 12.8). Several proembryos belonging to individual embryo sacs with long suspensor develop but ultimately only one survives and attains maturity.

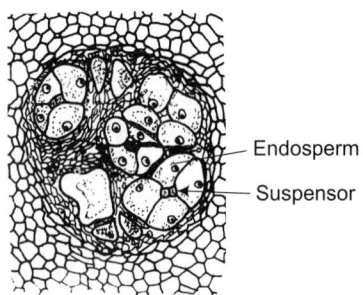

Fig. 10.12 *Tolypanthus involucratus*. Composite endosperm showing embryos.

(ii) Ruminate Endosperm

In certain plants the surface of the mature cellular endosperm shows a high degree of irregularity and unevenness assuming ruminate appearance. In fact, it is not a new type rather it belongs to any three normal types of endosperm. It is caused either by the activity of the seed coat or by the endosperm itself. Ruminate endosperm has been reported in 32 families of angiosperms. On the basis of morphology, Periasamy (1962) has distinguished seven types: *Passiflora, Annona* (Fig. 10.13A), *Myristica, Spigelia* (Fig. 10.13B), *Verbascum, Coccoloba* and *Elytraria*. In all of these types, except *Elytraria*, irregularity occurs in the growth of integument, which brings about rumination of endosperm. In *Elytraria*, the endosperm exhibits unequal peripheral activities during late stage of its development, and shows rumination.

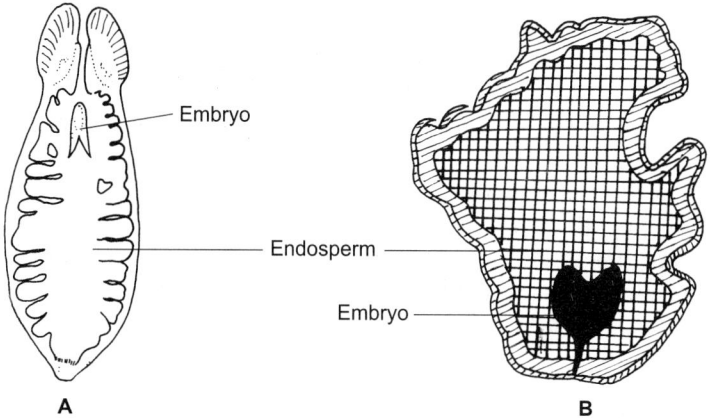

Fig. 10.13 Ruminate endospem. **A.** *Annona squamosa*. **B.** *Spigelia*.

(iii) Mosaic Endosperm

In some plants patches of two different colours appear in the tissues of endosperm, providing mosaic design. In maize, red and white patches of tissues are seen. The occurrence of such endosperm has also been reported in other plants, such as *Petunia, Lycopersicon* and *Acorus* etc. Many theories have been proposed to explain the mosaic endosperm but none of these has been satisfactory. The most convincing explanation is said to be either due to aberrant behaviour of the chromosome during mitosis or somatic mutation.

(iv) Xenia

The term xenia was described by Focke (1881) to denote the immediate or direct effect of pollen on the character of endosperm. It can be shown in *Zea mays*, where there are two sacs: one has white endosperm and the other has yellow endosperm. If pollen from the yellow endosperm sac is placed on the stigma of white endosperm sacs, one gets seeds with yellow endosperm, which grows into normal plants. Maize is the best example of xenia, where two races are found: one having yellow endosperm and the other with white endosperm. The genetics of xenia can be explained on the basis of single genes character.

Plant with yellow endosperm can be represented with dominant alleles (YY) and the plant with white endosperm is represented with recessive alleles (yy). If white variety is crossed with pollen from a yellow variety, one of the male gametes (Y) unites with the egg (y) and produces a hybrid embryo (Yy) which will behave as a heterozygote for yellow endosperm in the next generation. The second male gamete (Y) unites with the two polar nuclei (y, y) and produces the primary endosperm nucleus (Yyy). Since the latter has a factor for yellowness, the endosperm will naturally show the colour, although the ovule belongs to the white parents (Fig. 10.14). In the reciprocal cross, i.e., when the pollen from the white-grained variety (yy) is pollinated on a yellow-grained variety (YY), the grains are not white like those of the pollen parent, but yellow like those of the ovule parent. Here xenia would seem to be absent, this is merely due to the fact that yellowness is dominant over whiteness and therefore the male with a factor for whiteness has no effect on the colour (Fig. 10.14B).

(v) Metaxenia

The term metaxenia refers to the effect of the pollen grains on the tissues lying outside the embryo i.e. on the testa and fruit walls. The most important work in this connection is that of Swingle (1928), who found that in date palm *(Phoenix dactylifera)* the time of maturity of the fruit wall as well as their size can be made to vary according to the type of pollen used in fertilization. Regarding the nature of mechanism which enables this to take place, he suggests that possibly the embryo or the endosperm or both secrete hormones, or substances analogous to them which diffuse out into the wall of the seed and fruit and exert a specific effect on them, which in turn is influenced by pollen grains that fertilize the ovules.

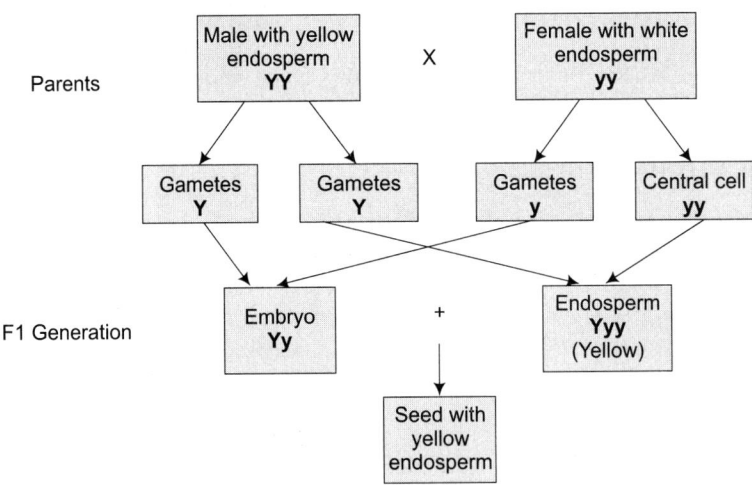

Fig. 10.14 Effect of yellow pollens on the colour of endosperm (xenia).

STRUCTURE OF ENDOSPERM

The cells of the endosperm are usually thin walled, isodiametric, and devoid of pits and store large amount of food materials. The starch along with other substances like oil and proteins are also accumulated in these cells. On account of heavy deposition, the nucleus becomes disorganized and deformed.

Usually, the endosperm is non-chlorophyllous. However, in some members of the family Amaryllidaceae, such as in *Crinum* seed coat as well as fruit wall is absorbed during seed development consequently endosperm is exposed to sunlight and becomes suberized and protective in nature.

In Poaceae, the outermost layer of the endosperm is highly specialized and constitutes **aleurone layer** (Fig. 10.15A). In barely, the aleurone layer is 3-4 layered thick, while in maize it single layered thick. The aleurone cells are characterized by the presence of thick walls and non-vacuolated cytoplasm. They are inter-connected by plasmodesmata. The most conspicuous structures of these cells are aleurone grains (10.15B).

Aleurone grains are surrounded by single unit membrane associated with spherosomes (Fig. 10.15C). The main components of the aleurone grains are proteins, phytin,

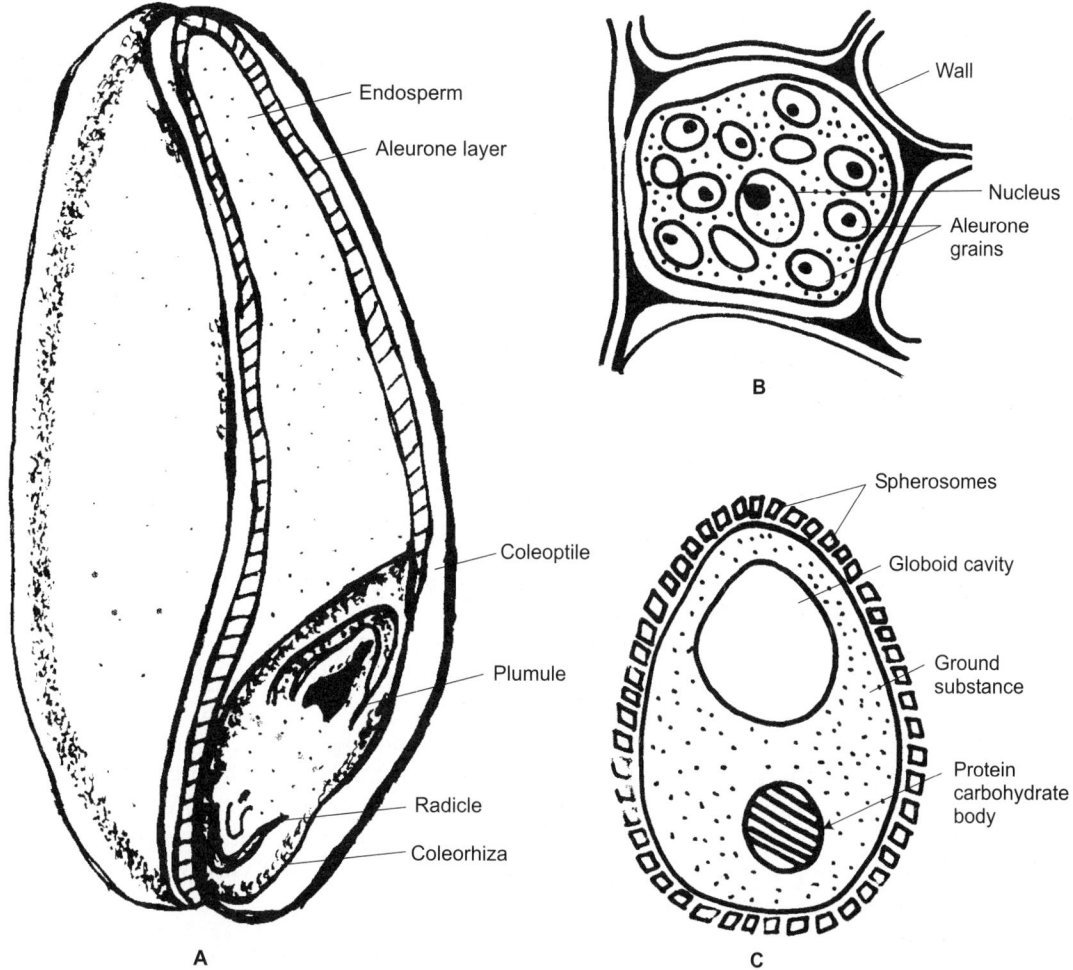

Fig. 10.15 A.L.S. of barley seed showing aleurone layer. **B.** Aleurone cells of barley, **C.** Single aleurone grain enlarged.

phospholipids and some carbohydrates. Structurally, the aleurone grains possess two types of inclusions besides the ground substances, namely: (i) Globoids—present within the globoidal cavity which contain phytin and lipids and (ii) Protein-carbohydrate body. During germination, the cells of the aleurone layer secrete certain hydrolytic enzymes like amylase and protease, which change the stored food materials of the endosperm so as to make them available to the germinating embryo. A phytohormone, gibberellins have been shown to induce *de novo* synthesis of these enzymes.

In most of the cases with the growth of the embryo sac, the nucellus gets used up and it is the endosperm which provides nutrition to the germinating seeds. However, in Piperaceae and Scitamineae, the endosperm is not the storage region rather it is the perisperm which is a persistent thin layer of nucellus and provides nutrition in germinating seeds.

CYTOLOGY OF ENDOSPERM

In majority of plants, the endosperm is initially triploid because it is derived from the fusion products of three haploid nuclei: one from the male gametophyte and two from the female gametophyte. In fact, the number of nuclei contributed by the male gamete is constant throughout the angiosperms but the number of nuclei contributed by the female gametophyte varies with the type of embryo sac. In *Oenothera*, it is just one and the endosperm is 2n, whereas in *Peperomia* it is eight and the endosperm is 9n.

The endosperm tissue is well known for polyploidization of its cells during development. According to Erbrich (1965), ploidy of the endosperm in *Thesium alpinum* is up to 385n. The highest ploidy, however, is reported in *Arum masculatum*, which is 2457n. According to Kapoor (1962), endomitosis and nuclear fusion are some of the methods of polyploidization of endosperm cells. He further reported the occurrence of various mitotic irregularities such as chromosome bridges, lagging chromosomes, spontaneous breakage of chromosomes and fragmentation of nuclei which are quite common in endosperm tissues. The size of nuclei and the number of nucleoli per nucleus is also variable.

MORPHOLOGICAL NATURE OF ENDOSPERM

In Gymnosperms, the endosperm develops as a result of free nuclear divisions of the megaspore, thus it is distinctly haploid and a pre-fertilized product. The nature of endosperm in angiosperms had been a matter of great discussion since the middle of nineteenth century and has been interpreted variously by the morphologists from time to time. Commonly, in angiosperms it develops from primary endosperm nucleus, which is a product of triple fusion, i.e. two polar nuclei usually and one male gamete. However, in angiosperms there are number of examples, where more than two polar nuclei are taking part in second fertilization. The fact, that the endosperm exhibits different types of development during its early stages and shows great diversity in its nature as well as in the number of nuclei taking part has brought about much discussion and varied interpretations. Some of the important interpretations are given on next page:

(i) Hofmeister's View (1858, 1859 and 1861)

He was of the view that the endosperm in angiosperms represents the vegetative tissues of the female gametophyte, just like that of gymnosperms and is haploid in nature. The only difference is that in angiosperm the growth of the endosperm remains arrested till the penetration of the pollen tube into the ovule. The syngamy and the triple fusion were not discovered till that time.

(ii) Le Monnier's View (1887)

After the discovery of syngamy by Strasburger (1884), Le Monnier stated that the fusion of two polar nuclei, generally in the central region of the embryo sac, is to be considered as a sexual process. Therefore, endosperm is a second embryo modified to serve as the nutritive tissue. He supported his view with the fact that the male nucleus in many cases enters into the organization of the primary endopserm nucleus. Thus the endosperm is the second embryo, i.e. diploid and a sporophyte.

(iii) Sargant's View (1900)

Sargant strongly supported Nawaschin in considering triple fusion as true fertilization. She is of the opinion that out of the three nuclei taking part in the triple fusion, one polar nucleus coming from the micropylar end is a true female unit akin to the egg nucleus. The male fusing with the secondary nucleus is identical to the male nucleus fusing with the egg. But the polar nucleus, coming from the chalzal end with its vegetative character and usually with rudiment and increased number of chromosomes is a disturbing factor. Thus, resultant primary endosperm nucleus, instead of developing a normal embryo, gives rise to living nutritive tissues termed as endosperm. The true embryo thus survives without any sort of interference. Sargant considered the endosperm as a degenerative embryo. She believes triple fusion is a sexual union, whose normal result has been interfered with the presence of a non-sexual nucleus (polar nucleus of the chalazal end).

(iv) Strasburger's View (1900)

He did not agree with the view of Sargant. Strasburger concludes that the fusion of the secondary nucleus (two fused polar nuclei) with the male nucleus is not a real fertilization. According to him, the true fertilization is a definite union of the parental qualities, resulting in an embryo. The triple fusion is a vegetative fertilization which results as a growth stimulus. He considers endosperm as a reduced female gametophyte with little or no reserve food materials. The endosperm is thus a belated gametophyte tissue.

(v) Nemec's View (1910)

He was of the opinion that fertilization of the secondary nucleus brings about a stimulus for the endosperm development and formation of nutritive tissues. This tissue is physiologically harmonious with the embryo. The hybridity of the endosperm, according to the views of

Thomas and Nemec, is a systematic procedure of making its composition fit for the use of the developing plants. If there were no such adjustment of the composition of endosperm, the hybrid embryo would be left with no other alternative but to rely upon food materials obtainable to it from the parent plant.

The most acceptable recent view regarding the morphological nature of the endosperm is that, it is neither haploid (gametophyte) nor diploid (sporophyte) but is altogether a new structure of complex morphological nature, characteristic to angiosperms. It should be considered as an undifferentiated tissue of polyploidy in nature, the composition of which has been adjusted according to the nutritive need of the developing embryo.

11 The Embryo

The zygote represents the first cell generation of the sporophyte, which is a fusion product of male and female gametes and undergoes a predetermined division to form the embryo. The embryo is a polarized structure, which could develop into a complete plant.

The zygote undergoes a period of rest after syngamy during which the vacuoles disappear and zygote starts shrinking. In cotton, the zygote is reduced to half the original size of the egg within 24 hours after pollination. The resting period of the zygote varies with different species and is to some extent depend upon the environmental conditions. In general, the primary endosperm nucleus divides first and the zygotic nucleus divides afterwards. In *Theobroma cacao*, the zygote divides 14-15 days after fertilization. In *Colchicum autumnale*, the zygote divides 4-5 months after winter. However, there are examples, which show short resting period, as in *Crepis capillaries*, the first division of zygote takes place 5-10 hours after fertilization, while in *Oryza sativa* zygote divides after 6 hours.

The zygote, which is ready to divide, shows high degree of polarization. The micropylar part is vacuolated and the chalazal part shows prominent nucleus. A marked increase occurs in the density of cytoplasm organelles, such as mitochondria, dictyosomes and plastids. Endoplasmic reticulum becomes more extensive and the density of ribosomes increases indicating intense metabolic activities.

A typical dicotyledonous embryo has two cotyledons attached laterally to an embryonal axis. The portion of the embryonal axis above the level of cotyledons is called **epicotyl** and the axis below is called **hypocotyl**. The epicotyl end in plumule and the hypocotyl ends in radicle which subsequently grow into shoot and root systems, respectively. The monocotyledonous embryo has only one cotyledon and the embryonal axis shows plumule surrounded by coleoptile, while the radicle is surrounded by coleorhiza.

EMBRYOGENY

In the majority of angiosperms, the zygote divides by transverse wall forming a small apical cell or terminal cell (conventionally designated as *ca*) towards the interior of the embryo sac and facing the chalazal end and a basal cell (conventionally *cb*) towards the micropyle.

11.2 Plant Embryology: Classical and Experimental

Rarely, the division of the zygote may be vertical or oblique (Loranthaceae, Piperaceae). From the 2-celled stage of the zygote till the differentiation of cotyledons, the sporophyte is called **proembryo**.

(i) Embryogeny in Dicotyledons

Soueges, the famous French embryologist is of the view that the mode of origin of four-celled proembryo and the contribution made by each of these cells to the adult embryo serve as most important means in classifying the embryonal types. Later Schnarf (1929), Johansen (1945) and Maheshwari (1950) have recognized five chief types of embryo among the dicotyledons, which can be easily distinguished from one another as shown below (Fig. 11.1):

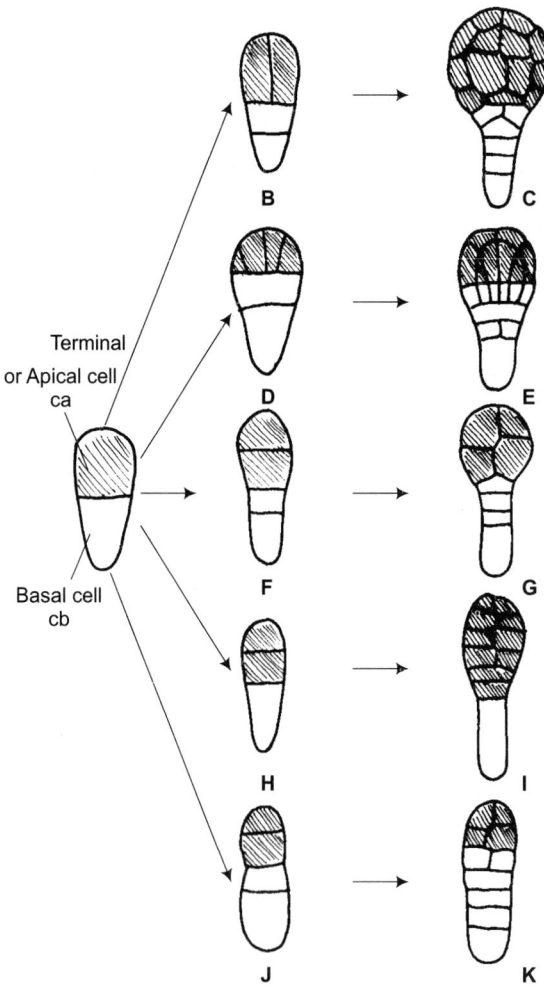

Fig. 11.1 Diagrammatic representation of the mode of embryo development in dicotyledons. **A.** Zygote showing transverse division **B-C.** Crucifer or Onagrad type. **D-E.** Asterad type. **F-G.** Solanad type. **H-I.** Caryophyllad type. **J-K.** Chenopodiad type.

I. **The terminal cell of two-celled proembryo divides by longitudinal wall:**
  1. The basal cell plays only a minor role or none in the subsequent developments of the embryo .... *Crucifer type*
  2. The basal cell and the apical cell, both contribute in the development of embryo ....*Asterad type*.

II. **The terminal cell of the two-celled proembryo divides by transverse wall:**
  (a) The basal cell plays only a minor role or none in the subsequent development of the embryo:
      (i) The basal cell usually forms suspensor of two or more cells .. *Solanad type*.
      (ii) The basal cell undergoes no further divisions and the suspensor, if present, is always derived from the terminal cell .... *Caryophyllad type*.
  (b) (i) The basal cell and the terminal cell, both contribute to the development of the embryo .... *Chenopodiad type*.

Abbreviation: Some of the abbreviations have been frequently used while describing the parts of proembryo and embryo proper which are found of general acceptance in embryological studies, listed as under:

  ca, terminal cell of the proembryo
  cb, basal cell of the proembryo
  cc, upper daughter cell of *ca*
  cd, lower daughter cell of *ca*
  ce, upper daughter cell of *cc*
  ci, upper daughter cell of *cb*
  cm, lower daughter cell of *cb*
  e, epiphysis
  f, lower daughter cell of *m*
  h, hypophysis
  iec, initials of root cortex
  l, lower daughter cell of quadrant of terminal cell
  l', upper daughter cell of quadrant of terminal cell
  n, upper daughter cell *ci*
  n', lower daughter cell *ci*
  o, lower daughter cell of *n'*
  p, upper daughter cell of *n'*
  pco, cotyledonary region
  pe, periblem
  ph, hypocotyl
  pl, plerome
  q, quadrant
  s, suspensor

(a) **Crucifer type or Onagrad type:** This type of embryogeny was first described by Hanstein (1870) in *Capsella bursa-pastoris* (Fig. 11.2 A-Q) of Brassicaceae family.

11.4 Plant Embryology: Classical and Experimental

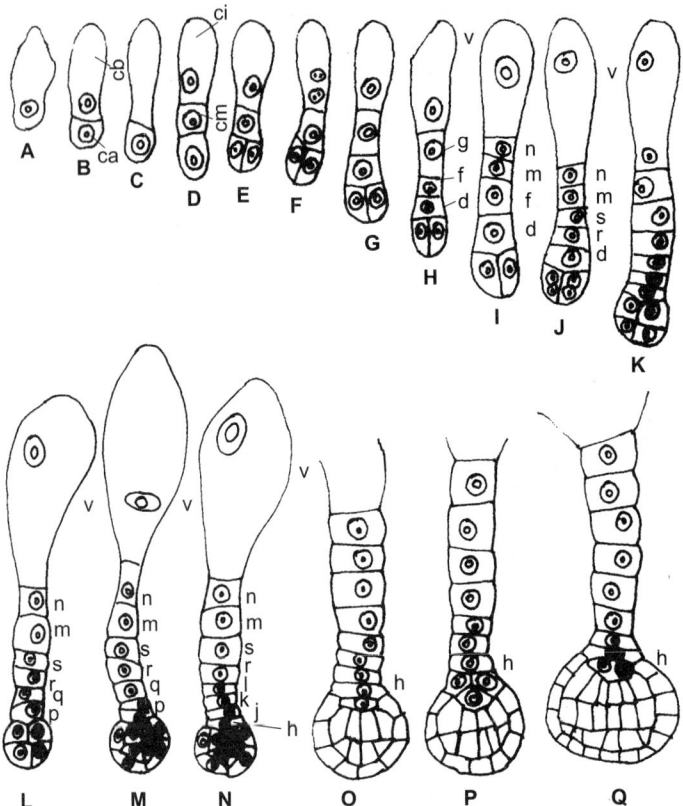

Fig. 11.2 Successive stages of the development of proembryo in *Capsella bursa-pastoris*.

The first division of the zygote is transverse resulting in basal cell *cb* and a terminal cell *ca*. The former divides transversely and the latter divides longitudinally resulting 4 cells ⊥-shaped proembryo. The two terminal cells then divides by a vertical wall lying at right angle to the first, so as to result in a quadrant stage. A transverse wall in a quadrant stage results an octant. Of this the lower four are destined to give rise to the stem and cotyledons and the upper four to the hypocotyls. A periclinal wall is laid down at this stage to form a peripheral dermatogen, while the inner one undergoes further divisions to form periblem and plerome.

In the meantime, the two upper cells *ci* and *cm* of the 4-celled proembryo divide to form a row of 6-10 suspensor cells, of which the uppermost cell *v* becomes swollen and vesicular, probably functions as haustorium. The lowermost cell *h* functions as the hypophysis. Although, initially it is similar to other cells of the suspensor, but subsequently it becomes somewhat rounded at the lower end and divides transversely to form two daughter cells, each of which undergoes two vertical divisions at right angle to each other, resulting 8 cells; the lower 4 form the initials of the root cortex and the upper 4 give rise to the root cap and the root epidermis.

At the same time, further divisions take place in the embryo proper, especially at two points in the lower tier, which are destined to form cotyledons. At this stage the embryo appears more or less cordate-shaped in longitudinal sections. The hypocotyl as well as the cotyledons soon elongate in size, mostly by transverse divisions of their cells.

During further development, the ovule becomes curved, like a horse-shoe and the growing cotyledons also conform to this shape for spacial reasons (Fig. 11.3 A-D).

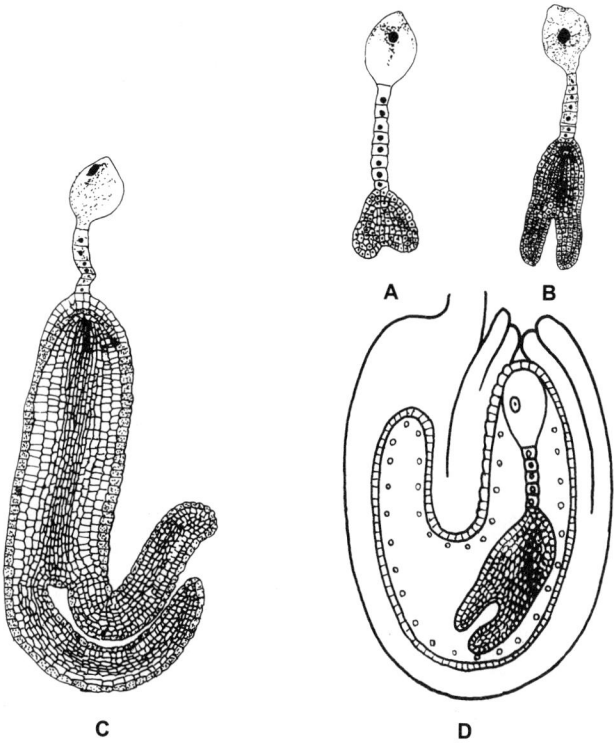

Fig. 11.3 Advanced stages in the development of proembryo in *C. bursa-pastoris*.

(b) **Asterad Type:** Caramo (1915) and Souèges (1920) have studied this type of embryo in various members of the family Asteraceae. However, *Lactuca sativa* may be used as an illustration of the Asterad type (Fig. 11.4 A-G).

The 4-celled proembryo consists of two juxtaposed cells derived from the terminal cell *ca* and two superposed cells *ci* and **m** derived from the basal cell *cb*. In the following stage, the terminal tier divides by vertical walls at right angle to the first one, thus forming quadrant *q*, the middle tier shows two juxtaposed cells by vertical division and *ci* divides transversely to form two daughter cells *n* and *n'*. The four cells of the quadrant divide to

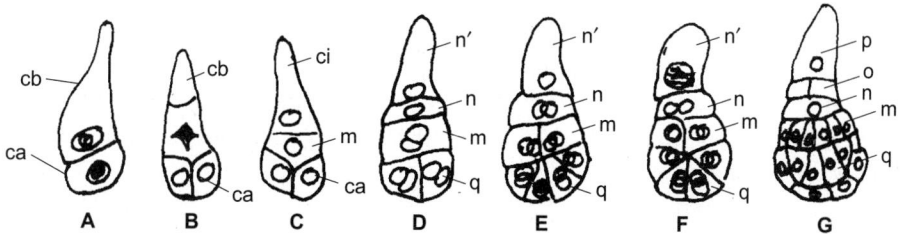

Fig. 11.4. Successive stages of the development of embryo in *Lactuca sativa*.

form the octant stage, the walls are laid down diagonally, the two cells of the tier *m* undergo a vertical division to give rise 4 cells, lying directly above the octant; *n* also divides by vertical wall and *n'* divides by transverse wall to form *o* and *p*. At the same time tangential walls are laid down in the tiers *q* and *m* to cut off an outer layer of dermatogen cells from the inner cells, which undergo further divisions to give rise periblem and plerome. Regarding further development, the cell *p* gives rise to a suspensor consisting variable number of cells; *o* to the root cap and dermatogen of the root; *n* to the remaining part of the root cap; *m* to the hypocotyledonary region; and *q* to the cotyledons and stem tip.

(c) **Solanad Type:** Souges (1920, 1922) and Bhaduri (1936) have studied a number of species of the Solanaceae, of which *Nicotiana* (Fig. 11.5 A-H) may be described as an example.

Here, both the terminal cell and basal cell of the two-celled proembryo divide by transverse wall only to form 4-celled stage. The 4 tiers may be designated from below upwards as *l, l', m,* and *ci*. Now the *l* and *l'* divide by vertical walls at right angle to each other to give rise to an octant, while *m* and *ci* divide transversely to produce *d, f, n* and *n'*. By subsequent divisions, the tier *l* gives rise to cotyledonary portion, *l'* to the hypocotyl and to the periblem and plerome of the root and *d* to the root tip. The remaining cells *f, n, n'* produces the suspensor.

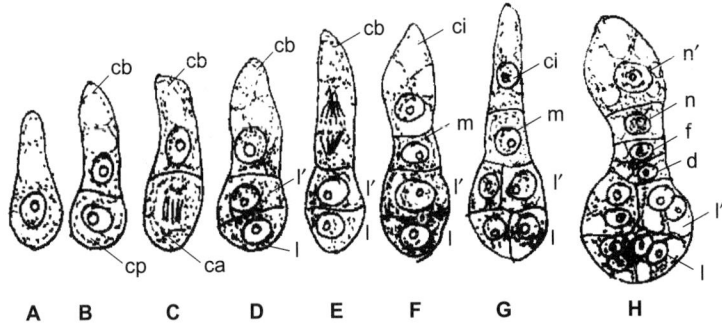

Fig. 11.5 Successive stages of the development of embryo in *Nicotiana*.

(d) **Caryophyllad Type:** This type has been described in *Sagina procumbens* (Fig. 11.6 A-H) by Soueges (1924), which is the most typical type.

The basal cell *cb* remains undivided and form a large vesicular structure, which does not take any further part in the development of the embryo. The terminal cell *ca* undergoes transverse divisions to form a row of four cells designated as *l, l', m* and *ci*. Of these, each of the three lower cells divide by a vertical wall and the upper cell *ci* divides by transverse wall. The embryo now comprises five tiers (excluding *cb*), namely *l, l, m, n* and *n'*. The next division is also vertical (at right angle to the first) in *l, l', m* and results in the formation three quadrants; *n'* also divides by transverse wall to give rise to *o* and *p*. Of the six tiers formed in this way *l* is destined to give rise to the stem tip, *l'* to the cotyledons; *m* to the hypocotyls; *n* to the root cap and *o* and *p* to a short suspensor, which abuts on the large cell *cb*.

However, variation may be found in the embryogenesis of *Saxiflora granulate, Androsaemum officinale* etc.

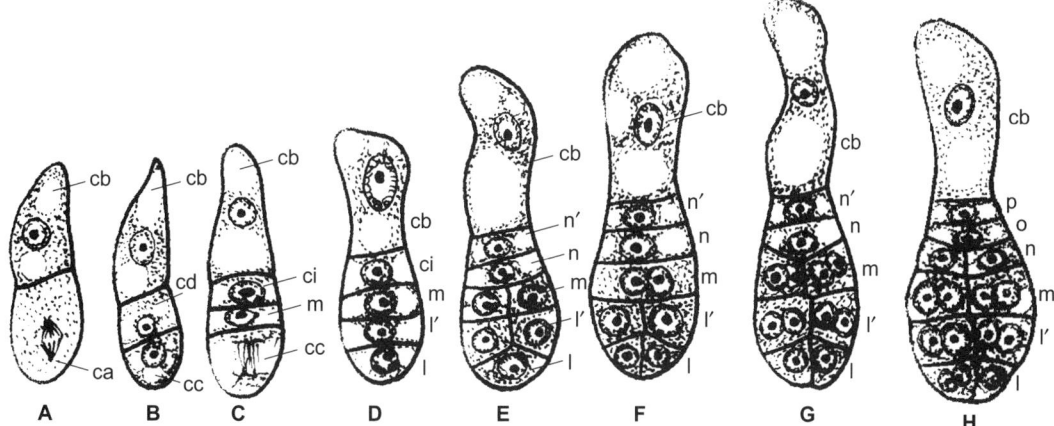

Fig. 11.6 Successive stages of the development of embryo in Sagina procumbens.

(e) **Chenopodiad Type:** *Chenopodium bonus-henricus* (Soueges, 1920) has been cited as a typical example of this group, which shows 4-celled proembryo, similar to the *Nicotiana*, consisting of the cells l, l', m and ci arising by the transverse divisions of ca and cb (Fig. 11.7 A-H). During the course of further development l, l' and m becomes segmented into 4 cells each by laying down of two vertical walls oriented at right angle to each other, while ci divides transversely to form n and n' and then the four cells h, k, o, p. During further development the tier l gives rise to the cotyledons, l' to the lower part of the hypocotyl and m to the upper part of the hypocotyls. The cells originating from ci form the suspensor except the last cell h, which functions a hypophysis and contribute to the root tip. Variation has been reported in *Beta vulgaris*, *Myosotis hispida* etc.

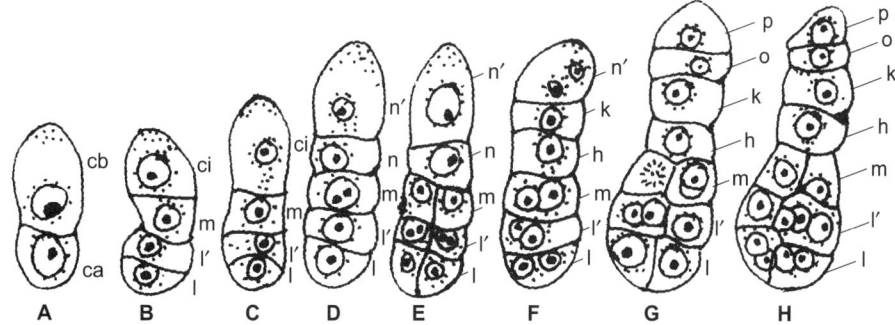

Fig. 11.7 Successive stages of the development of embryo in *Chenopodium bonus-henricus*.

The embryogeny of the *Drosera rotundifolia* is interesting because it shows resemblance with Asterad, Solanad, Chenopodium and Caryophyllad types in one way or other.

(ii) Embryogeny in Monocotyledons

There is no essential difference between the embryogenesis of monocotyledons and dicotyledons regarding the early cell divisions of the proembryo. In monocotyledons one half of the terminal cells and its derivatives have retarded growth, whereas the other half grows rapidly to form one cotyledon. As a result of this asymmetrical growth, the shoot apex occupies a lateral position in somewhat cylindrical embryo and the cotyledon is terminal.

According to Laksmanan (1972), the major differences between dicot and the monocot arise due to disparity in the number and the position of the cells of the terminal quadrant which contribute to the formation of cotyledon and epicotyl (11.8A). In monocotyledons the number of cells involved in the formation cotyledon is variable; two adjacent cells in Amaryllidaceae, Hydrocharitaceae and Potamogetonaceae (Fig. 11.8B), 3 cells in Iridaceae, Pontederiaceae, Sparganiaceae (Fig. 11.8C) and it is practically all the 4 cells in Philydraceae (Fig. 11.8D). However, in dicotyledons derivatives of the opposite cells of the terminal quadrant gives rise to two cotyledons (11.8E).

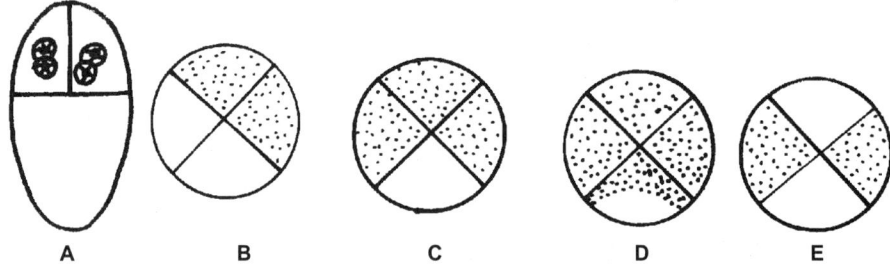

Fig. 11.8 Diagrammatic representation of the derivation of cotyledons in monocots and dicots. **A.** Proembryo at quadrant stage. Stippled portion representing progenitors of cotyledons. **B-D.** Development of cotyledons in monocot. **E.** Development of cotyledons in dicots (After Laksmanan, 1972).

Soueges (1923) has described the embryogeny of *Luzula forsteri*, a member of the family Juncaceae (Fig. 11.9A-I). The terminal cell **ca** of the two-celled proembryo divides by longitudinal wall to produce two juxtaposed cells and a little later the basal cell **cb** divides by transverse wall to produce the two cells **ci** and **m**. In the next stage, two cells of the tier **ca** undergo another vertical division at right angle to the first to give rise to quadrant **q**; cell **m** also divides by longitudinal wall to give rise to two juxtaposed cells and **ci** divides transversely to form two cells **n** and **n'**. By further divisions, the quadrant becomes partitioned into two portions **l** and **l'**, of which **l** gives rise to the lower part of the single cotyledon and **l'** to its upper part and to hypocotyl and plumule. Of the remaining tiers, **m** gives rise to the periblem and part of root cap; **n** to the remaining part of the root cap; and **n'** to the short suspensor.

ULTRASTRUCTURE AND HISTOCHEMICAL STUDIES OF THE ORGANOGENETIC PART OF EMBRYO

Except in the Asterad and Chenopodiad types of development, after the first division of the zygote, the small terminal cell at the chalazal end produces the organogenetic part of the embryo and the large cell at the micropylar end gives rise to the suspensor or become vesicular in nature. Thus, in most angiosperms, polarity displayed by the zygote profoundly affects the fate of daughter cells formed, as an unequal partitioning of the cytoplasm materials of the cells elicits different modes of nuclear expression.

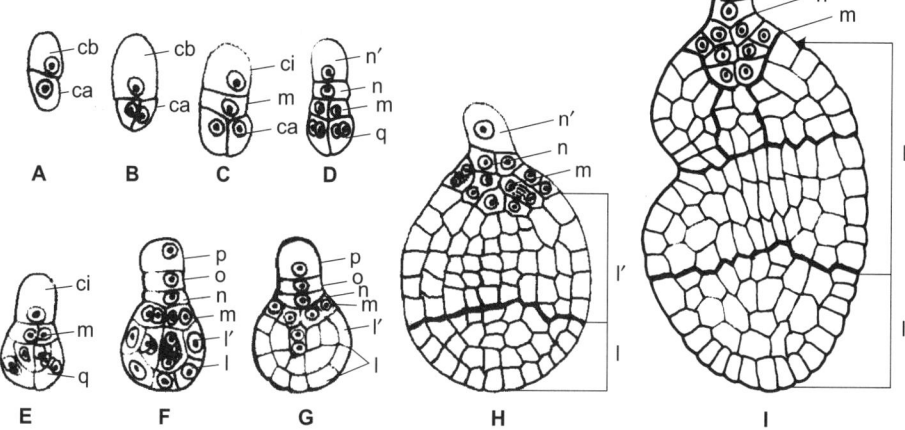

Fig. 11.9 Successive stages of the development of embryo in *Luzula forsteri*.

Ultrastructure and histochemical studies have shown that the marked degree of asymmetry displayed by the two-celled embryo is complemented by the subtle variations in the distribution of the organelles and the concentration of macromolecules. For example, in cotton(Jensen, 1964), *Capsella* (Schulz and Jensen, 1968) and *Quercus* (Singh and Mongensen, 1975) the terminal cells of the two-cell embryo is similar to the chalazal part of the zygote in that it has a dense cytoplasm enriched with organelles, whereas the basal cell is relatively poor of organelles.

(i) Mature Embryo

(i) **Dicotyledonous Embryo:** A typical dicotyledonous embryo consists of an embryonal axis having two broad cotyledons. The portion of the embryonal axis lying above the level of cotyledons origin is termed as epicotyl, which terminates into a plumule or stem tip. The cylindrical portion below the level of cotyledons is called hypocotyl, which terminates at the lower end in radicle and root tip. The root meristem is covered by well-defined root cap (Fig. 11.10).

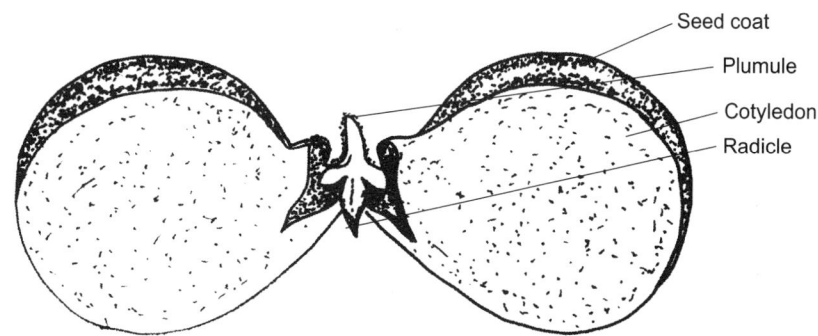

Fig. 11.10 Dicot seed (gram). Embryo with unfolded cotyledons, showing radical and plumule.

(ii) **Monocotyledonous Embryo:** The embryo of monocotyledons possesses only one cotyledon. It has single cotyledon in the form of scutellum which appears to be laterally attached to the embryonal axis. At its lower end, the embryonal axis has the radicle and root cap enclosed in an undifferentiated part of the embryo called coleorhiza. On one side of the coleorhiza is given out a small outgrowth, called the epiblast. The upper portion of the embryonal axis is called epicotyl. It has plumule, which changes into shoot apex and a few leaf primordia enclosed within hollow foliar structure called coleoptile (Fig. 11.11).

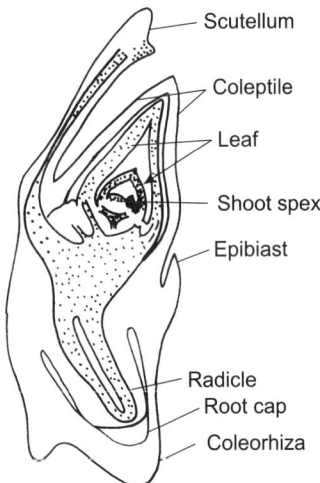

Fig. 11.11 Monocot seed (wheat). Longitudinal section of the embryo showing growing points and other parts.

(ii) Suspensor

The suspensor is an ephemeral structure found at the radicular end of the proembryo. The suspensor grows much faster than the embryo during early stage of embryo development and usually attains its maximum size at globular or early heart-shaped stage. In a mature seed only remnants of the suspensor may be seen. The suspensor anchors the embryo to the embryo sac and pushes it deep into the endosperm so that the embryo lies into nutritionally favourable environment. The suspensor shows great variation with regards to its size, shape and the number of its component cells. The variation is usually related mainly to its chief function i.e. nutrition of embryo.

The longest suspensors have been observed in the members of Loranthaceae family. Leguminosae is known to display a range of suspensor morphologically matching that of Orchidaceae. According to Lersten (1983), who has collected much of the information on suspensor in this family, the form of the suspensor may be (1) spheroid, accompanied by various degree of swellings of the cells (*Lotus*); (2) long slender, biseriate and filamentous (*Cicer*); (3) uniseriate with grossly inflated cell (*Ononis*) or (4) massive, elongated and two to six cells wide (*Phaseolus*). In the sub-family Genisteae, two striking different forms of suspensor have evolved; in *Genista monisperma*, it appears as an enormous spherical structure virtually engulfing the small proembryo, whereas in the different species of *Lupinus* it ranges from a uniseriate filamentous to a biseriate column of short, broad cells (Fig. 11.12 A-F).

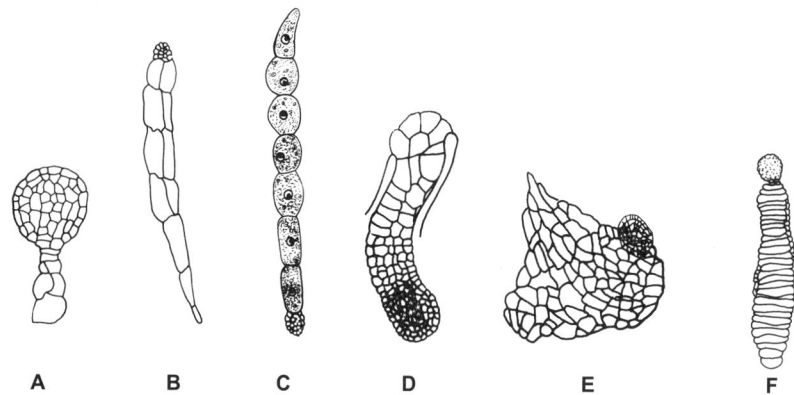

Fig. 11.12 Papilionaceous members showing different types of suspensor. **A.** *Lotus corniculatus*. **B.** *Cicer arientinum*. **C.** *Ononis fructicosa*. **D.** *Phaseolus multiflorus*. **E.** *Genista monisperma*. **F.** *Lupinus pilosus*.

(a) **Suspensor Haustoria:** A most interesting modification of the suspensor is found in *Tropaeolum majus*, in which it attains the several millimeters in length at globular stage of the embryo (Fig. 11.13A). At one end the suspensor is attached to the embryo by a rosette elongated cells, and at the micropylar end it forms a cellular mass from which two multicellular branches arise. One penetrates the integument near the micropyle and grows around the ovule in the cells of carpel (chalazal haustorium). The other branch (placental haustorium) traverses through the integument and funiculus to reach its final destination in the vascular bundle of the placenta (Walker, 1947). However, in several other plants the suspensor cells show a pronounced increase in size or give rise to prominent haustorial structures that penetrate between the cells of endosperm and encroach upon the surrounding tissues of the ovule. The suspensor haustoria of some members of Rubiaceae have been well known since the days of Hofmeister (1885). Detailed studies later revealed that many lateral haustoria are produced which penetrate into the endospermal tissues with their swollen cells (Fig. 11.13B). In *Sedum acre* the basal cell forms an aggressive haustorium whose branches penetrate the seed coat (Fig. 11.13C). The suspensor haustoria of *Myriophyllum* bear a remarkable resemblance to synergids. Here, the two-celled proembryo consists of large basal cell and a much smaller terminal cell. The former divides longitudinally to form two daughter cells which enlarge to such an extent as to occupy the entire space of in the micropylar part of embryo sac (Fig. 11.13D).

(b) **Ultra-structure of Suspensor:** At the globular stage of the embryo certain cells of the suspensor develop lateral wall projections which increase in number and complexities at heart-shaped stage (Fig. 11.14). The presence of wall projections is a feature which a basal cell shares with the suspensor cells. Soon after the formation of wall projections the suspensor cells are accompanied with increase in the density of ribosomes, plastids, mitochondria, dictyosomes, and smooth ER, which continue until the cotyledonary stage.

The transverse wall of the suspensor cells, except the one separating the suspensor from the hypophysis, bears plasmodesmata. The hypophysis cells and its derivatives are also connected through plasmodesmata but there is no such connection between the lens-shaped cell and the embryo proper.

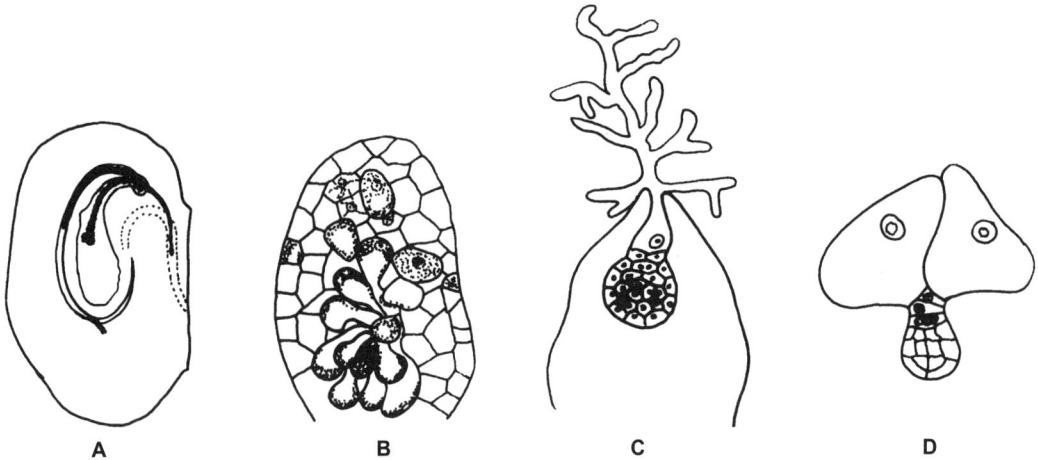

Fig. 11.13 Different types of suspensor haustoria: **A**. *Tropaeolum majus*. **B**. *Asperula*. **C**. *Sedum acre* **D**. *Myriophyllum alterniflorum* (Synergid-like).

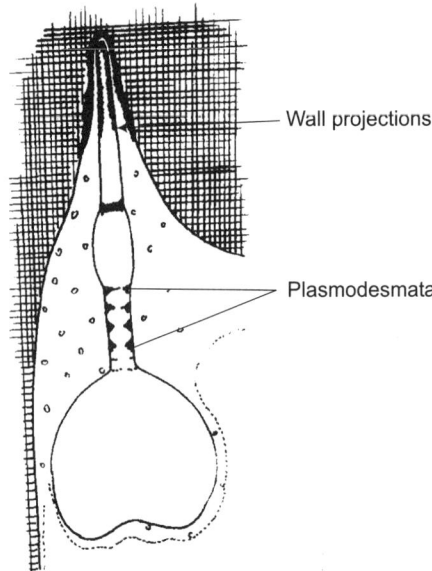

Fig. 11.14 *Diplotaxis* erucoides. Diagrmmatic representation of embryo showing the presence of plasmodesmata and wall projection in the micropylar part of the endosperm and in the basal cell of suspensor. (After Simoncioli, 1974)

(c) **Cytology of Suspensor:** The suspensor is polarized from the early proembryo stage onwards with respect to the mitotic activities. Cell division in the basal cell usually ceases by the proembryo stage but they continue in the chalazal region. In *Phaseolus coccineus*, the suspensor attains its maximum number of about 200 cells at the heart-shaped stage; at the cotyledonary stage it reaches to its maximum size.

Polyteny in suspensor cell is achieved due to endoreduplication of genome. On the basis of nuclear volumes or Feulgen assay the highest values for nuclear DNA content are 8192C for *P. coccineus* (Brady, 1973) and 4,096 for *P.vulgaris, P. hysterinus* and *P.multiflorus* (Nagl, 1974). The lowest values are 16C for *Sophora flavescens* and 32C for *Geranium phaeum* and *Tunica saxifragra* (Nagl, 1978).

Nuclear DNA suggests a progressive increase in the level of endoreduplication beginning with low degree in the cells at the junction between the suspensor and embryo proper, medium degree in the cells in the neck region of the suspensor, and very high degree of endoreduplication in the basal cell of the suspensor.

(d) **Role of Suspensor:**

   (i) *Absorption of Nutrients:* Presence of wall projections and increased number of ER, dictyosomes and mitochondria in close proximity suggests the role of wall ingrowth in the exchange of nutrients (Yeung and Clutter, 1979). However, the solute absorption theory was first articulated by Schulz and Jensen (1969). This theory envisages that metabolites from endosperm or from the surrounding diploid cells of the ovule are delivered to the embryo through the suspensor, the wall invaginations facilitates the transfer by increasing the surface area of absorption. The plasmodesmatal criss-crossing in the embryo-suspensor complex may serve to maintain an open communication for the flow of solutes between the suspensor and embryo. The absorption of the nutrient has further been consolidated with experiment done by Yeung (1980), who demonstrated that if ^{14}C-sucrose is administered through pod or isolated embryo, much of the radioactivity appear in the suspensor and in the suspensor end of the embryo.

   (ii) *Biochemical Indices of Suspensor Activity:* The most distinctive cytochemical changes occur in the suspensor during its programmed cell death. The presence of specialized plastids in suspensor cells of *Stellaria media, Phseolus coccineus* and *P. vulgaris* has provoked the idea on the role of acid phosphatase activity in the plastids during autolysis of suspensor. Other enzymes, such as alkaline phosphatase and acetylesterase are also active in the process of autolysis. Enzyme activity is maximal at cotyledonary stage. The lysed materials of the organelles are the good source of energy for use by the maturing embryo; an increase in the pool of free amino acids in the degenerative suspensor of *Tropaeolum* provides circumstantial evidence of the nature of the substances transmitted to the embryo (Singh *et al.*, 1980).

The suspensor cells produce GA, cytokinins and auxins. Evidence for the activity of GA is the outcome of the work done by Alpi *et al.*, (1975). Their study showed that in heart-shaped embryos of *Phaseolus coccineus*, the total amount of GA-like substances present in the suspensor is many times greater than in the embryo proper. An abrupt change occurs in the cotyledon- stage embryos when there is decrease in the GA content of the suspensor and a concomitant increase in the content of this hormone in the embryo proper. Changes in the cytokinin status of suspensor also bear certain similarities to the change pattern seen in GA activity (Lorenzi *et al.*, 1978). As the status of auxin is concerned, the suspensor contains the higher amount of auxins than does the embryo proper (Przybillok and Nagl, 1977). High concentrations of phytohormones in the suspensor has also been supported by cultural studies as there is spontaneous growth of callus from the suspensor of *Phaseolus coccineus* cultured in a growth medium lacking growth hormones. High concentration of phytohormones in suspensor suggests a definite role of growth hormones in embryogenesis. However, recently it has been suggested that GA in the suspensor enhancse or maintain the protein level in the embryo (Brady and Walthall, 1985).

NUTRITION OF THE EMBRYO

The young proembryo derives its nutrition from the ovular tissues with the help of suspensor. As the embryo develops, its suspensor degenerates. Later, the chief source of nutrition of the embryo inside the developing seed is the endosperm. Another feature of endosperm, which enables it to nourish the embryo, is its capacity to develop haustorial structures. Apart from the endosperm, developing embryo derives its nutrients even from the nucellar tissues, such as pseudo-embryo sac, perisperm and chalazosperm.

(i) Pseudo-embryo Sac

It is the exclusive feature of the family Podostemonaceae. The nucellar cells below the 4-celled stage of proembryo breaks down forming a large cavity called pseudo-embryo sac (Fig. 11.15A). However, pseudo-embryo sacs develop as early as when the megaspore mother cell is under going meiosis for e.g., *Dicraea, Hydrobryum, Zeylanidium*. In two genera reported by Mukkada (1962), namely *Indotristicha* and *Terniola*, the development of the pseudo-embryo sac is a post-fertilization event.

(ii) Perisperm

Usually nucellar tissue is consumed by the developing embryos. However, there are certain families *viz.*, Cannaceae, Capparidaceae, Piperaceae etc., where the nucellar tissues surrounding the embryo sac persists in the seed and becomes densely packed with food materials called perisperm. It functions as the main storage tissue and the endosperm is the intermediate tissue in between the perisperm and the embryo and functions for conduction of food from perisperm to the embryo (Fig. 11.15 B).

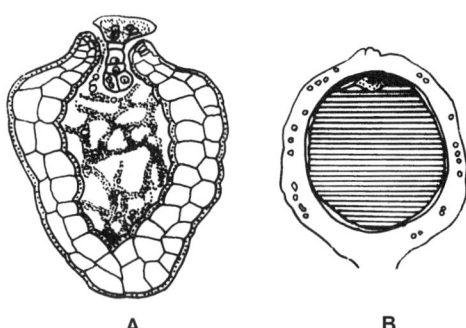

Fig. 11.15. **A.** *Indotristicha ramosessima* showing pseudo-embryo sac (After Chopra and Mukkada, 1966). **B.** *Piper nigrum*. Fruit showing massive perisperm (After Kanta, 1962).

(iii) Chalazosperm

In *Cyanastrum* (Fries, 1919; Nietsch, 1941) the nucellus and the endosperm disappear during the development of seed but the cells in the chalazal part of the ovule, lying just above the vascular bundle, divide actively and form a very prominent tissue, which although loose and possessed of many air spaces, soon becomes filled with fat and starch and serve as a substitute for the endosperm. Fries, who gave the name "**chalazosperm**", suggest that it may also functions as a sort of food body designed to facilitate the distribution of the seed by animals.

UNCLASSIFIED AND ABNORMAL EMBRYOS

There are several plants, whose embryos are not conforming to any of the described types. The following variations have been found:

(i) Treub (1885) reported that in *Macrosolen cochichinensis*, the first wall in zygote is not transverse but vertical.

(ii) Soueges (1937) described the embryogeny of *Scabiosa succesa*, where the first wall of the zygote is oblique.

(iii) Earle (1938) described *Cimicifuga*, where well-differentiated suspensor is lacking.

(iv) In several members of Poaceae also there is no regular pattern of cell divisions in the development of the embryo.

UNORGANIZED AND REDUCED EMBRYOS

Irrespective of the mode of development, a mature embryo generally possesses an embryonic root (radicle), an embryonic shoot (plumule) and one or two cotyledons. However, some groups of plants are characterized by the presence of reduced embryos, lacking differentiation of the organs as in Balanophoriaceae, Rafflesiaceae, Gentianaceae, Burmanniaceae and Orchidaceae, therefore, appear to be associated with some degree of parasitic or saprophytic mode of life.

Reduced embryos without the usual organization of radicle, plumule and cotyledons have also been described in *Ranunculus ficaria*, *Corydales cava* and a few other plants although these are not characterized by a parasitic or saprophytic habit.

12

Polyembryony

Polyembryony has been defined as occurrence of more than one embryo in a seed. The phenomenon first discovered by Leeuwenhoek (1719), which subsequently invited attention because of its potential application in horticulture. Except for a few taxa (*Citrus, Mangifera*), polyembryony occurs as an abnormal feature in angiosperm, whereas it of common occurrence in gymnosperm.

Gustafsson (1946) has reviewed the classification of polyembryony proposed by Ernst (1918) and Schnarf (1929) as follow:

(i) **True Polyembryony**: (a) embryos arising within an embryo sac either by cleavage of the egg or from the synergids, antipodals or endosperm; and (b) embryos arising from the tissues lying outside the embryo sac i.e., the cells of the nucellus or the integuments, although ultimately they come to lie within the embryo sac (adventive embryony).

(ii) **False polyembryony**: It includes the fusion of two or more nucelli or the development of two or more embryo sac within the same ovule.

Polyembryony can be broadly classified into: (i) Simple and (ii) Multiple, depending upon the presence or absence of one or more embryo sacs in the same ovule (Laksmanan and Ambegaokar, 1984). In simple polyembryony the embryos develop within a single embryo sac but in multiple polyembryony accessory embryos are produced from two or more embryo sacs in the same ovule (Fig. 12.1).

TYPES OF POLYEMBRYONY

True Polyembryony

True polyembryony in angiosperms may arise in the following manner:
(a) Cleavage polyembryony.
(b) Formation of embryos by cells of the embryo sac other than egg
(c) Activation of some sporophytic cells of the ovule (Adventive Polyembryony).

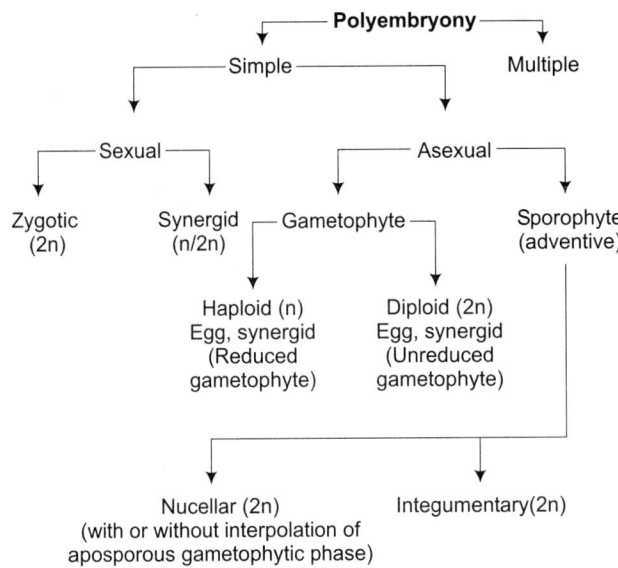

Fig. 12.1 Schematic representation of polyembryony in angiosperms (After Laksmanan and Ambegaokar, 1984)

(a) Cleavage Polyembryony

The simplest method of an increase in the number of embryo is cleavage of the zygote or the proembryo into two or more units. It is very common in gymnosperms but it is less frequent in angiosperms. Jeffrey (1895), described cleavage polyembryony in *Erythronium americanum*, where synergids degenerate and disappear after fertilization and zygote divides to form a small group of cells, irregularly arranged. This embryogenic mass gradually increases in volume and produce two, three or rarely four embryos from its lower end which eventually functions as an independent embryo.

Following Jeffrey's discovery, a similar proliferation of embryonic cells was reported in several other plants. It is only in the family Orchidaceae, where cleavage polyembryony seems to be of very common occurrence. In *Eulophia epidendraea*, Swamy (1943) has described the following variations:

  (i) Cells of the embryonic mass lying towards the chalazal end grow simultaneously into multiple embryos (Fig. 12.2A).
  (ii) The proembryo gives out small buds or outgrowths which function as embryos (Fig. 12.2B).
  (iii) The filamentous proembryo becomes branched and each of the branches grows into an embryo (Fig. 12.2C).

Kausik and Subramanium (1946) described an embryo sac of *Isotoma longiflora*, in which an additional embryo seems to have budded out from a suspensor cell. Crete (1938) reported one or more embryos arising from the cells of suspensor in *Lobelia syphilitica* (Fig. 12.3). However, suspensor polyembryony (about six) has been found to be a common feature in the genus *Exocarpus*, a member of Santalaceae.

Polyembryony **12.3**

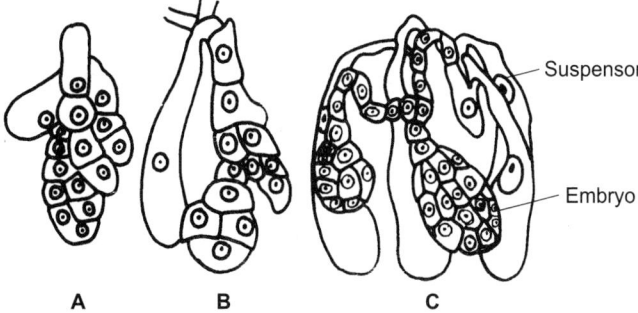

Fig. 12.2 Cleavage polyembryony in *Eulophia epidendrea*. **A.** Zygote divided to form group of cells, three of which have divided to form independent embryos. **B.** Bud arising rom the right side of embryo. **C.** Two embryos arising by the splitting of a single embryo; large vacuolated cells belong to suspensor (After Swamy, 1943).

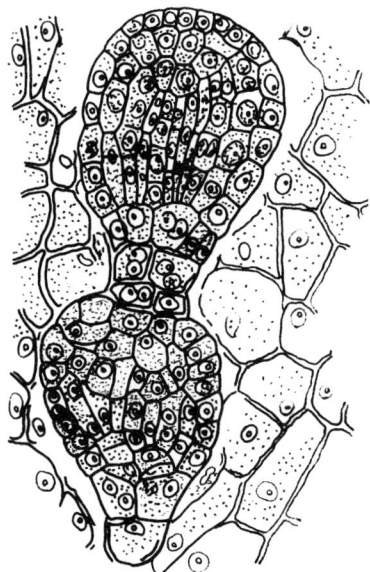

Fig. 12.3 *Lobelia syphilitica*, twin embryos; the basal has arisen by proliferation of the suspensor (After Crete, 1938).

(b) Formation of Embryos by Cells of the Embryo Sac other than the Egg

(i) *Embryo arising from synergids:* Besides the zygotic embryo produced from the fertilized egg, embryos may also be produced from other parts of the embryo sac. The most common source is the synergids, which frequently becomes egg-like and may be fertilized by sperm from normal pollen tube or an additional pollen tube or develops without fertilization and accordingly embryo may be diploid or haploid. If only two sperms enter an ovule, twin embryos are produced as a result of fertilization of egg and one of the synergids by the two male gametes. There would be no triple fusion in these cases and consequently no endosperm. Under such condition the twin embryos invariably become aborted.

In *Aristolochia bracteata* (Johri and Bhatnagar, 1955) the fertilization of the egg or one or both synergids is accompanied by the formation of endosperm. This is because of the entry of more than one pollen tube into the embryo sac, or presence of additional sperm in the pollen tube. However, *Pelliphylum peltatum* (Lebegue, 1952) seems to be an exception, as the twin embryos are said to develop in the absence of endosperm.

Synergids may also develop without fertilization but such embryos contain the gametophytic number of chromosome. Haploid-diploid embryos have been observed in several instances. *Bergenia delavayi* is one of the examples where twin proembryos are seen, of which the smaller seems to be derived from an unfertilized synergid (Fig. 12.4).

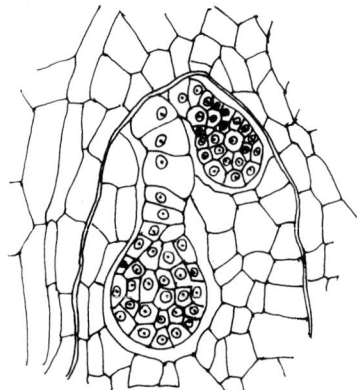

Fig. 12.4. *Bergenia delavayi*. Twin proembryos, of which smaller on the right is derived from unfertilized synergids (After Lebegee, 1949).

(ii) ***Embryos arising from the antipodals:*** Usually the antipodals are ephemeral and show sign of degeneration before or after fertilization. However, Shattuck (1905) noted that *Ulmus americanum* presented an egg-like appearance of antipodals. Ekdahl (1941) has confirmed this finding in *U. glabra* (Fig. 12.5). Subsequently, several scientists working on different plants have reported embryos from antipodals. These embryos develop to some extent but fail to develop into an adult viable embryo.

(iii) ***Embryos arising from endosperm:*** There have been a few reports of embryos arising from the cells or nuclei of the endosperm. Reinvestigation of most of the cases has, however, yielded negative results. Embryos may be wrongly considered to have originated from the endosperm, either because of their displaced position in the interior of the embryo sac or because of adventive embryony which make close association with the endosperm after the destruction of the nucellar tissues from which they have originated. However, *in vitro* studies have established that embryo-like structures, called embryoids (non-zygotic) can be developed from endosperm culture (Nag and Johri, 1971).

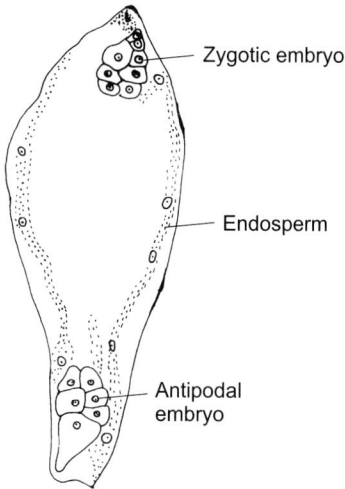

Fig. 12.5 *Ulmus grabra*, embryo sac showing zygotic embryo at the micropylar end and antipodals at chalazal end (After Ekdahl, 1941).

(c) Activation of Some Sporophytic Cells of the Ovule (Adventive Polyembryony)

(i) **Embryos arising from cells outside the embryo sac:** The embryos arising from the maternal tissues (nucellus and integuments) are called adventive embryos. Members of the family Rutaceae, Anacardiaceae, Cactaceae, Myrtaceae and Orchidaceae have marked tendency for nucellar polyembryony. In various species of *Citrus*, nucellar embryos are of great horticultural importance, as the plantlets they produce are more vigorous, virus- free and endowed with well developed tap root system in comparison with the shoot cuttings of the mother plants. Moreover, the seedlings are more uniform than those obtained through the seeds. In *Mangifera indica*, polyembryonal seeds contain as many as 50 embryos. It has been found that favoured cells grow at the expense of their neighbours and proliferate into the embryo sac cavity where they complete their further development. The adventive embryos may be distuinghed from the sexual embryo by their lateral position, irregular shape and lack of suspensor.

According to Esen and Soost (1977), pollination and fertilization are not essential for initiation of nucellar embryo in the ovule of certain cultivars of *Citrus sinensis* and *C. reticulata* but subsequent development is clearly dependent upon double fertilization and endosperm development.

Swamy (1940) described that the orchid, *Spiranthes cernua* showed the embryo sac, where both the male and female gametophytic tissues are functionless. The cells of the inner integument give rise to adventive embryos of which 2-6 may mature within the seeds, whereas in *Zeuxine*, they originate from the nucellar epidermis. Formation of adventive embryos from the cells of the integument is rare but it has been reported in *Melanopodium divaricatum* (Devi and Pullaiah, 1976)

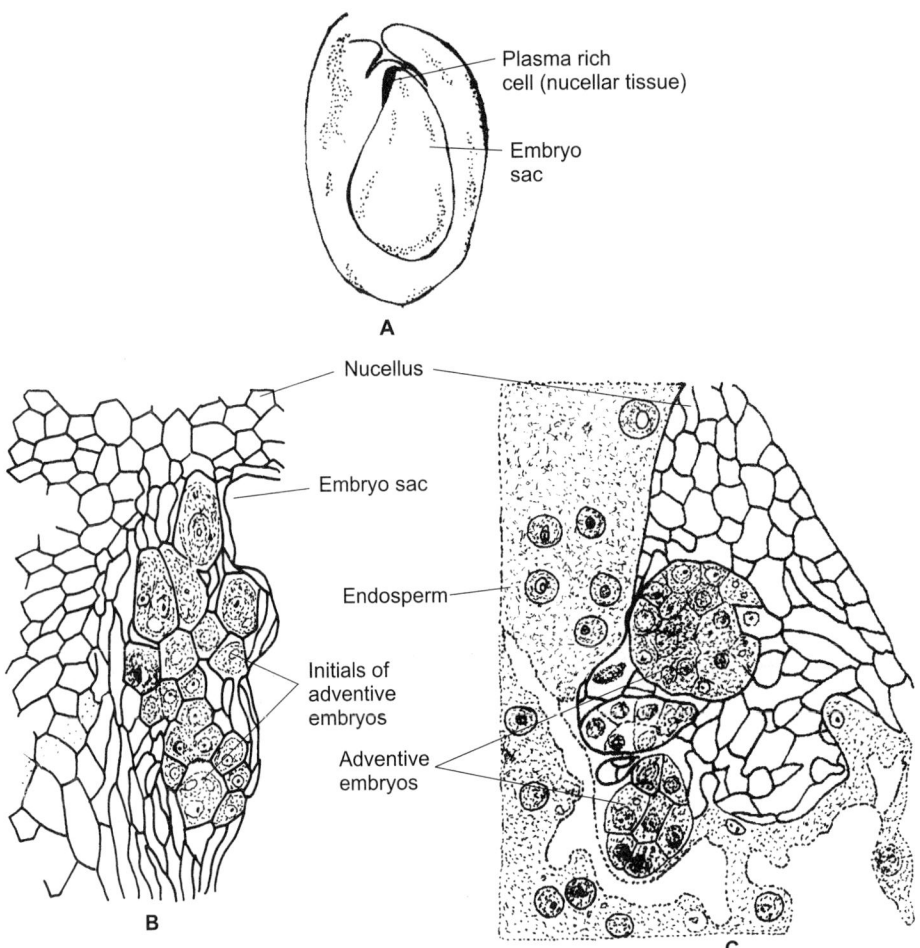

Fig. 12.6 Nucellar polyembryony in *Mangifera*. **A, B.** *M. indica* var. *olour* **A.** L.S. ovule showing position of plasma rich cells. **B.** Part of nucellus (plasma rich) from **A** to show details. **C.** An enlarged view to show embryonic masses protruding into embryo sac. (After Sachar and Chopra, 1957)

Although, both nucellar and integumentary embryos have their origin outside the embryo sac but during later stage of their development, they come to lie inside the embryo sac and are suitably nourished by the endosperm. In *Mangifera* species adventive polyembryony has been observed, where nucellar tissues divide to form multiple embryos which later protrude into embryo sac (Fig. 12.6A-C).

As a rule nucellar embryos originate either from the cells situated at the micropylar end of the embryo sac or from those on its sides. In *Trillium undulatum* (Swamy, 1948b), however, they are said to originate from the cells of the chalazal end of the embryo sac.

In unitegmic, tenuinucellate ovules, the innermost layer the integument differentiates as endothelium. Sometimes some of the cells may become distinct and show cytoplasm and divide repeatedly to form several proembryos. This feature has been observed in *Melampodium divaricatum* (Fig. 12.7) and *Carthamus tinctorius* (Maheshwari, Devi and Pullaiha, 1976, 1977). These proembryos never undergo further development.

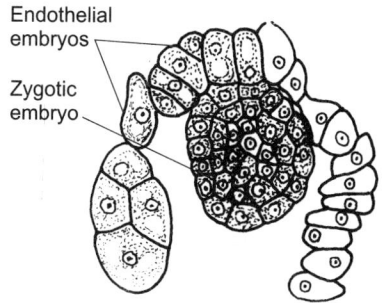

Fig. 12.7 *Melampodium divaricatum*, endothelial polyembryony (After Maheshwari and Pullaiah, 1976).

FALSE POLYEMBRYONY

The polyembryonic condition may also results from the fusion of two or more ovules or the occurrence of multiple embryo sacs within single ovule.

Multiple embryo sacs may arise from; (i) derivatives of the same megaspore mother cell, (ii) derivatives of two or more megaspore mother cell and (iii) nucellar cells by apospory.

Solntseva (1957) observed two or more embryo sacs arising in the nucelli of *Fragaria grandiflora*, which are said to fuse to form single structure containing several embryos. In *Malus hupehensis*, the megaspore fails to develop into a functional embryo sac; instead some of the somatic cells give rise to typical 8-nucleate embryo sac which often fuses with one another. Polyembryony occurs as the result of parthenogenetic development of more than one egg cell.

Many members of the Loranthaceae are characterized by the absence of clearly demarcated ovules. There is a group of sporogenous cells from which several embryo sacs arise and even extended to stylar canal. After fertilization biseriate proembryo grow downwards and enter the ovarian tissue surrounded by a composite mass of endosperm cells formed by the fusion of the several embryo sacs (Fig. 12.8). Usually a single embryo matures while the rest degenerate (Maheshwari *et al.*, 1957). However, sometimes two embryos are found as in *Dendrophthoe neelgherrensis* and *Lepeostegeres gemmiflorus* (Dixit, 1958).

SOME SPECIAL CASES

In some plants, multiple embryos are produced by the simultaneous operation of more than one methods named above. One of the examples is *Allium odorum*, reported by Tretjakow (1895), where both antipodal and synergid embryos were recorded in one-third to one-half of the ovules of this species. Later, Haberlandt (1923, 1925) reported that even in castrated flowers, there is an increase in the size of the ovules accompanied by the production of embryos from several sources: egg, synergids, antipodals and the cells of inner integument. He found diploid number of chromosomes in all the embryos.

Modilewski (1925, 1930, and 1931) reported some of the plants, where two types of embryo sacs are found: haploid and diploid. In the diploid embryo sacs only the polar nuclei are fertilized, resulting in pentaploid endosperm whereas embryos arising from the

12.8 Plant Embryology: Classical and Experimental

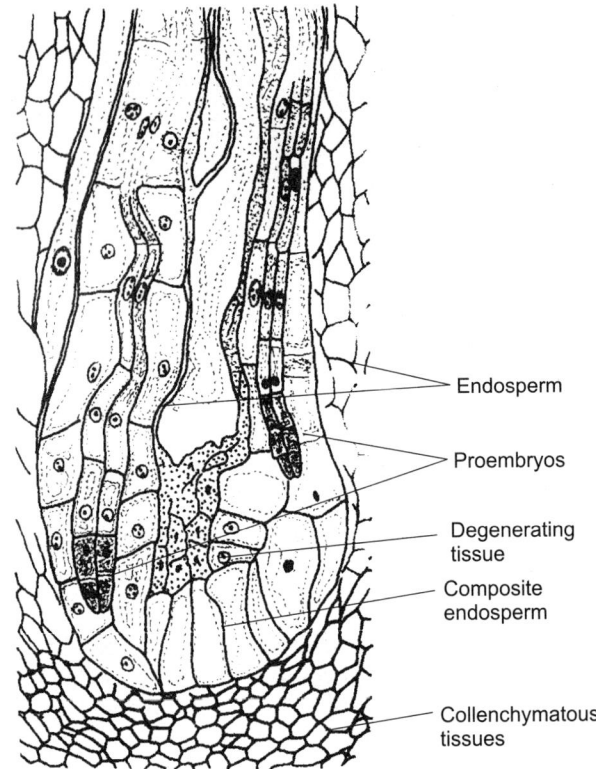

Fig. 12.8 False polyembryony in *Dendrophthoe neelgherrensis*. L.S. showing two proembryos embedded in the composite mass of endosperm (After Narayana, 1954).

unfertilized egg and antipodals are diploid. However, in haploid embryo sac viable embryos are formed only after the fertilization.

In some species of *Elatostema* (Fagerlind, 1944), synergids, antipodals and nucellar cells, all take part in the production supernumerary embryos. Additional embryos may also be formed owing to the presence of multiple embryo sacs in the same ovule. Sometimes the wall separating the embryo sacs dissolve, resulting in the formation of common cavity and the developing embryos come to lie very close to one another, they show various degree of fusion (Fig. 12.9A-G).

TWINS AND TRIPLETS

The frequency with which multiple embryos occur in a seed is often so low that it is not so easy to trace their exact developmental studies. One can only speculate on their possible source, based on cytological and genetical studies of the seedlings. Twin embryos usually show the following combinations: diploid- diploid, haploid- diploid, diploid-triploid and haploid- triploid.

Radoph (1936) found two embryos arranged parallel to each other in some kernel of *zea mays*. The seedlings produced by such kernels were found to be genetically identical and he attributes such a condition to the cleavage of a zygotic embryo.

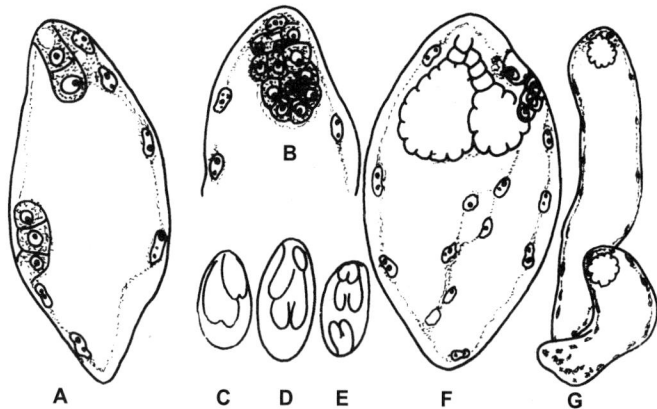

Fig. 12.9 Polyembryony in *Elatostema*. **A.** *E. eurhynchum*; embryo sac with two embryos, one arising laterally. **B.** Upper part of embryo sac showing three embryos, two of which are lying somewhat horizontally. **C, D**, older ovules, showing two embryos. **E.** Ovule showing two embryos, at micropylar end and one at chalazal end. **F.** *Elatostema acuminatum*. Compound embryo sac formed by the fusion of two sacs, showing well developed embryo (left) and poor developed embryo (right). **G.** *E. pedunculosum*; two adjacent embryo sacs each with normally developed embryo at micropylar end. (After Fagerlind, 1944).

Greenshield (1951) made a cytological studies of 55 pairs of twins of *Medicago sativa* and showed that all of them had 2n = 32 chromosomes. He concludes that twinning is either due to cleavage of the normal zygote or to fertilization of twin embryo sacs. Cytological similar diploid seedlings may also be produced by the fertilization of more than one of the embryo sac. As all the nuclei of the monosporic embryo sac are genotypically alike, they would give rise to identical embryos if fertilized by the male gametes discharged by the same pollen tube. Hagerup (1947) indicated such possibility in some orchids. In addition, nuclear budding can also give rise to identical twins, triplets, etc. Further a haploid cell of embryo can also give rise to a diploid embryo by endoreduplication of the chromosomes.

Haploid-diploid twins have been reported in *L. usitassimum*, *Oryza sativa*, *Solanum tuberosum*, *Triticum vulgare* etc. There are two views regarding haploid-diploid twins: (i) the diploid member is derived from the fertilized egg and the haploid member from the unfertilized cell of the same embryo sac or different embryo sac in the same ovule, (ii) the fertilization of the egg in one embryo sac might stimulate a parthenogenetic development of the egg of a neighbouring embryo sac.

Cameron (1949) found a high degree of diploid-diploid twins in *Nicotiana tabacum* which were fused and suggested that these may have originated by the cleavage of the zygote embryo. Separate diploid-diploid twins are also presumed to have resulted from the fertilization of the egg and one synergid.

Diploid-triploid has been recorded in *Triticum vulgare*, *Secale cereale* and a few other plants. As the authentic information about the origin of embryo from the endosperm is lacking altogether, therefore, it seems more likely that the triploid embryo originated either by the fertilization of an unreduced (aposporic) embryo sac or by the fusion of a cell of haploid embryo sac with two male gametes or by one unreduced gamete.

12.10 Plant Embryology: Classical and Experimental

Randall and Rick (1945) found 405 multiple seedlings in *Asparagus officinalis*. Out of these 97% were twins, 11 were triplets and 1 was quadruplet. Diploids (2n = 20) were the most frequent but a few showed variations, such as 30 (3n), 21 (2n + 1) and 40 (4n). Twins seedlings, showing similar morphological features usually had the same chromosome number. In Haploid–diploid pairs, the haploid member was always much smaller than the diploid but in others the size of the seedlings seldom gave any clue to their origin of chromosome number.

EXPERIMENTAL INDUCTION OF POLYEMBRYONY

In nature, formation of embryo is restricted to the ovular tissues. However, in 1960's it has been established that by providing suitable conditions, any cell of the plant body can be stimulated to give rise to viable embryo, called '**embryoid**', non-zygotic embryo which resembles the normal embryo structurally and can grow into complete plantlet. Embryoids have been obtained from *in vitro* cultures of different types of organs, tissues or cells, such as zygotic embryos, nucellus, root segments, stem segments, leaves, fruit tissues, ovary, ovules and anthers. So far, somatic embryogenesis has been reported in over 102 species including many important crop plants (Rangaswamy, 1986).

Butter cup (*Ranunculus sceleratus*) and carrot (*Daucus carota*) are very good examples of embryogenesis. All the parts of these plants give rise to embryoids in tissue culture. Of the various factors, genotype of the donor plant is very important although response of the explants is also determined by culture medium (Fig. 12.10 A-D). Detailed discussion can be seen in Chapter-21.

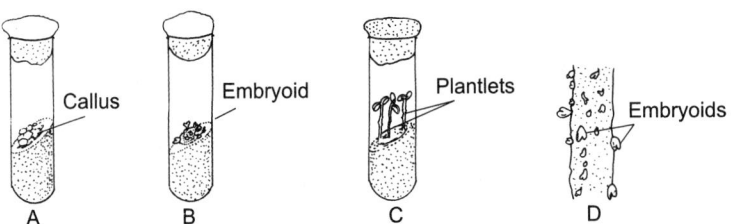

Fig. 12.10 Embryoid formation in floral buds of *Ranunculus sceleratus*. **A.** Callus formation. **B.** Six-week-old culture with numerous embryoids. **C.** Plantlet formation through embryoids; note numerous embryoids developed at the hypocotyledonary region. **D.** Portion of the hypocotyls enlarged to show somatic embryoids (After Nataraja, 1968).

CAUSES OF POLYEMBRYONY

Various theories have been proposed to explain the occurrence of polyembryony but none has been proved to be satisfactory so far.

Haberlandt (1921, 1922), proposed the '**Necrohormone Theory**', according to which degenerating nucellar tissues secrete some substances which stimulate adjacent cells to divide and to develop into adventive embryos. He succeeded in inducing polyembryony in *Oenothera* by pricking the ovule by fine needle. Subsequent workers could not confirm his findings.

Leroy (1947) tried to show genetic control of polyembryony in *Mangifera indica*. According to him the polyembryonate forms of mangoes are due to the presence of one or more recessive genes. Mango forms which originate from the secondary center of origin such as China, Philippines and a few other countries show polyembryonate forms. However, in India which is primary center of origin of mangoes, contains monoembryonic forms carrying dominant genes.

Frusato *et al.* (1957) showed that embryo number in citrus seeds may be influenced by following factors:

(i) Age of the trees-increasing in older trees.
(ii) Fruit sets-being higher in the years of higher fruit set.
(iii) Nutritional status of the plant.
(iv) Orientation of the branches of the trees- being higher on **Northern** than on **Southern** branches.

ROLE OF POLYEMBRYONY IN PLANT BREEDING AND HORTICULTURE

Polyembryony is of considerable interest to the plant breeders as well as to the horticulturists.

(1) Adventive embryos have been successfully employed in the propagation of trees like *Mangifera* and *Citrus*. Such embryos not only inherit the character of maternal parent but also provide a genetically uniform plantation. In *Citrus* they have been used as orchard stock upon which grafts from other types can be made for better root system (the nucellar seedlings have tap roots, therefore, develop better root system than do cuttings. The latter have only small lateral root system).

(2) The nucellar seedling show a restoration of the vigour lost after repeated vegetative propagation through cuttings.

(3) The multiple embryos within a seed can be haploid, diploid or triploid. Haploids are of immense use in plant breeding because homozygous diploids can be obtained by doubling the chromosome numbers by treating them with colchicine. Such haploids and diploids could be utilized in plant improvement programme.

(4) Adventive embryos are useful in agriculture and horticulture because they are genetically true copy of the parents and are generally disease free. The seedlings derived from such embryos are more vigorous than those produced from the cuttings.

(5) Through *in vitro* studies large number of embryoids could be produced. Thus this feature can be employed for large-scale propagation of selected genotypes.

(6) Somatic embryos can be converted into "synthetic seeds" which has great potential. (Discussed in Chapter-22).

13 Apomixis

Meiosis and syngamy are two main characteristics of sexual reproduction. Meiosis results production of haploid gametophytes, which bear male and female gametes. Syngamy results in the restoration of the diploid sporophytic generation. Thus the haploid and diploid phases alternate with each other in the life cycle of plants. In many plants, however, normal sexual reproduction is replaced by asexual reproduction. This phenomenon of substitution of sexual process by asexual method is known as **apomixis** and the plants showing this phenomenon are called apomictic plants. According to Wrinkler (1908), the term apomixis refers to the substitution of sexual reproduction by any such method which does not involve meiosis and syngamy.

Apomixis has been reported in more than 300 plant species belonging to 35 families (Khush *et al.*, 1994) but is prevalent in the families Asteraceae, Poaceae, Rosaceae and Rutaceae. As a rule apomixis is more common in polyploids than diploids. Within Asteraceae, apomixis has been reported in more than 125 species.

TYPES OF APOMIXIS

According to Maheshwari (1950), apomixis can be classified into following four types:
(1) Non-recurrent apomixis, (2) Vegetative reproduction, (3) Recurrent apomixis and (4) Adventive embryony or sporophytic budding.

Non-recurrent apomixis has been proposed by Maheshwari (1950). In this type, the megaspore mother cell undergoes usual meiosis resulting in haploid embryo sac. The new embryo may then arise from the egg (haploid parthenogenesis) or from some other cell of the gametophyte (haploid apogamy). However, according to the present concept (Heslop-Harrison, 1972), the non-recurrent apomixis does not fall within the purview of the apomixis, as it involves meiosis. A true apomixis eliminates both meiosis and syngamy. Presently, apomixes is classified into following two main types (1) Vegetative reproduction, (2) Agamospermy (Fig. 13.1).

13.2 Plant Embryology: Classical and Experimental

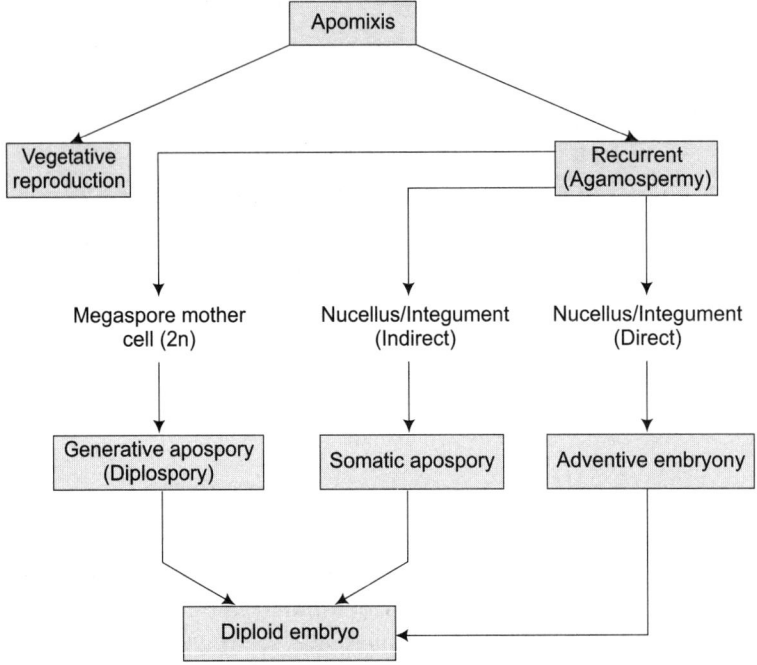

Fig. 13.1 Scheme showing different types of Apomixis.

(i) Vegetative Reproduction

Vegetative reproduction includes propagation by parts other than seed such as bulbils, bulbs, runners, suckers etc. These propagules are formed by sporophytes only. Gustafsson (1946) has described three types of vegetative reproduction in higher plants:

(i) The propagules are formed outside the floral regions and despite the occurrence of functional sex organs, no fertilization or seed setting takes place. This may be seen in *Elodea canadensis* and *Agave americana, Globba* (Fig. 13.2).

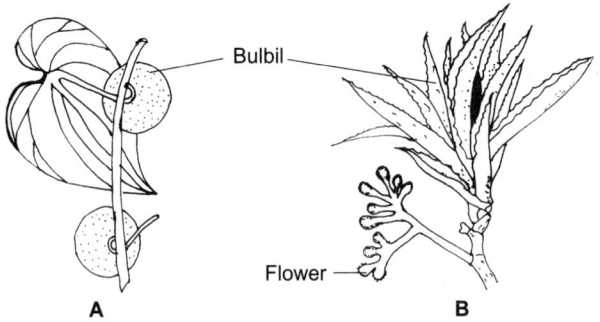

Fig. 13.2 Vegetative propagule (bulbil). **A.** *Globba* **B.** *Agave*.

(ii) The propagules are formed outside the floral parts; the plant itself is sterile as it is seen in *Fritillaria imperialis* and *Lilium bulbiferum*. They produce bulbils or bulblets as a means of reproduction.

(iii) The propagules are formed on the floral branches either in addition to flowers or in place of them. This phenomenon is called vegetative vivipary in contrast to normal vivipary as in mangroves. Vegetative vivipary is quite common in grasses, for example, *Deschcampsia, Festuca, Poa* and *Lilium*.

(ii) Recurrent Apomixis (Agamospermy)

In this category, the embryo sac either arises from a cell of the archesporium (**generative apospory**), also called **Diplospory** or from some other part of the nucellus (**somatic apospory**). The embryos may either arise from the egg (diploid parthenogenesis) or from some other cell of the embryo sac (diploid apogamy).

(i) Generative Apospory (Diplospory)

In diplospory, a diploid embryo sac is formed from megaspore mother cell without a regular meiotic division. Diplospory may be divided into following three types depending upon the type of division of the megaspore mother cell.

(a) *Taraxacum Type*: The division of the megaspore mother cell begins like meiotic prophase. Some degree of pairing of chromosomes also takes place. The dissociation of chromosome occurs with the formation of restitution nucleus. The megaspore mother cell with restitution nucleus undergoes mitotic division to form two cells, of which one degenerates and the other forms the embryo sac. This type of development can be cited in *Taraxacum albidum* (Fig. 13.3), which is polyploidy. In

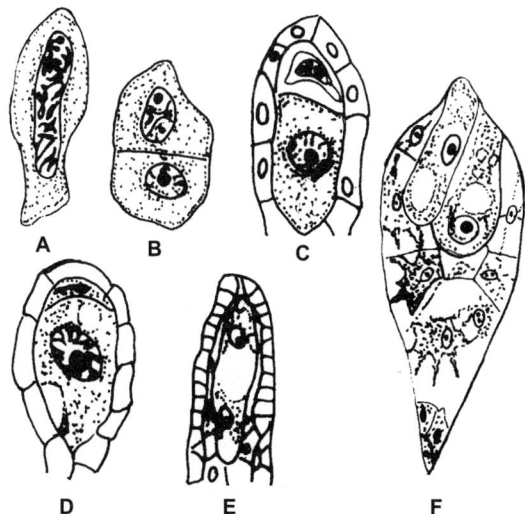

Fig. 13.3 Stages in the formation of diplosporous embryo sac in *Taraxacum albidum*. **A**. Megaspore mother cell with restitution nucleus. **B**. Dyad stage **C, D**. Upper cell of dyad is degenerating and lower is functional. **E**. Two-celled embryo sac. **F**. Mature embryo sac; endosperm development has started (After Osawa, 1913).

this species, the diploid ones show normal meiotic divisions. Usually it is the chalazal dyad cell which functions but in *T. laevigatum*, it is frequently the upper micropylar cell.

(b) ***Ixeris Type***: In this type also there is semihetertypic division followed by restitution nucleus. The basic chromosome number in this genus is 7. Species with the diploid chromosome number (2n = 14) reproduce normally by the asexual method but *I. dentata*, which is a triploid (2n = 21) is an apomictic species. During the division of megaspore mother cell there is no evidence of any pairing of the chromosomes and finally restitution nucleus is formed which contains unreduced number of chromosomes. Further divisions are entirely mitotic and lead to formation of an eight- nucleate embryo sac, organized in the usual manner (Fig. 13.4).

(b) ***Eupatorium Type***: Holmgren (1919) made a very detailed study of the embryology of the several species of the genus *Eupatorium*. In this case, the nucleus of the megaspore mother cell directly undergoes the normal mitotic divisions (three nuclear divisions) to form unreduced 8-celled embryo sac, as in *Eupatorium glandulosum* (Fig. 13.5).

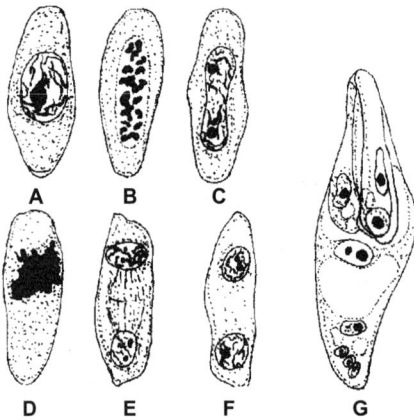

Fig. 13.4 Stages in the formation of diploporous embryosac in *Ixeris dentata*. **A.** Megaspore mother cell with prophase nucleus. **B.** Later stage; meiosis with univalents only. **C.** Restitution nucleus. **D.** Mitosis in restitution nucleus. **E.** Binucleate embryo sac. **F.** Mature embryo sac (After Okabe, 193).

(c) ***Allium Type***: In this type of diplospory premeiotic chromosome doubling is the cause of unreduced embryo sac formation (Kojima and Nagato, 1992). Chromosome number in MMC is doubled by the endomitosis and the ensuing mitosis results in the chalazal dyad cell results in 8-nucleate embryo sac. *A.nutans* and *A. odorum* show Allium type of diplospory.

(ii) Somatic Apospory

Here somatic cell of the sporophyte, usually belonging to the nucellus or chalaza functions as the initial of the embryo sac. Somatic apospory is frequently preceded by degeneration of the

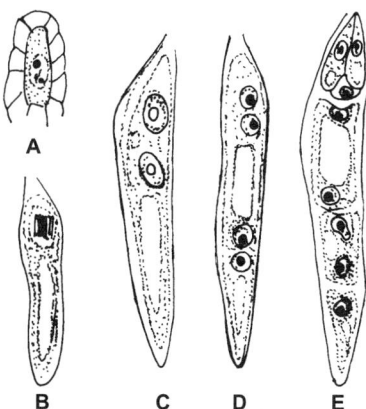

Fig. 13.5 Development of diplosporous embryo sac in *Eupatorium glandulosum*. **A.** Megaspore mother cell. **B.** Megaspore mother cell with its nucleus in late anaphase of mitosis. **C-E**. Two, four and eight nucleate embryo sacs, respectively. (After Holmgren, 1919).

legitimate megaspore mother cell. Sometimes both the aposporous and the normal embryo sac develop together. The following two types are distinguished:

(a) **Hieracium Type**: Rosenberg (1907) for the first time described the occurrence of somatic apospory in three species of *Hieracium viz., H. excellens* (Fig. 13.6A-E), *H. flagellare* (Fig. 13.6F-H) and *H. aurantiacum*.

The megaspore mother cell undergoes normal meiosis and forms tetrad. At this stage somatic cell situated at the chalazal end becomes enlarged and vacuolated. This cell gradually increases in volume encroaching upon the megaspores and finally crushing them. The aposporic embryo sac, arising from it, has unreduced chromosome number and is able to function without fertilization. In *H. excellence*, the normal and reduced embryo sac as well as the aposporic and unreduced embryo sac sometimes develops simultaneously but this is rare in other two species. *H. aurantiacum* is peculiar in that the the aposporic embryo sac usually originates from a cell of nucellar epidermis.

Aposporic embryo sac has also been reported in *Artemisia, Crepis, Hypericum, Malus, Parthenium, Poa, Ranunculus* and some other species.

(b) **Panicum Type**: Warmke (1954) found that one or more nucellar cells develop into an aposporous embryo sac. After vacuolation there are two divisions producing four nuclei all of which remain in the micropylar region. Wall formation occurs at this stage giving rise to two synergids, one egg and one polar nucleus. Somatic apospory is very common in other members of the Panicoidae.

(iii) Adventive Embryony

In adventive embryony, there is no gametophytic generation. The embryo originates from the diploid cells of the ovule lying outside the embryo sac and belonging either to nucellus or integument.

A common feature of the process is that the cells concerned in such development become richly cytoplasmic and actively divide to form small group of cells which eventually pushes their way into the embryo sac and grow further to form a true embryo.

13.6 Plant Embryology: Classical and Experimental

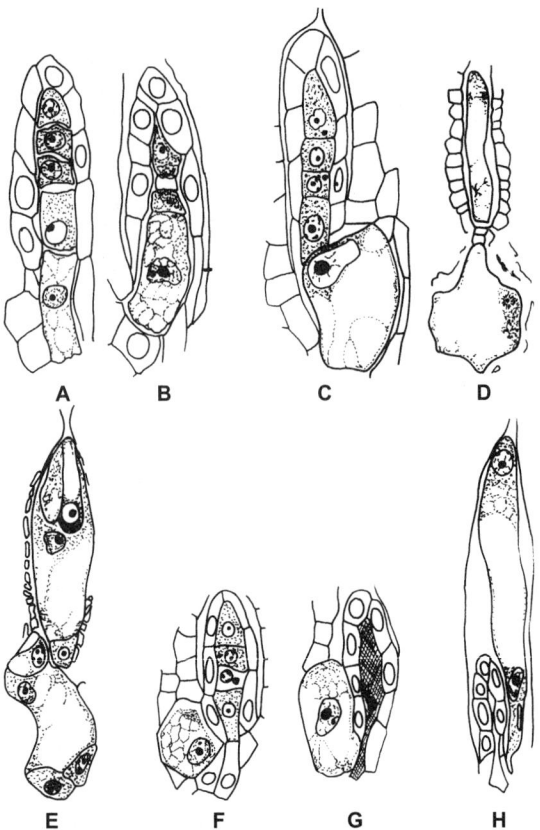

Fig. 13.6 Development of embryo sac in *Hieracium excellens* (**A-E**) and *H. flagellare* (**F-H**). **A**. Nucellus showing tetrad of megaspores; one enlarged cell lying below chalazal megaspore. **B**. Megaspore tetrads in the process of degeneration; chalazal cell showing increase in size. **C**. Megaspore tetrad and large nucellar cell destined to give rise to embryo sac. **D**. Normal and aposporic embryo sacs growing simultaneously. **E**. Two fully developed embryo sacs, lower is of aposporic origin. **F-H**. Some stages in the development of aposporic embryo sac; progressive degeneration of megaspore tetrad. (After Rosenberg, 1907).

Adventive embryony usually leads to form polyembryony. Besides the well-known example of citrus (Fig. 13.7 A-C), adventive embrony is known to occur in Buxaceae, Cactaceae, Euphorbiaceae, Myrtaceae and Orchidaceae. It has been observed that when nucellus is intact, the adventive embryos originate from the nucellar cells but when it becomes disorganized the cells of the integument take over this function.

Adventive embryony is although independent of pollination and fertilization, however, there are a few reports of induction by one or both of the factors. In orchids, like *Nigritella nigra* (Fig. 13.8 A-E), Afzelius (1928) reported that neither pollination nor fertilization is essential but the occurrence of pollen tube in the ovary seems to accelerate the tendency towards the production of adventive embryos. *Eugenia jambos* (Pijl, 1934) is a very interesting example, where adventive embryos may originate quite independent of pollination but do not attain their full maturity unless fertilization takes place.

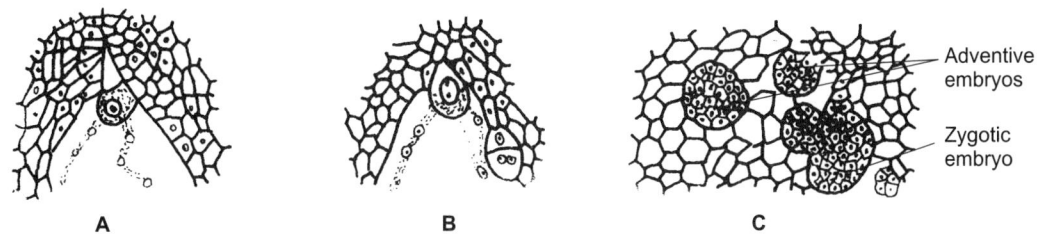

Fig. 13.7 Development of adventive embryos in *Citrus trifoliate*. **A**. Micropylar portion of embryosac showing fertilized egg, pollen tube and endosperm nuclei; some of the nucellar cells have become prominent. **B**. Same at more advanced stage. **C**. Upper part of the embryo sac showing several embryos lying in endosperm; only zygote embryo has suspensor (After Osawa, 1922).

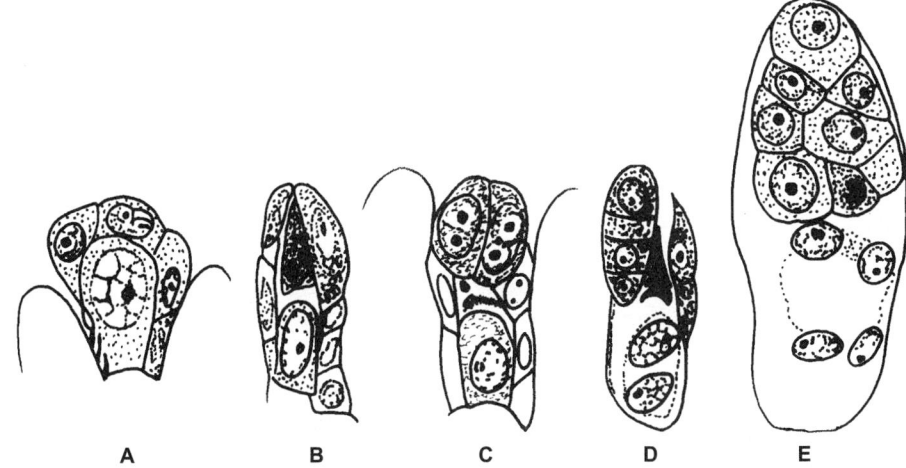

Fig. 13.8 Development of adventive embryogeny in *Nigritella nigra*. **A**. Megaspore mother cell. **B**. Micropylar dyad cell in the process of degeneration. **C**. Functional megaspore with degenerating megaspores; enlargement of two cells of nucellar epidermis. **D**. Two nucleate embryo sac with young adventive embryos arising from cells of nucellar epidermis. **E**. Large nucellar embryo lying at the apex of 4-nucleate embryo sac (After Afzelius, 1928).

In all the cases of adventive embryony, there is formation of endosperm, whether it develops as a result of triple fusion or without it, except in *Opuntia aurantiaca* (for further detail see Chapter-12).

ORGANIZATION OF EMBRYO SAC

By whichever methods apomictic embryo sac arises, it is usually 8-nucleate. Sometimes fewer than 8 nuclei occur as in *Ochna serrulata* or more than 8 as in *Elatostema euryhnchum*. A more common feature is the disturbed polarity and lack of proper organization of the various elements of the embryo sac.

EMBRYOGENESIS

Here we will discuss the various mode of embryo development in apomictic embryo sac. When the development of egg is associated with pollination it may be eugamous, semigamous or pseudogamous. On the contrary, when pollination is excluded the embryo is formed parthenogenetically.

  (i) *Eugamy*: Reports of the occurrence of the normal fertilization in apomictic egg can be cited in a number of plants. A few species are, *Parthenium, Leontodon hispidus, Poa, Malus, Rubus* and *Potentilla*.

  (ii) *Semigamy*: Sometimes the male nucleus penetrates into the egg but does not fuse with its nucleus. Subsequently, both nuclei divide independently. According to the position of the male nucleus inside the egg, a variable part of the embryo possesses nuclei of pure male origin. Frequently the sperm derivatives constitute the suspensor of the embryo. This phenomenon was first discovered by Battaglia (1945-1947) in *Rudbeckia speciosa* and *R. laciniata. R. sullivantii* is of particular interest because here an egg-like antipodal cell develop by semigamy (antipodal semigamy).

  (iii) *Unreduced Pseudogamy*: Here the sperm sooner or later degenerates either inside or outside the embryo sac. Battaglia (1947) favours a concept of pseudogamy, ranging from mere stimulation by pollen tube to the degeneration of the male nucleus in the egg cytoplasm. However, egg can develop by pseudogamy and the development of the endosperm usually takes place only after the fusion of the secondary nucleus with one sperm. Unreduced pseudogamy is comparatively widespread and is know in *Parthenium argentatum, Ranunculus auricomus, Malus seiboldi* and some species of *Potentilla* and *Rubus*.

  (iv) *Unreduced Parthenogenesis*: This implies the total absence of any stimulus referable to male gametophyte. In *Potentilla*, parthenogenetic embryos that arise in emasculated flowers are considerably reduced in size and are imperfectly differentiated. There are many examples of unreduced parthenogenesis, such as *Alnus rugosa, Antennaria, Crepis, Eupatorium, Taraxacum, Potentilla, Elatostema* sp. etc.

GENETICS OF APOMIXIS

Apomictic plants are generally hybrids or polyploids; as a consequence there is irregular meiosis. *Allium carinatum* is predominantly an apomictic species, where replacement of flowers by bulbils is under the control of genes. Recessive genes have been attributed for apomictic character.

Early hypothesis assumed several genes were involved for apomictic behaviour of plants. Experiments on *Parthenium argentatum* have revealed that three pairs (AABBcc) of genes determine the breeding behaviour of individual plant of this species. In homozygous condition gene *a* forms unreduced eggs, while *b* prevents fertilization and *c* promote egg development without fertilization. Therefore, only those plants with *aabbcc* genetic constitution would be apomictic. However, plants with *AABB cc* genotype have normal sexual behaviour because the gene *c* has no effect in the presence of *A* and *B*. However, Valle (1995) suggests a simple monogenic control for apomixis.Experimental studies on the genus *Ranunculus*, Nogler (1984) concluded that apospory is due to a single dominant gene **A**. The wild allele a$^+$ does not contribute to apospory but may function in normal embryo sac development. The sexual parent *R. cassubacfolius* (2n = 16) is homozygous for the wild allele (a$^+$a$^+$), whereas the aposporous parent *R. megacarpus* (2n = 32) is heterozygous (a$^+$a$^+$ AA).

In facultative apomictics, environmental conditions seem to play an important role in the shift from sexual mode of reproduction to apomixis as in *Deschampsia caespitosa*, which behaves as sexually reproducing plant in Sweden but shows vegetative vivipary when shifted to California.

SIGNIFICANCE OF APOMIXIS

As apomixis does not involve meiosis, there is no segregation and recombination of chromosome. Thus it is useful in preserving desirable characters for indefinite period. However, importance of meiosis in variation can not be ignored. In obligate apomictic species this advantage is enjoyed for a long time at the cost of long-term evolutionary flexibility, which is advanced mainly through the sexual reproduction. On the contrary, in facultative apomictic species, where sexual and apomictic plants occur together, the phenomenon is of great significance.

14 The Seed

Seed may be defined as ripened ovule, which develops after fertilization. All spermatophyta (i.e. gymnosperms and angiosperms) are characterized by the presence of seeds. In gymnosperms the seeds are naked because megasporophylls do not form ovary, called naked seeded plants, while angiosperms seeds are enclosed within the ovary, therefore, called covered or closed-seeded plants. The angiospermic seeds may either have two or one cotyledon and accordingly called dicotyledons or monocotyledons.

A typical seed has embryo, endosperm and seed coat. The embryo is diploid and is an outcome of fusion of male gamete and egg, whereas the endosperm is a post-fertilized product, develops as a result of fusion of secondary nucleus and a male gamete, usually triploid (Fig. 14.1).

	Ovary structure	Fertilization	Seed structure
Embryo sac in the ovule			
	Two polar nuclei	+♂gamete → Triple fusion	Endosperm (Food reserve)
	Egg Synergids	+♂gamete → Zygote	Embryo
Integuments of ovule wall			Seed coat (testa + tegmen)
	Ovary wall		Fruit wall (pericarp)

Fig. 14.1 Scheme in the events of seed formation.

14.2 Plant Embryology: Classical and Experimental

The nucellus is consumed during the development of seeds; however, it may persists as perisperm in some seeds, for example *Eriospermum carpense, Orygia decumbens*. In seeds, food is stored either in the cotyledons or in a special storage tissue, the endosperm. In legumes, for example, the food is mainly stored in cotyledonds and there is no endosperm. Absence of endosperm in seeds suggests that it is consumed completely by the developing embryo. Such seeds are known as **non-endospermic or exalbuminous**. In some dicotyledons (e.g., castor; Fig. 14.2) and monocotyledons (e.g., cereals; Fig. 14.3, and grasses) the food is mainly stored in endosperm. Such seeds are known as **endospermic or albuminous**. Food materials stored in the seeds include carbohydrates, proteins and fats. Different types of seeds accumulate different types of food. For example, Seeds of rice, wheat and maize etc. are rich in carbohydrates. Seeds of castor, sunflower, soybean, peanut etc. are rich in fats and seeds of legumes are rich in protein contents.

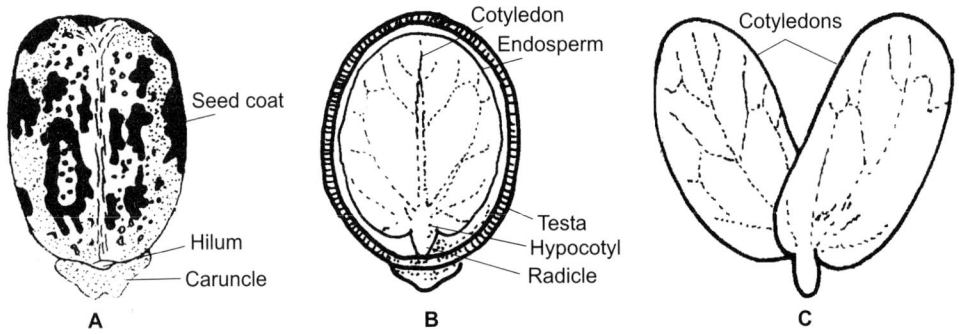

Fig. 14.2 Seed of castor (*Ricinus communis*). **A.** Protuberance at the base of seed the caruncle. **B.** L.S. showing embryo, cotyledon and endosperm. **C.** Embryo with cotyledon opened out.

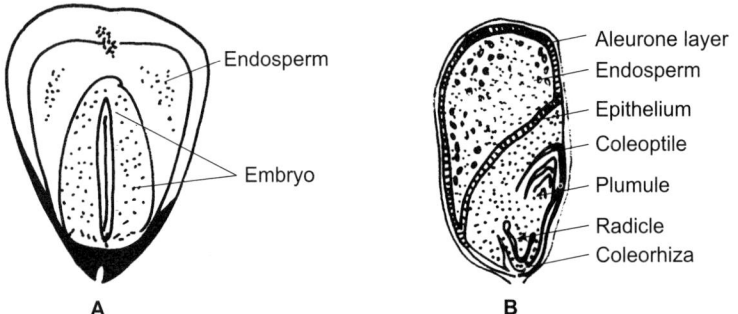

Fig. 14.3 Grain of maize (*Zea mays*). **A.** Entire grains showing embryo. **B.** L.S. of seed showing endosperm and scutellum consisting embryonal axis.

There are variations in the seed size, shape, colour and surface. The seeds sizes vary from tiny dusty particles as is found in some orchids to a large like double-coconut (*Lodoicea maldivica*).

As the seeds get detached from the funicle, a scar is left on its surface, called hilum. In family Fabaceae the funicular scar persists as hilum and slightly below is the micropyle as a minute pore. Seed coat has two layers: outer **testa** and the inner one is called **tegmen**. Inside the seed coat there are two massive cotyledons attached laterally to the embryonal axis. A portion of the embryonal axis projects beyond the cotyledons, the pointed end of which is the embryonic shoot (**plumule**). The other end of the axis is the embryonic root (**radicle**). The portion between the plumule and the point of attachment of the cotyledons to the axis is known as **epicotyl**, whereas the portion of the axis between the radicle and the cotyledons is called **hypocotyl.**

PARTS OF SEED

(i) Seed Coat

As the ovule develops into seed, the integument matures into seed coat. In bitegmic ovules the seed coat is developed either from both the integument or the inner integument is lost and the seed coat is formed from the outer integument alone. The seed coat morphology provides valuable information about the taxonomical identification and relationship because seed characters are considered to be very conservative (Corner, 1961; Tobe *et al.*, 1987). Different species of *Abrus* may be identified on the basis of their seed colour. Seeds of *A. precatorius* are scarlet and jet-black; reddish brown and mottled in *A. laevigatus* and olive green in *A. fructiculosus*. The surface of the seeds varies from highly polished to markedly roughened or sculptured. During formation seed coats, the integuments undergo significant histological changes.

In the Cucurbitaceae the ovules are bitegmic but the outer integument alone forms the seed coat, the inner integument degenerates. Seed coat development in *Luffa* has been described by Singh (1971), in an unfertilized ovule, the mature embryo sac shows 10-15 layered outer integument, while the inner integument is only 2-3 layered thick. However, in mature seed, the seed coat is differentiated into five layers. In *Gossipium* sp., both the integuments contribute in the formation of seed coats.

In Acanthaceae, the ovules are unitegmic. In *Andrographis*, *Elytraria* and *Haplanthus* the integument is completely digested; therefore the seeds are devoid of seed coat. In many parasitic angiosperms also (e.g., members of Santalaceae and Loranthaceae) the seeds are naked because integuments are altogether absent.

(ii) Some Special Structures

During development of seed, some special structures arise from the various parts of the ovules. They include caruncle, operculum, aril and elaisome.

(a) Caruncle

This is fleshy, whitish structure present on the micropylar end of the seed. It arises due to proliferation of cells at the tip of outer integument, on the side of the funiculus or all round the micropyle. It is very common in Euphorbiaceae (Fig. 14.2 A). Two functions have been assigned to the caruncle:

(i) Absorption of water due to its hygroscopic nature, therefore, facilitates germination process and (ii) invites insects because of sugary nature, thus seed dispersal is brought about.

(b) Operculum

It is a plug-like structure in the micropylar region of the seed, which gets detached during germination. The ontogeny of the operculum is variable. It arises either from the nucellar apex, endosome, exosome or the hilar region of the seed. The operculum is generally present in monocotyledonous families, such as Araceae, Commelinaceae, Marantaceae, Musaceae, Zingiberaceae and Lemnaceae (Fig. 14.4A) but is also met in some dicotyledonous families like, Bignoniaceae, Droseraceae, Melastomaceae and Nymphaeaceae. The operculum facilitates germination and provides protection to the micropylar region of the seed.

(c) Aril

Arils are fleshy seed appendages often with vivid colours to attract animals. It arises from the funiculus or testa or both. Aril is regarded as the third integument and generally surrounds the ovules more or less completely in post-fertilization stage e.g., *Asphodelus, Cardiospermum, Lichi, Lomantia, Myristica fragrans, Trianthema* (Fig. 14.4B) etc.

(d) Elaiosome

Sernander (1906) introduced the ecological term elaiosome for all the fleshy and edible parts of seeds which are dispersed by ants (Fig. 14.4C). The elaiosome arise as an outgrowth of raphe or hilum (e.g., *Trillium ovatum*). Elaiosome is rich in nutritive substaces like proteins, lipids, starch, and vitamins. Three functions have been ascribed to the seed coats:

(a) It protects the embryo and endosperm from desiccation, mechanical injury, unfavourable temperatures and attack by bacteria, fungi and insects.

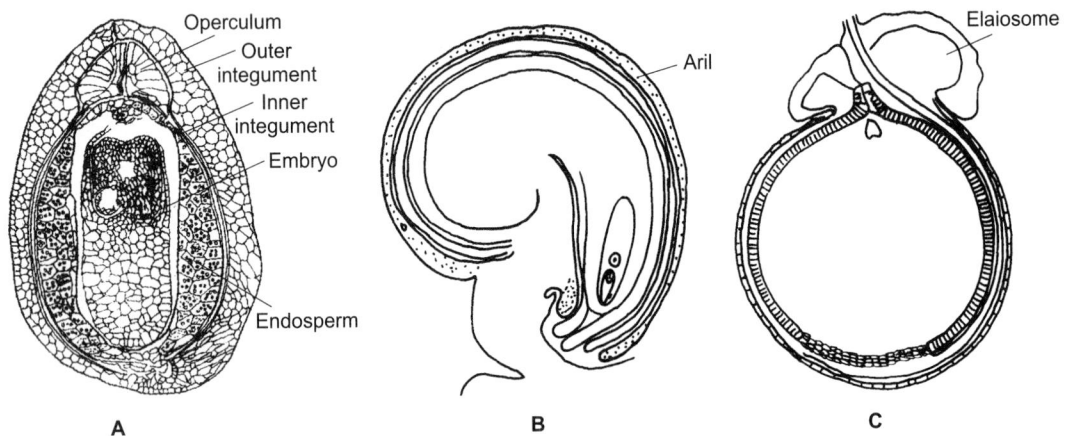

Fig. 14.4 **A.** L.S. of mature seed of *Lemna pausicosta*, showing operculum. **B.** Ovule showing aril in *Trianthema monogyna*. **C.** *Trillium ovatum*, seed with elaiosome.

(b) It helps in the dispersal of seeds by developing special structures such as wings, fleshy and brightly coloured hairs and air filled cavities.
(c) It may provide nutrition to the developing embryo.

LABYRINTH SEEDS

Seeds that show an irregular internal structures when cut in any planes are called labyrinth seed (Van Heel, 1970). The seed coat first encroaches on the endosperm and later the cotyledons, forming the labyrinth structure. The labyrinth becomes more pronounced if the lobing of the cotyledons occur in combination with the intrusion of the testa between the folds and lobes e.g., *Kingiodendron pinnatum* (Fig. 14.5), *Harnandia peltata* etc.

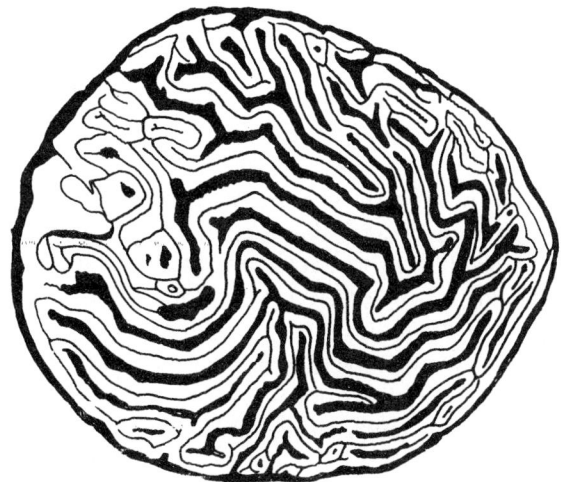

Fig. 14.5 *Kingiodentron pinnatum*, Median Longitudinal section of a labyrinth seed (After Van Heel, 1970).

SEED DISPERSAL

Seed dissemination to various distances from the parental plant is called seed dispersal. Seed dispersal evolved as a mechanism to avoid overcrowding of seedlings in order to overcome competitions, interbreeding, vulnerability to animals and epidemic attack of pathogens during the course of evolution.

Seeds cannot move independently from one place to another. Their movement or transport is brought about by external agencies. The seeds are dispersed either along with the fruit or the fruit remain attached to the parental plant and burst to expose the seeds, which are then carried to various localities through one or the other agencies. Depending upon the agencies the dispersal is categorized as following on next page:

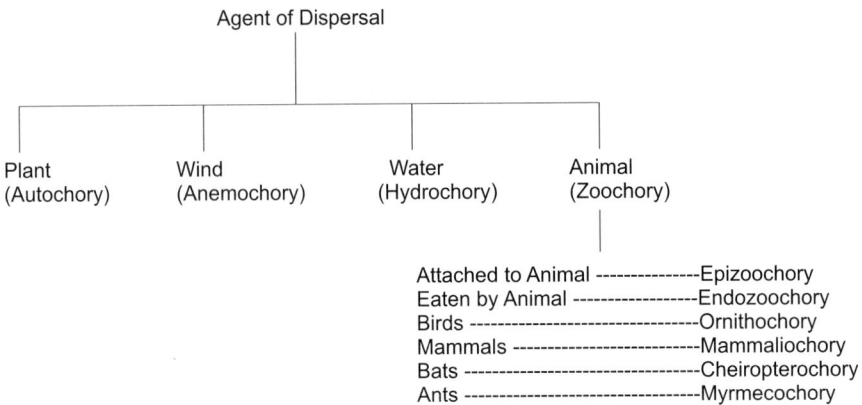

(i) Autochory

In plants showing mechanical dispersal of seeds, the fruits remain attached on the parental plant for long period. The ejection of the seeds is brought about by the desiccation or the turgidity of the cells of fruit wall.

In balsam (*Impatiens parviflora*), fruit is pentacarpellary capsule. The mature fruit, when touched, burst open with a force and the carpels rolls up immediately like a spring (Fig. 14.6A). In this process seeds are thrown to several feet. The pods of camel's foot climber (*Bauhinia vahlii*) burst violently with a cracking sound scattering the seeds away (Fig. 14.6B).

In squirting cucumber (*Ecballium elaterum*) the seeds are ejected while the fruit is still soft and succulent. Here the pedicel remains attached to the fruits like a stopper. As the fruit matures, the turgor pressure within the fruit causes it to burst open at the point of attachment of the pedicel. Consequently, the seeds along with the mucilagenous mass come out like fountain.

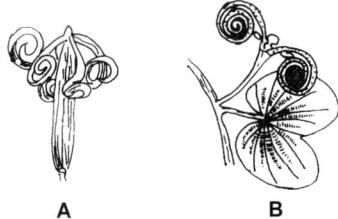

Fig. 14.6 Seeds dispersal by bursting of fruits. **A.** *Impatiens parviflora*. **B.** *Bauhinia vahli*.

(ii) Anemochory

The wind is probably the most impotant agency of seed dispersal in nature. The fruit and seeds show different kinds of adaptations to be carried away by the slightest blow of wind. Some seeds of orchids are so thin and light that they are called **'dust seeds'** (more than 10,000 seeds are produced per capsule and each seed weigh about 0.004gm) and blown away by wind to distant places. Some seeds (e.g., *Jacaranda, Moringa, Lagerstroemia, Pinus*) and fruits (e.g., *Acer, Holoptelia, Terminalia, Shorea*; Fig. 14.7A) have wings, which enable them to float in the air.

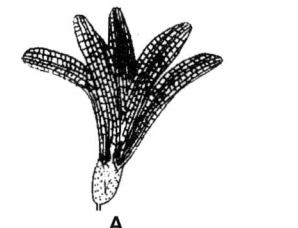 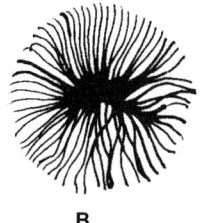

Fig. 14.7 **A**. Winged fruits of sal (*Shorea robusta*). **B**. Hairy seed of cotton (*Gossipium* sp.).

Seeds of certain plants possess hairs which also help to float in air and eventually carry them to a long distance. In cotton (*Gossipium*; Fig. 14.7 B) hairs are the outgrowth from the seed coat and occur all along the surface. In some plants tuft of hairs is present at one (e.g., *Calotropis*, *Asclepias*) or both the ends of seed (*Alstonia*).

In the members of family Asteraceae sepals are modified into pappus, which acts like a parachute and help in dispersal. In *Aristolochia*, *Papaver*, *Argemone mexicana*, *Nigella* etc. 'censer mechanism' is operative, where the fruit is capsule and on maturity splits but remains attached to the parental plant. With the swinging of the fruits with slightest blow of wind seeds are scattered.

(iii) Hydrochory

Fruits and seeds, which show water dispersal generally, develop some kind of floating device and protective covering which make them water resistant.

In *Lodoicea* (double coconut) and *Cocos nucifera* (coconut) the fibrous mesocarp functions as a protective as well as floating device. In *Nelumbo nucifera* (lotus), the seeds are embedded in the spongy thalamus, which help the fruits to float on water (Fig. 14.8).

(iv) Zoochory

There are two methods mainly by which animals bring about the dispersal of seeds. Some fruits are eaten by the animals and the seeds passed out with the excreta unharmed (**endozoochory**), for examples banyan (*Ficus benghalensis*) and peepal (*Ficus religiosa*).

Fig. 14.8 *Nelumbo nucifera*: Seeds are embedded in fleshy thalamus.

Fig. 14.9 Fruit of *Xanthium* showing hooks.

Alternatively, the fruits or the seeds may be carried by the animals, externally sticking to their body by hooked fruits or seeds (**epizoochory**), for eamples, *Achyranthes, Aristida, Xanthium* (Fig. 14.9) and *Tribulus* etc.

Unlike the birds, which are attracted by the colour of the fruits, bats are attracted by the rancid odour and they feed upon the fleshy part only, seeds are pressed out (**cheiropterochory**). They feed upon the fleshy and juicy parts of fruits e.g., some members of the families Anacardiaceae, Meliaceae and Sapotaceae etc.

Dispersal of seeds by ants is describes as **myrmecochory**. The ants like oily bodies or the oily appendages of the fruits of *Anemone nemeros, A. hepatica* and many species of *Trillium*. After consuming the oily substances, the seeds are either dropped on the way or out of their nests.

Mankind through their agricultural or commercial practices brings about the widest dissemination of both fruits and seeds.

SEED DORMANCY

Nearly all land plants pass through a phase of dormancy at some stage in their life cycle, either as spores in lower plants such as thallophyta, bryophytes and ferns, or as seeds in case of higher plants. Apart from spores and seeds, perennial plants show different types of resting organs, such as the winter resting buds of trees, tubers, rhizome, bulbs, corms. Usually the phase of dormancy coincides with the period of unfavourable climatic conditions, usually low temperature, high temperature or drought. Thus, many resting organs show adaptation to adverse climatic conditions either in the form of special structures or attaining a physiological state, which confers resistance to frost or heat than is shown during the active-growing phase of the plants. However, there are instances where despite all favourable environmental conditions growth is ceased.

Inability of the viable seeds to germinate is known as seed dormancy. The fully dormant condition of the seeds or the other organs is not attained suddenly, but is gradually developed over a period. However, the seeds of many species, including the crop plants, corn (*Zea mays*), pea (*Pisum sativum*) and bean (*Phaseolus aureus*) germinate almost immediately after harvesting, when placed under suitable conditions. However, there are large numbers of species, which show different types of innate dormancy, the main features of which are summarized as follow:

(i) Impermeable Seed Coats

One of the most common factors associated with seed dormancy is the presence of a hard seed coat. The hard seed coat may be responsible for dormancy by preventing the absorption of water as in the members of Papilionaceae (e.g., *Lupinus arboreus, Melilotus, Trigonella*); by preventing gaseous exchange as in *Xanthium* and; by mechanically restricting the growth of the embryo as in the *Amaranthus retroflexus*.

The barriers can be overcome, depending upon the nature; either by mechanical scarification, such as shaking the seeds with some abrasive materials (e.g., sand) or scratching or nicking the coat with a knife; or by chemical scarification such as immersing the seeds into strong sulphuric acid or into organic solvents (alcohol, acetone).

In nature, the action of acids and enzymatic conditions of the digestive tract of birds and other animals or by the action of fungus and other microorganism or by abrupt change of the surrounding temperature accomplish the process. In some grasses and legumes the seeds would germinate only after these get fire treatment. Their seeds have high degree of tolerance to temperature due to fire. Common examples are *Themeda, Heteropogon, Andropogon, Rheus, Tephrosia* and *Stipa*.

(ii) Immature Embryo

At the time of release of seeds or fruit from the parent plant the embryo are still not fully developed. Dormancy due to immature embryos may be found in the members of Orchidaceae and Orobancheae as well as some *Fraxinus* and *Ranunculus* species. Dormancy in immature embryos can only be overcome by allowing the embryo to complete its development within the seed in an environment favourable to germinate.

(iii) After-ripening

A large number of plants produce seeds that do not germinate immediately but do so after a period of time under normal conditions for germination. They gradually emerge from dormancy over a period ranging from a few weeks to several months. Many common cereals show this type of dormancy, e.g., barley, oat and wheat.

After-ripening occurs for some species during dry storage. For others, moisture and low temperature are necessary process called **stratification**. Natural stratification occurs when seeds shed in the fall are covered with cold soil, debris and snow. Many workers are of the opinion that after-ripening is a period of rest, however, several studies have revealed that considerable physiological activities occur during the so-called after-ripening or dormant period.

Usually this type of seed dormancy may be removed either by storing them at 35-40°C for 2-4 days or by removing the seed coats, e.g., in cereal grains and *Avena fatua*.

(iv) Inhibitors

Seed dormancy may also be due to the presence of inhibitory substances in the seed. Inhibitors of germination and seedling growth have been isolated from different parts of a seed *viz*. pericarp, testa, endosperm and embryo. Phenolic compounds in general act as naturally occurring inhibitors of germination. Of these compounds, coumarin and its derivatives are regarded as the most important germination inhibitors. Abscisic acid (ABA) is another common inhibitor in plant tissues; caffeic acid and ferulic acid fulfill the same function in the fruit pulp of the tomato.

Of the growth hormones, gibberellic acid and cytokinins have significant role in breaking the dormancy and promoting the germination. Inhibitor-induced dormancy can be removed by leaching out of the inhibitors and germination can be increased.

CONDITIONS FOR GERMINATION

The process by which the dormant embryo of the seed resumes active growth and is expressed with the emergence of plumule and radicle is called germination. As we have seen that barrier mechanism can either be present in the embryo itself or in all the layers that surround it. If they are removed the embryo is able to germinate. However, it germinates only if certain external conditions in ambience are congenial. These conditions are water, oxygen, temperature and light.

(i) Water

In dormant seeds food materials are stored in the concentrated form and therefore, have low physiological activities. In the presence of water the macromolecules are converted into forms that are utilized by the growing points. Water also serves as a medium where enzymatic function starts. Besides, the seed coat becomes soft after absorbing water and allows the growing embryo to come out of it. Usually moisture present in the soil is sufficient for germination but some seeds with hard seed coat or inhibitors require more water.

(ii) Oxygen

Energy is necessary for germination. It is made available in the form of ATP, which is derived from substrate chain and respiratory chain phosphorylation. Here, oxygen is a necessary prerequisite for the functioning of the respiratory chain and thus, for oxidative phosphorylation. For this reason the presence of oxygen is usually a condition needed for germination.

(iii) Temperature

A number of physiological processes occur within the seed during germination. Therefore, a suitable temperature is always required for germination. The range of optimum temperature varies greatly in different types of seeds. However, most of the seeds fail to germinate below 0°C and above 48°C.

(iv) Light

In numerous seeds, light is required for breaking the dormancy. These seeds are called **photoblastic** as opposed to the **non-photoblastic** seeds, which do not require light for dormancy release. Borthwick and Hendricks (1952) for the first time established the effects of red and far-red light in regulation of germination and involvement of photoreversible phytochrome pigments, while investigating the influence of light on lettuce seeds (*Lactuca sativa*). In lettuce seeds, red light induces gemination and far-red inhibits, thus P_{fr} is promotive and P_r appears to be suppressive.

$$P_r \underset{730 \text{ nm}}{\overset{660 \text{ nm}}{\rightleftarrows}} P_{fr}$$

TYPES OF GERMINATION

Under appropriate conditions, the seed absorb water from the soil through the micropyle. The first visible indication of germination is the swelling of the seed. It is accompanied with the softening of the seed coat. Absorption of water causes a series of physiological changes that result in the establishment of the seedling. Increased physiological activity is indicated by the higher rate of respiration of the germinating seedling.

Cell division start in the growing part of the embryo (radicle and plumule) when they get food materials. The radicle is the first part of the embryo to come out of the seed coat. It is positively geotropic and soon grows towards the soil irrespective of it initial orientation. With the further growth of embryo the seed coat ruptures allowing the plumule to come out and to form shoot system.

On the basis of the behaviour of the cotyledon, the germination may be of the following types:

(i) Epigeal Germination

In the seeds with epigeal germination the cotyledons are brought above the ground due to elongation of hypocotyls (the part of the embryonal axis between cotyledons and radicle). Here the cotyledons besides food storage, perform photosynthesis till the seedlings become independent. A few examples are cotton, castor (Fig. 14.10 A), papaya etc.

(ii) Hypogeal Germination

In hypogeal germination the cotyledons do not come out of the soil surface. In such seeds epicotyl (part of the embryonal axis between plumule and cotyledons) elongates pushing the plumule out of the soil. All monocotyledons show this type of germination. Among dicotyledons gram, pea (Fig. 14.10B), groundnut and mango etc. are common examples of hypogeal germination.

(iii) Vivipary Germination

It is special type of germination found in mangrove plants for e.g., *Rhizophora*, *Heritiera*, *Ceriops* and *Avicenia* etc. In this process embryo does not undergo any period of rest and the seeds germinate inside the fruit while it is still attached to the parental plant. The radicle elongates considerably and projects out of fruit. The lower part of the seed becomes thick and swollen (Fig. 14.10 C). Finally seedling breaks off the parental plant due to its increasing weight and gets embedded in the soil. Soon lateral roots develop and plant is established independently.

IMPORTANCE OF SEED

Seeds are valuable to mankind as well as to plants themselves. Seeds ensure the perpetuation of plants and also dissemination of plants to different areas. Especially, for annuals the seeds are the only means of multiplication and their perpetuation.

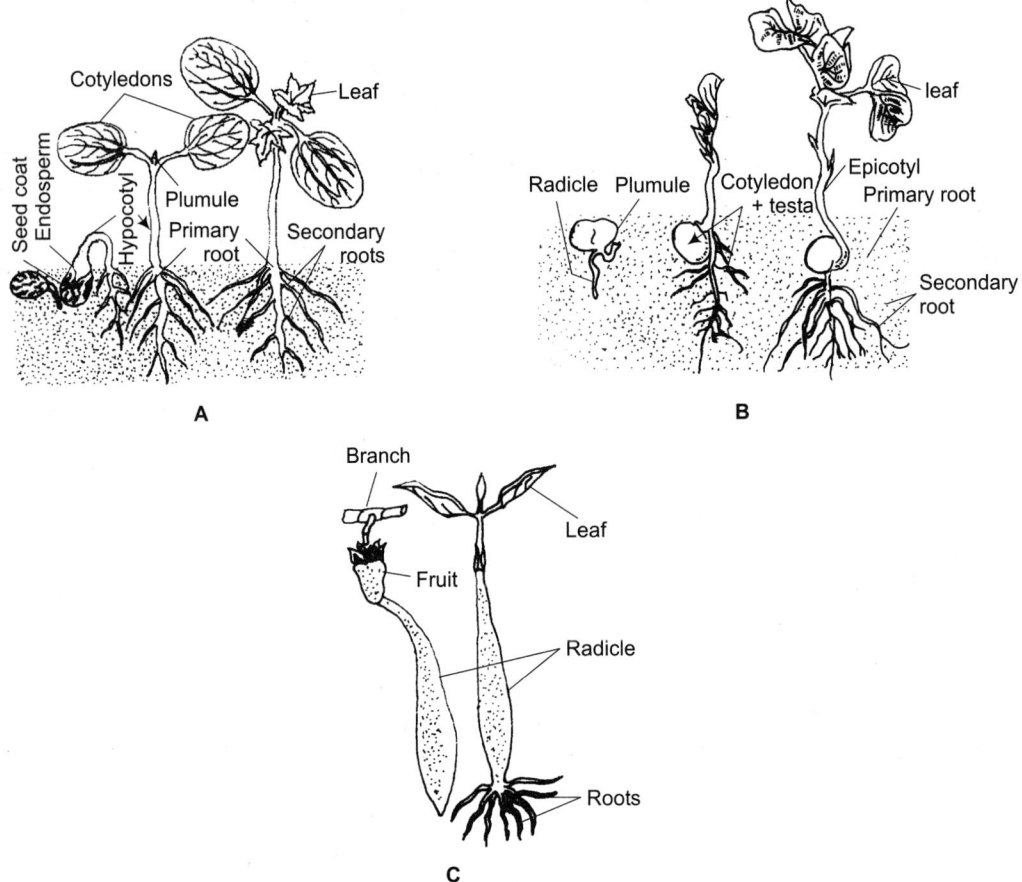

Fig. 14.10 Different types of germination. **A.** Epigeal germination in castor. **B.** Hypogeal germination in pea. **C.** Vivipary germination in *Rhizophora*.

The endosperm or cotyledons of the seeds with their rich reserves provide staple food to mankind and animals. The family Poaceae contributes more food seeds than any other family. Rice, wheat and maize meet the 90% food demand of global population; other edible seeds provided by Poaceae are oat, barley, sorghum, millet, rye etc. The second most important family is Fabaceae that provides pulses.

Seeds are also the sources of several other items of human use, such as fibres, oils, spices, beverages, medicines etc. The cotton fibre, which is derived from the seed coat of *Gossypium* species, has been in use far more than any other plant. Edible oils are obtained from the seeds of groundnut, coconut, mustard and sunflower. Coffee and cocoa made from *Coffea arabica* and *Theobroma cacao* seeds, respectively. Seeds of castor (*Ricinus communis*), Isabghul (*Plantago* sp.) and nux vomica (*Strychnos nuxvomica*) are used as medicines.

15 Embryology in Relation to Taxonomy

Modern embryology includes three main disciplines: descriptive, phylogenetic and the experimental embryology. The second or the phylogenetic embryology deals with the evaluation of various embryological data in determining the inter-relationships of the different orders and families with a view to improve the system of classifications.

The older system of classification is mainly based on external morphology, especially floral characters. Presently all such system, which are based on one aspect of plants tend to be artificial and may serve the general purpose only for which they are constructed. A natural system of classification is the ultimate goal of taxonomy. In view of attaining a true natural system of classification is recommended that it must be based on scrutinization of all the plant characters and a harmony of interpretation is obtained from all aspect of organs and tissues of the plants (Baily, 1949). Gross morphology, no doubt provided a foundation for framework of classification, yet embryology together with other aspects of plants is equally important.

Since the embryological studies were difficult, therefore, its development has been rather slow (Maheshwari, 1964). Drawing attention towards the fact, he stated that embryology involves tedious practice to obtain desired data about plants under investigation. The embryological techniques are complicated and time consuming. The time period for the collection of material is very long and often one has to wait for weeks and months together in order to obtain the desired stage. In spite of the above limitations the use of embryology in taxonomy has shown rapid progress recently. Much attention of the workers has been attracted towards the valuable embryological characters in deriving phylogenetic conclusions. Despite odds, the embryological characters are of great significance because of its conservative nature. However, we are now better equipped with embryological information covering almost all major groups of angiosperms to evaluate affinities and phylogenetic relationship (Kapil and Bhatnagar, 1991, 1994).

The role played by the embryology in taxonomy was for the first time highlighted by German embryologist Schnarf (1931). Susequently, P. Maheshwari (1950,1961) emphasized that the following embryological characters are of diagnostic values:

1. *Anther tapetum*: Whether it is of glandular or amoeboidal type.
2. *Quadripartition of microspore mother cell*: Whether it takes place by furrowing or by the formation of cell plate and whether the mode of division is successive or simultaneous.
3. *Development and organization of pollen grains*: Number and position of germ pores and furrows: adornment of the exine; place of formation of generative cell; number and shape of the nuclei in the pollen grains at the time of discharge from the anther.
4. *Development and structure of ovule*: Number of integuments and the alteration in structure in which they undergo during the formation of seed; presence or absence of vascular bundles in the integument; shape of the micropyle, whether it is formed by the inner integument, or the outer, or both; presence or absence of an obturator.
5. *Form and extent of nucellus*: Whether it is broad and massive or thin and ephemeral; presence or absence of hypostase; place of the origin of integuments, whether close to the apex of the nucellus or near its base; persistence or gradual disappearance of the nucellus during seed formation.
6. *Origin and extent of the sporogenous tissue in the ovule*: Nature of archesporium, whether it is one-celled or many- celled; presence or absence of wall layers; presence or absence of periclinal divisions in the cells of the nucellar epidermis.
7. *Megasporogenesis and development of the embryo sac*: Arrangement of the megaspores; position of the functioning megaspore; whether the embryo sac is monosporic, bisporic, or tetrasporic; number of the nuclear division intervening between the megaspore mother cell stage and the differentiation of egg.
8. *Form and organization of mature embryo sac*: Shape of the embryo sac and the number and distribution of the nuclei; persistence or early disappearance of the synergids and antipodal cells; increase in the number of the antipodal cells, if any; formation of the embryo sac caeca or haustoria.
9. *Fertilization*: Path of the entry of the pollen tube; interval between pollination and fertilization; any tendency towards branching of the pollen tubes during their course to the ovule.
10. *Endosperm*: Whether it is nuclear, cellular, or helobial type; orientation of the first wall in those cases in which it is cellular; presence or absence of haustoria and the manner in which they are formed if present; nature of the food reserve in endosperm cells; persistence or gradual disappearance of the endosperm in the mature seed.
11. *Embryo*: Relation of the proembryonal cells to the body regions of the embryo; form and organization of the mature embryo; presence or absence of suspensor haustoria.
12. *Certain abnormalities of development*: Parthenogenesis; apogamy; adventive embryony, polyembryony etc.

Although large number of floral characters are required for the identification of a family, however, a few families are so distinctive that they can be recognized on the basis of single character only, for e.g., a single pollen grain from microspore mother cell in Cyperaceae; Oenothra type of embryo sac in Onagraceae and composite endosperm in the Loranthaceae.

A comparative account of distribution of embryological characters in Dicotyledons and Monocotyledons is given in the Table 15.1:

Table 15.1 Comparative distribution of embryological characters (after Kapil and Bhatnagar, 1991)

Character	State	Dicot (%)	Monocot (%)
Anther wall	Basic type	100	-
	Dicot type	97	3
	Monocot type	60	40
Tapetum	Secretory	87	13
	Amoeboidal	38	62
Cytokinesis	Simultaneous	95	5
	Successive	13	87
Microspore tetrads	Tetrahedral	96	4
	Isobilateral	14	86
Pollen	Binucleate	84	16
	Trinucleate	63	37
Ovule	Bitegmic	70	30
	Unitegmic	100	-
	Crassinucellate	79	21
	Tenuincellate	89	11
Endosperm	Nuclear	83	17
	Cellular	97	3
	Helobial	17	83
Embryogeny	Asterad	79	21
	Onagrad	93	7
	Solanad	100	-
	Caryophyllad	55	45
	Chenopodiad	83	17
	Piperad	80	20

EMBRYOLOGY IN SOLVING TAXONOMICAL PROBLEMS

Cyperaceae

In angiosperms usually microspore mother cell gives rise to four microspores. In family Cyperaceae it has been noticed that instead of four only one microspore develops as pollen from the microspore mother cell. During meiosis in the megaspore mother cell, three nuclei are cut off, on one side and remain non-functional, whereas the remaining fourth one is functional and forms the generative cell. This phenomenon is so unique and constant amongst both Indian and overseas members of the family that it can be relied upon for identification. Further, simultaneous type of microspore formation numbering four is indicative of the phylogenetic correlation amongst *Juncaceae* and *Cyperaceae* and perhaps *Cyperaceae* seeks its origin from *Juncaceae*.

Epacridaceae

Except Cyperaceae the phenomenon of degeneration of three nuclei during the formation of microspore tetrad in the anther is rather rare. Family Epacridaceae resembles Cyperaceae in this respect. However, these families differ from each other in cellular stage of pollen at the time of shedding. Pollens are shed at two-celled stage in Epacridaceae and three-celled stage in Cyperaceae.

Onagraceae

In the family Onagraceae, the distinguishing feature is the presence of *Oenothera* type of embryo sac (Fig. 15.1). The megaspore mother cell on meiosis produces the megaspore tetrad. Generally, the micropylar microspore is functional developing into embryo sac. The chalazal megaspore remains non-functional cell at the base of embryo sac. In functional megaspore the nucleus divides by two repeated divisions resulting into four nuclei. All the nuclei remain located at the micropylar end of the embryo sac only. Its chalazal end is enucleate. Out of the micropylar nuclei three develop into egg apparatus and one functions as polar

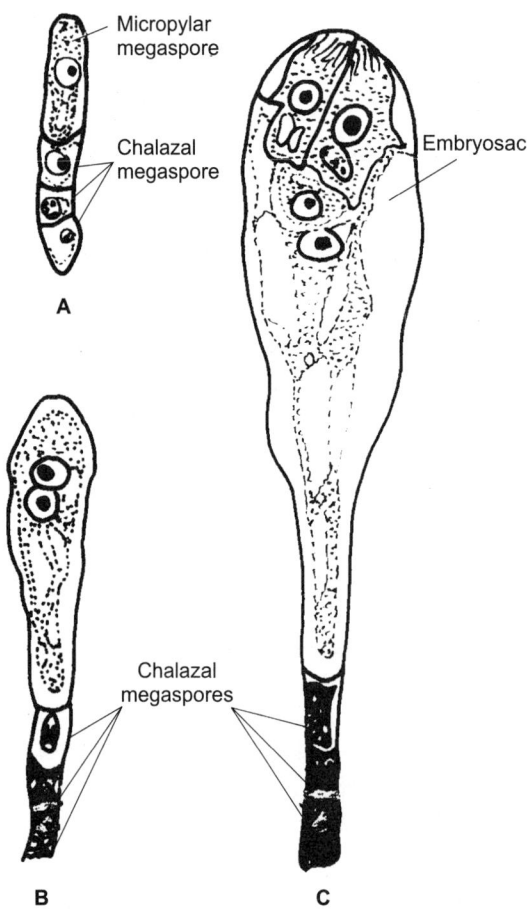

Fig. 15.1 Embryo sac development in *Oenothera acaulis*. **A.** Megasore tetrad; the functional enlarged micropylar megaspore. **B.** Two-nucleate megaspore. **C.** Mature embryo sac (After Pagni, 1958).

nucleus. Antipodals are absent. This type of monosporic embryosac is unique in this family and is predominant amongst all genera.

Earlier, genus *Trapa* was placed (Bentham and Hooker, 1883) in the family Onagraceae. Others (Raimann, 1898; Engler and Dials, 1936; Hutchinson, 1959) have included *Trapa* in the family Hydrocharitaceae, which has now been given a new name Trapaceae. Pulle (1938) and Ram (1956) have supported the earlier view to place it under family Trapaceae on the ground of embryological differences with Onagraceae. Eames (1953) on the basis of anatomical features has suggested that *Trapa* should be placed as a separate family Trapaceae rather than as a genus under the family Onagraceae.

Podostemaceae

Engler and Prantl (1931) placed Podostemaceae with Crassulaceae, Saxifragraceae, and Hydrostachyaceae. Hutchinson (1926) grouped them with Hydrostachyaceae in the order Podostemales. Subramanyam (1962) compared the embryological characters of Crassulaceae with those of Podostemaceae and Hydrostachyaceae and emphasized that the differences between these families are probably due to their special mode of life.

Family Podostemaceae constitutes mainly aquatic plants. The individuals have typical features of development of pseudoembryosac due to the disintegration of nucellar cells situated just at the base of embryo sac. Other embryological features of taxonomic importance are paired pollen grains, bisporic embryo sac, solanad type of embryogeny, absence of antipodals except in the genus *Dicraea* and lack of endosperm due to absence of triple fusion.

These features justifies separation of Podostemaceae from other two families, namely Crassulaceae and Saxafragaceae and placing together with Hydrostachyaceae in order Podostemales.

Ranunculaceae

The Family Ranunculaceae has long been considered a natural taxon. Recent embryological studies point out that there are outstanding differences amongst various members. In this connection special attention is drawn on the position of the genus *Paeonia*. Worsdell (1908) and Eames (1961) on the basis of floral and vascular anatomy pointed out that the study is indicative of the anomalous position of the genus in family Ranunculaceae. Therefore, Tyagi (1970) has recommended the placement of *Paeonia* either as a separate family or a sub-family under Ranunculaceae itself. Morphology and the chromosome number are also in support of the above view (Gregory, 1941). An observation on the behaviour of zygote nucleus that follows the course of repeated divisions results in coenocytic structure (Yakovlev and Yoffe, 1957). After considerable numbers of nuclei are developed, they start moving towards the periphery and arrange themselves around a large central vacuole. However, Murgai (1952) states that first division of the zygote is followed by wall formation resulting in the formation of two-celled stage, one cell being apical and the other is basal. The basal cell thus, formed becomes coenocytic due to repeated nuclear divisions. Just after this, wall formation around the nuclei starts from the periphery inwards. Few peripheral cells then proliferate and give rise to many embryo initials. Out of these embryo initials, one attains maturity and the rest degenerate. This type of embryo development is unique. It also happens to be distinguishing feature and supported by many workers (Fig. 15.2).

15.6 Plant Embryology: Classical and Experimental

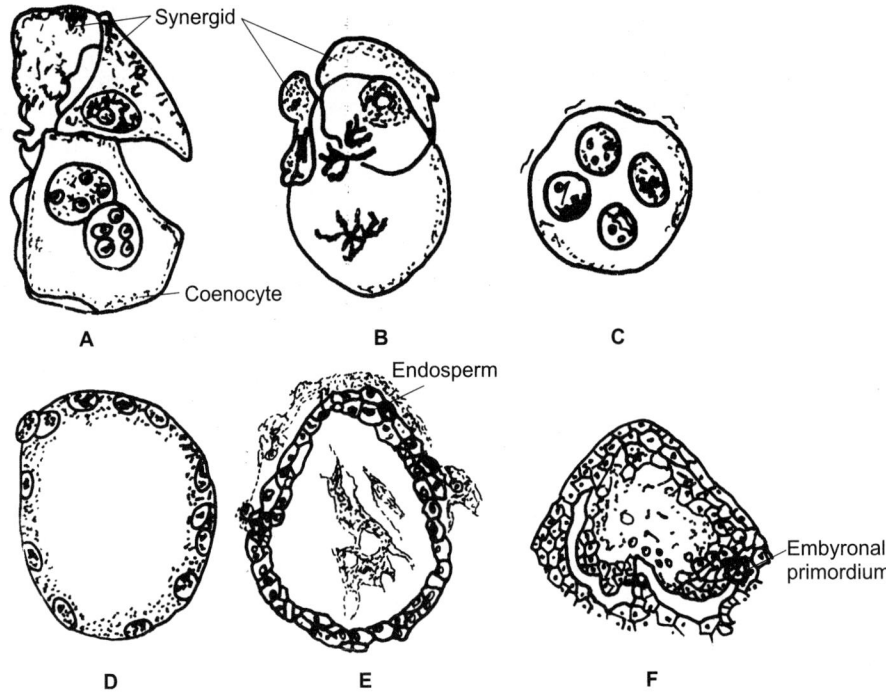

Fig. 15.2 Embryogeny in *Paeonia lactiflora*. **A**. Binucleate coenocyte. **B**. Division phase in the binucleate coenocytes. **C**. Four-nucleate stage. **D**. Multinucleate stage; all the nuclei are arranged peripherally. **E**. Wall formation. **F**. Embryo differentiation (After Yakolev, 1969).

Table 15.2 Comparison between *Paeonia* and *Ranunculus*

Character	Paeonia	Ranunculaceae
Stamen	Spirally arranged, Centifugal	Spirally arranged Centripetal
Anther	Multilayered endothecium, mostly 2-layered tapetum	One-layered endothecium and one-layered tapetum
Pollen exine	Reticulate pitted exine Large and elongated generative cell	Granular, papillate or smooth Lenticular generative cell
Female archesporium	Multicelled, many megaspore mother cells function	Uni- or multicelled, one Megaspore mother cell functions
Antipodal cells	Persistent, not polyploid	Persistent (ephemeral in *Adonis*), nuclei one or more than one, polyploid
Embryogeny	Unique (free nuclear or first division of zygote is transverse	Onagrad or rarely Solanad type
Seed	Arillate	Non-arillate
Fruit	Follicle	Achene

The above dissimilarities strongly support the removal of Paeonia from the Ranunculaceae to a separate family, the Paeoniaceae.

Lemnaceae

Lawrence (1945, 1952) has suggested shifting of Lemnaceae to the order Helobiales. On the contrary embryological studies of the representatives of the family reveal absolutely different picture. The form and development of proembryo and endosperm and presence of integumentary operculum bring Lemnaceae and Araceae much more closely as compared to Helobiales.

Santalaceae

Studies on the genus *Exocarpus* have recently revealed interesting facts. Gagnepain and Boureau (1947), made suggestions for removal of the genus from Santalaceae on the basis of presence of well developed pollen chamber; an articulate pedicel resembling the genera *Podocarpus* and *Acmopyle*; occurrence of naked ovule and subsequently, its placement in a separate family Exocarpaceae somewhat near Taxaceae in gymnosperms. Lam (1948) said that the simple nature of ovule is suggestive of its correlation with *Salix* as well as *Casuarina*, members of Protangiospermae.

Ram (1959) on the basis of embryogeny of *Exocarpus cupressiformis*, *E. strictus* and *E. sparteus* has clearly shown that beyond doubt *Exocarpus* is a member of angiosperms, belongs to Santalaceae only, and not of gymnosperms. His statement is supported by the resemblances existing between *Exocarpus* and angiosperms with respect to the structure of flower, development and structure of anther, endothecium in the anther being fibrous, presence of distinct middle layers and tapetum, release of pollen grain at two-celled stage, an archesporial cell functioning as megaspore, Polygonum type of embryo sac, first division of zygote being transverse, cellular endosperm with chalazal haustoria and the derivation of pericarp from the ovary wall (Fig. 15.3).

Campanulaceae

Genus *Pentaphragma* has been assigned position under the family Campanulaceae by well-known taxonomists (Bentham and Hooker, 1876 and Engler and Prantl, 1889). However, on the basis of vegetative characters Airy Shaw (1941, 1954) has pointed out its relationship with various other families. The vegetative structure resembles Begoniaceae, Cucurbitaceae, Gesneriaceae and Rubiaceae. The nature of inflorescence brings it nearer to Boraginaceae and Hydrophyllaceae. Anyway, none of the above features strongly advocates its placement in any of the families mentioned above. Embryological investigations of *Pentaphragma horsfieldii* by Kapil and Vijayaraghvan (1962, 1965) revealed distinctive features of different parts, such as ovule (unitegmic and tenuinucellate); endothelium (restricted to chalazal portion of the embryo sac); embryo sac (Polygonum type); endosperm (cellular type, micropylar haustorium) and Solanad type of embryogeny remarkably reflects similarities with Campanulaceae and not with Boraginaceae. Therefore, on the basis of embryological characters its retention in Campanulaceae has been recommended.

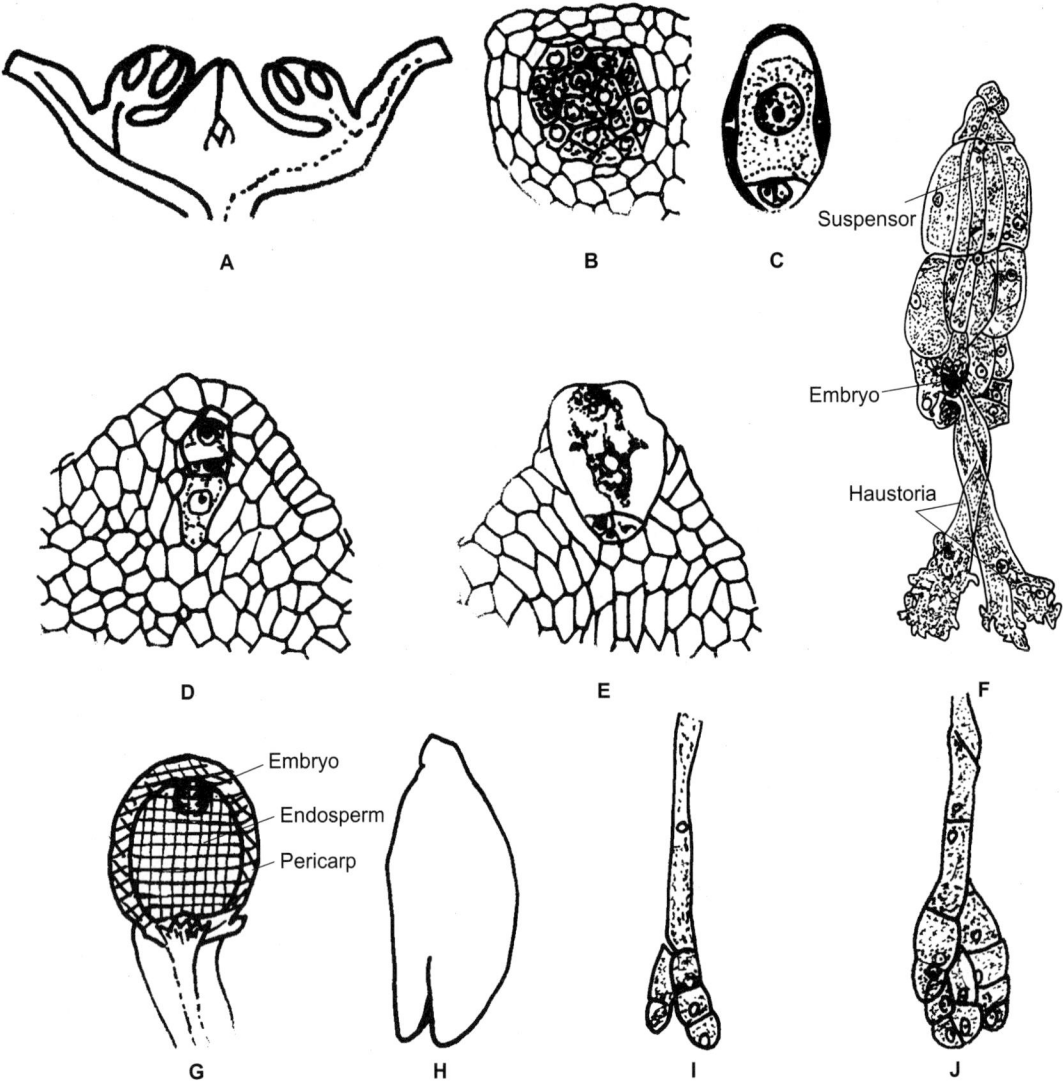

Fig. 15.3 A - I. *Exocarpus sparteus* and J. *E. cupressiformis*. **A**. L.S. of a flower. **B**. T.S. of anther lobe. **C**. 2-celled pollen grain. **D, E**. L.S. of placenta showing linear tetrad and 8-nucleate embryo sac, respectively. **F**. Endosperm with 2-celled chalazal haustorium. **G**. L.S. of mature fruit. **H**. Dicotyledonous embryo **I, J**. Supernumerary embryos formed from suspensor. (After Ram, 1959)

Butomaceae

This family includes the genus *Butomus* along with other three genera. *Butomus* has *Polygonum* type of embryo sac and other three genera are characterised by *Allium* type of embryo sacs, which is more like Alismataceae members. In addition to the above feature *Butomus* differs from other genera in characters like absence of laticiferous ducts, sessile and linear leaves, persistent petals, anatropous ovule and straight embryo. On the contrary, other genera have laticiferous ducts, petiolate leaves, lamina expanded, petals caducous, campylotropous ovule and curved embryos. The cytological evidences also confirm that *Butomus* has distinctive features. It, therefore, becomes clear that there is valid ground for retention of *Butomus* under the family Butomaceae. The other three genera since resemble more to Alismataceae, therefore, they may suitably be shifted from Butomaceae, either to Alismataceae or some other position nearby it.

Loranthaceae

The family Loranthaceae includes two sub-families *Loranthoideae* and *Viscoideae*. Their structure and development of embryo shows wide differences. Therefore, Maheshwari (1964) based on suggestions made earlier by Miers, is also of the opinion that Loranthaceae be reatained for sub-family Loranthoideae. The sub-family *Viscoideae* may also be raised to the status of a family Viscaceae. These two sub-families namely *Loranthoideae* and *Viscoideae* differ from each other in floral structure, mode of development of endosperm, embryo sac and orientation of vascular supply with respect to viscid layer in the fruit. *Loranthoideae* has tri-radiate pollen and the pores are situated at the extremities, embryosac is monosporic and of *Polygonum* type, endosperm is composite, the first division of zygote is vertical, early ontogeny is biseriate and in the fruits viscid layer is outside the vascular supply. On the contrary, in sub-family Viscoideae pollen grains are spherical; embryo sac is bisporic *Allium* type straight or 'U' shaped, endosperm non-composite and in fruits viscid layer is inside the vascular supply.

CONCLUSION

In view of some of the above examples regarding various families, one has to accept that embryology certainly provides some important facts to support status of taxa in the system of classification. However, it provides only additional knowledge and does not advocate construction of any classification solely on embryological basis.

PART B

Experimental Embryology

Chapter 16. Technique of Tissue Culture
Chapter 17. Somatic Embryogenesis
Chapter 18. Haploid Production
Chapter 19. *In Vitro* Pollination and Fertilization
Chapter 20. Embryo Culture
Chapter 21. Culture of Different Parts of Pistil and Seed (Organ Microculture)
Chapter 22. Synthetic Seeds

16 Technique of Tissue Culture

The term tissue culture, as it is used in plant sciences, refers to artificial culture/glass culture or *in vitro* culture of plant parts, whether single cell, tissue or organ under aseptic condition. Each type of culture requires slightly different methods; however, the principle of culture is the same.

The underlying principle in plant tissue culture is very simple. Firstly, a plant part from the intact and its tissue are isolated. Secondly, it is necessary to provide the explant with an appropriate environment in which it can express its intrinsic or induced potential. Finally, the above procedures must be carried out in aseptic ambience.

CULTURE MEDIUM

The greatest progress towards the development of culture/nutritional media for plant cell grown in culture took place in 1960s and 1970s. The basic cultural requirements of cultured plant cells are very similar to those utilized by *in vivo* grown plants. However, nutritional composition varies according to cells, tissues, organs and also with reference to particular plant species.

Excised plant tissues and organs will grow only *in vitro* on suitable artificial prepared medium, which is known as culture medium. From time to time many workers (White; Gamborg; Nitsch and Nitsch; Murashige and Skoog etc.) have proposed the composition of nutrient medium for growth of the plant tissues (Table No. 1). But no single medium is universally applicable for optimum growth of all the plant tissues. Consequently, the most suitable medium for the growth of particular tissue is determined by trial and error. The proposed culture medium has often been modified to stimulate the growth of particular plant material.

A culture medium is composed of inorganic salts, an iron source, amino acids, vitamins and phytohormones. Inorganic salts are supplied as major salts and minor salts. In most of the media, iron is chelated as ferric-sodium ethylene-amine tetra-acetate (Fe-EDTA). In this state iron is gradually released into the culture medium and it is utilized by the living cells.

16.4 Plant Embryology: Classical and Experimental

The carbon source is supplied usually in the form of sucrose. The plant cells also require organic nutrients for proper growth and development. There is an absolute requirement for vitamin B1 (thiamine). Growth is also improved by the addition of nicotinic acid and vitamin B6 (pyridoxine). Some media contain pantothenic acid, biotin, folic acid etc. The amino acids included in the media are; glycine, aspargine, tyrosine, arginine and cysteine.

Table 16.1 Nutritional components of media often used for plant cell culture

Constituents	Media (amount in mg/l)				
	White's	MS	B_5	Nitsch's	N_6
Inorganic					
NH_4NO_3	–	1650	–	720	–
KNO_3	80	1900	2527.5	950	2830
$CaCl_2 \cdot 2H_2O$	–	440	150	–	166
$MgSO_4 \cdot 7H_2O$	750	370	246.5	185	185
KH_2PO_4	–	170	–	68	400
$(NH_4)_2SO_4$	–	–	134	–	463
$Ca(NO_3)_2 \cdot 4H_2O$	300	–	–	–	–
Na_2SO_4	200	–	–	–	–
$NaH_2PO_4 \cdot H_2O$	19	–	150	–	–
H_3BO_3	1.5	6.2	3	10	1.6
$MnSO_4 \cdot 4H_2O$	5	22.3	–	25	4.4
$MnSO_4 \cdot H_2O$	–	–	10	–	–
$ZnSO_4 \cdot 7H_2O$	3	8.6	2	10	1.5
MoO_3	0.001	–	–	–	–
$CuSO_4 \cdot 5H_2O$	0.01	0.025	0.025	0.025	–
$CoCl_2 \cdot 6H_2O$	–	0.025	0.025	0.025	–
$Fe_2(SO_4)_3$	2.5	–	–	–	–
$FeSO_4 \cdot 7H_2O$	–	27.8	–	27.8	27.8
$Na_2EDTA \cdot 2H_2O$	–	37.3	–	37.3	37.3
Sequestrene 300Fe	–	–	28	–	–
Organic					
Inositol	–	100	100	100	–
Nicotinic acid	0.05	0.5	1	5	0.5
Pyridoxine HCl	0.01	0.5	1	0.5	0.5
Thiamine HCl	0.01	0.5	10	0.5	1
Glycine	3	2	–	–	–
Folic acid	–	–	–	0.5	–
Biotin	–	–	–	0.05	–
Sucrose	20,000	30,000	20,000	20,000	–
pH	5.8	5.8	5.5	5.8	5.8

White	(White, 1953)
MS	(Murashige and Skoog, 1962)
B_5	(Gamborg et al., 1968)
Nitsch	(Nitsch and Nitsch, 1969)
N_6	(Chu, 1978)

Phytohormones usually incorporated into the medium are either auxins or cytokinins. The most common auxins are indole-3- acetic acid (IAA), indole-3- butyric acid (IBA), naphthalene-1-acetic acid (NAA), 2,4-dichlorophenoxy acetic acid (2,4-D) etc. and the cytokinins are 6-furfurylaminopurine (kinetin), 6- benzoylaminopurine (BAP), 6- isopetenylaminopurine (IPA) and zeatin.

In order to enhance growth, sometimes complex additives, such as coconut milk, casein hydrolysate, yeast extract, malt extracts etc. are used.

Agar, a seaweed derivative, is the most popular solidifying agent. It is polysaccharide with high molecular weight and has the capability of gelling the media. Medium may either be solid when agar-agar is added or liquid when the medium is devoid of agar-agar. The former medium is used for callus culture, while latter is used for cell suspension culture.

Medium Preparation and Sterilization Procedure

It is not possible to weigh and mix all the constituents just before the preparation of the medium; moreover, a few constituents are used in very small quantity. So it is convenient to prepare the stock solutions of macro-salts, micro-salts, iron source, vitamins, amino acids and phytohormones. The widely preferred medium is Murashige and Skoog's medium (1962). All the stocks are stored in refrigerator and are checked periodically to ensure the stocks are not contaminated.

As per the experimental design stocks are mixed, and phytohormones are added together with the agar and sucrose. The pH of the medium ranges from 5.0 to 6.0 with the help of N/10 HCl or N/10 NaOH, though the 5.8 is usual practice. The pH higher than 6.0 gives a fairly hard medium and a pH below 5.0 do not allow satisfactory gelling of the agar. The final volume is attained with the addition of distilled water. The prepared medium is dispensed into culture test tubes (15 × 2.5 cm) measuring 10-15 ml and closed with cotton plugs.

The culture medium and the instruments to be used for inoculation are autoclaved for 15-20 minutes at 121°C (1.06 kg/cm^2) in order to get a sterile medium. The plant materials to be inoculated are also surface sterilized with the help of chemical agents such as ethyl alcohol (70%), $HgCl_2$ (0.01%), NaOCl, Cl_2-water etc. and finally washed with autoclaved distilled water to get a sterile plant material.

Inoculation and Incubation of Culture

Inoculation is performed under Laminar Air Flow cabinet near the spirit lamp adopting all the precautions. Cultures are placed in culture room, where the physical conditions such as temperature (25 ± 3°C), photoperiod (16 h light and 8 h dark), light intensity (1500-3000 lux) and relative humidity (60%) are maintained. After a week or so observation starts, results are recorded and photographed (Fig. 16.1).

16.6 Plant Embryology: Classical and Experimental

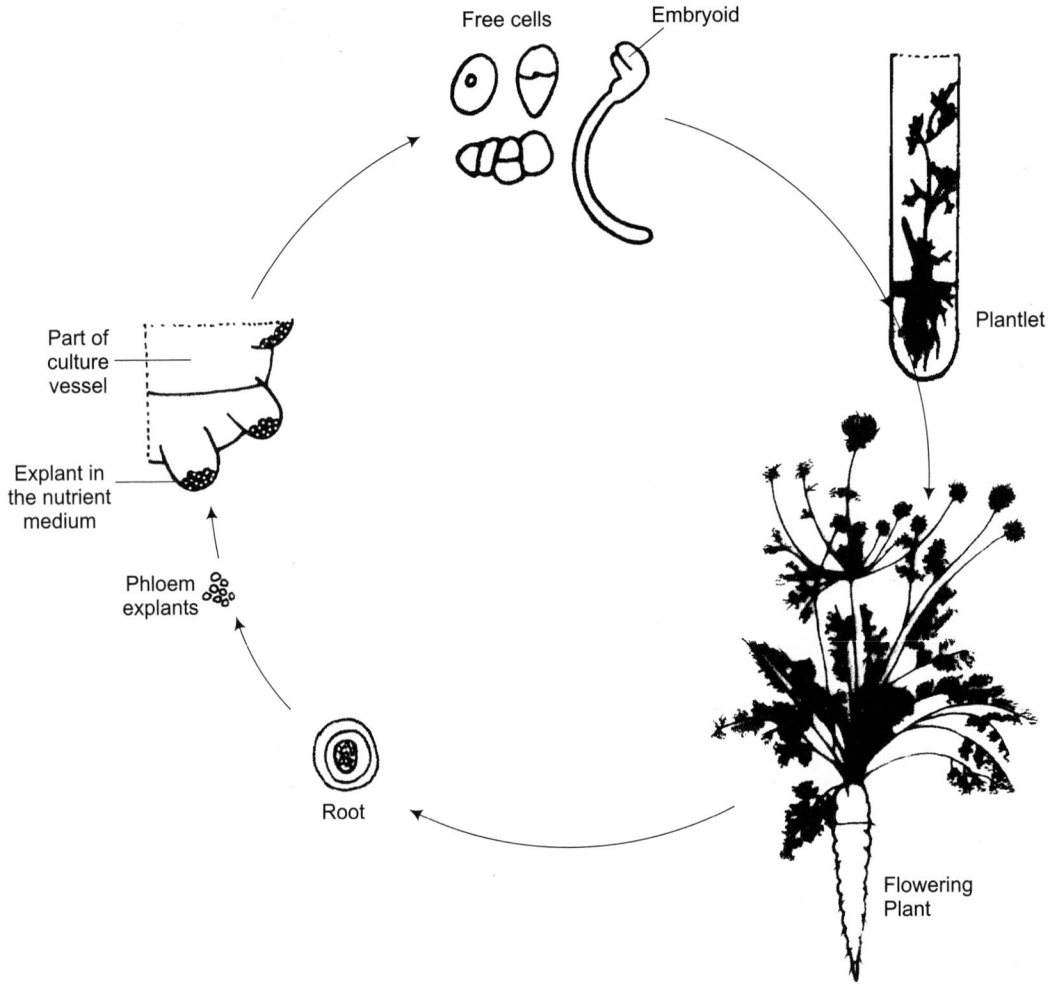

Fig. 16.1 Diagram showing events leading to regeneration of carrot plant.

Observation

From third day onwards observation is made and results are noted periodically (weekly). The results are observed in terms of germination, hypertrophy, callusing and regeneration.

Greenhouse

Greenhouse is required to grow regenerated plants for growing plants to maturity. This facility is required as a transitional step of taking plant from culture containers present in the controlled room to the field. Thus in the greenhouse, plants are acclimatized and hardened before being transferred to the field conditions. The plants are grown in the greenhouse to develop adequate root system and leaf structures to withstand the field environment.

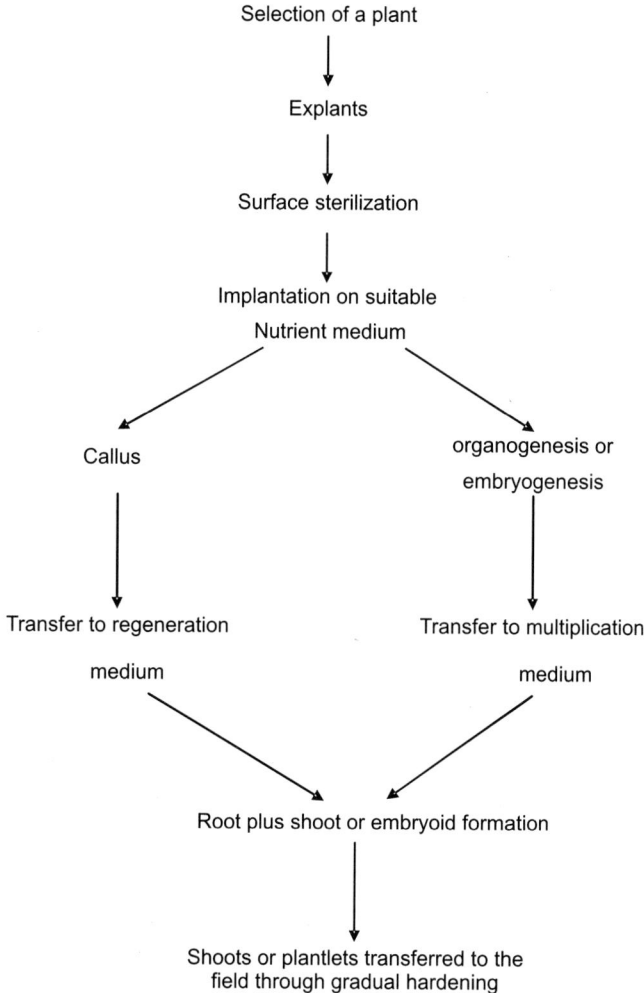

Fig. 16.2 Schematic events of tissue culture.

The whole event of tissue culture, starting from selection of plant, implantation of explant, callusing, embryogenesis or organogenesis and field transfer, has been depicted in Fig. 16.2.

17 Somatic Embryogenesis

In plant tissue culture, the totipotency of the living cells has been well proven i.e. each living cell is a blue print of a complete plant, as it has all the ingredients necessary for the development of plant. Usually it is the zygote which develops into embryo, through embryogenesis that subsequently grows into a complete plant. However, in tissue cultures the non-zygotic or somatic embryos (**embryoid**) are formed very frequently. The embryoids are polarized structure and the developmental behaviour is similar to the zygotic embryos. This pathway is called **somatic embryogenesis.**

The initiation of embryo from somatic tissues in plant tissue culture was first recognized by Steward (1958) and Reinert (1958, 1959) from the tap root culture of *Daucus carota*. Since then, reports started pouring in every now and then from different parts of the world. Tisserat *et al.* (1979) reported somatic embryogenesis in 32 families, 81 genera and 132 species.

In addition to somatic embryos from sporophytic cells, embryos have been arising from generative cells, such as in the classical work of Guha and Maheshwari (1964) with *Datura innoxia* microspores and Nitsch (1969) with *Nicotiana tabacum* microspores. Triploid embryos have also been observed in endosperm culture of *Santalum album* (Lakshmi Sita *et al.*, 1980). However, in nature there is no instance of *ex-ovule* embryo development.

Adventive or asexual embryogenesis is the well-known natural phenomenon, where cells of the nucellus or inner integument may develop into embryos; members of Rutaceae and especially *Citrus* species offer good examples. Cells within the embryo sac proper, such as the synergids or antipodal may also develop into embryos. The proembryo, embryo or its suspensor may also give rise to multiple embryos. In addition, there are examples of embryos arising from endospermal cells as in the case of *Brachiaria setigera* (Muniyamma, 1977).

Each non- zygotic embryo or embryoid is a bipolar structure, consisting plumular and radicular ends, thus resembles closely with their zygotic counterpart. In carrot, each developing embryoid passes through three sequential stages of embryo formation, such as globular stage, heart-shaped stage and torpedo stage. It is the torpedo stage which is bipolar structure and gives rise to a complete plant.

TYPES OF SOMATIC EMBRYOGENESIS

According to Sharp *et al.* (1982), somatic embryogenesis is initiated by two types of cells i.e. **'Pre-embryogenic Determined Cells' (PEDCs)** or by **'Induced embryonic determined cells (IEDCs)'**. The former type of cells are found in embryonic tissues of the ovule, whereas the latter type of cells need specific phytohormones to enter into embryogenic state, such as 2,4-D. These cells are differentiated generally in microspores and callus cultures. Somatic embryogenesis may be initiated in two different ways:

(i) In some cultures somatic embryogenesis occur directly, without entering into callus, as the cells are PEDCs, which are already programmed for embryogenesis. For instance, somatic embryos have been developed directly from leaf mesophyll cells of orchard grass (*Dactylis glomerata*) when leaf explant is cultured on SH (Schenk and Hilderbrandt, 1972) medium supplemented with 30μM 2-Methyl-3, 6-dichlorobenzoic acid (dicamba). Plant formation occurred on the same basal medium (Conger *et al.* 1983).

(ii) In the second or indirect type of somatic embryogenesis, explants initiate callus formation in the medium containing 2,4-D. This induced callus consists IEDCs which grow into embryoid on transfer to basal medium.

Indirect embryogenesis is very common. In rice, induction of callus takes place on MS medium supplemented with 2,4-D (2.5 ppm). Embryoids are formed in large number, when callus is transferred to basal MS medium i.e. without any phytohormone. In *Ranunculus sceleratus* various floral parts (including anthers) as well as somatic tissues proliferate to form callus on a medium cotaining coconut milk (10%) with or without IAA. Within 3 weeks numerous embryos appear on the callus (Konar and Nataraja, 1969). The embryoids originate from the peripheral as well as deep-seated cells of callus or embryo differentiation also occurs in suspension culture raised from these calli. Embryo germinates when these embryoids are isolated and cultured individually on a fresh semi- solid medium (Fig. 17.1). A feature of special interest is the development of a fresh crop of embryos (5-50) from the stem surface of these plantlets (Konar and Nataraja, 1965, 1969).

The embryogenic cells are generally characterized by dense cytoplasmic contents, large starch grains and relatively a large nucleus with darkly stained nucleolus.

TECHNIQUE

In *Daucus carota* any part of the plant, such as embryo, hypocotyl, young root, tap root, petiole, peduncle (taken at any time of development) have successfully produced somatic embryos. For some species, however, only certain parts of the plant body may respond in culture. This is commonly met in monocotyledons especially in Poaceae. For these plants actively dividing cells and immature embryo is extremely useful. Floral or reproductive tissues, in general, have proven to be an excellent source of embryogenic material, e.g., ovules of *Carica papaya*, and *Citrus* species.

The plant material carrot provides a classical example of somatic embryogenesis. The protocol is described below:

Cambium tissue is obtained from surface sterilized taproot. Following aseptic technique, explants are placed on semi-solid MS medium supplemented with 2, 4-D (2 mg/l). Cultures

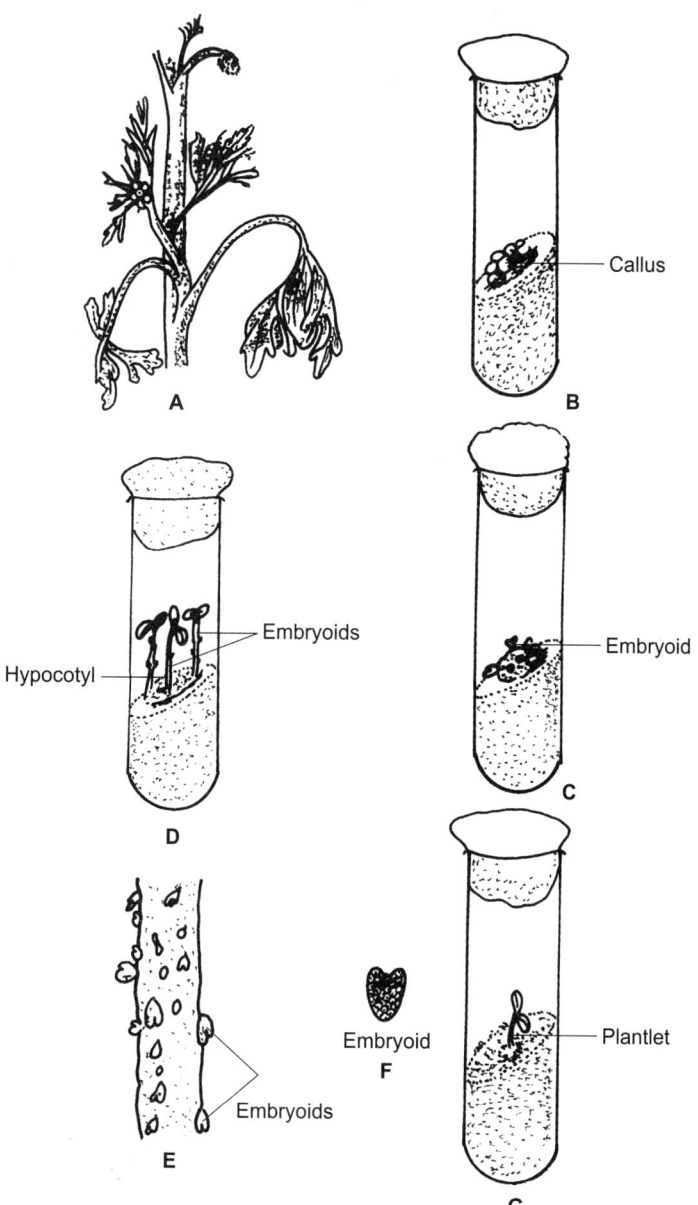

Fig. 17.1 Embryogeny in *Ranunculus scleratus*. **A.** Mother plant. **B.** Callus induction. **C.** Embryoid formation. **D.** Plantlets regenerated from embryoids on hypocotyls. **E.** Enlarged view of hypocotyl to show embryoids. **F.** Single embryoid. **G.** A Regenerated plantlet.

are incubated in dark under controlled temperature and humidity conditions. Profuse callus is produced within 10-15 days. After 4 weeks about 0.25 gm callus is transferred to 250 ml Erlenmeyer flask containing 25-30 ml liquid medium (same medium without agar). Flasks are placed on horizontal gyratory shaker with 125-150 rpm at 25°C.

Cell suspension is either transferred to semi-solid or liquid medium (without phytohormone) for embryoid formation which could be developed into complete plantlets, ready for transfer to the field through hardening process.

FACTORS AFFECTING SOMATIC EMBRYOGENESIS

(i) Genotype

Genotype of the donor plant has significant effect on somatic embryogenesis. Three cultivars of *Zea mays* produced cultures capable of shoot regeneration through organogenesis and only one cultivar, A 188 readily produced somatic embryos. Out of 500 varieties of rice screened by Kamiya *et al.* (1988), 19 showed 65-100% embryogenesis, 41 showed 35-64% embryogenesis and the remaining 440 cultivars were less efficient.

(ii) Explants

For *Daucus carota*, any part of the plant body taken at any time of their developmental stages has successfully produced somatic embryos in culture. However, for some species, only certain regions of the plant body may respond to culture. In Poaceae, actively dividing cells seem to respond more readily. Immature embryos have been proved to be extremely useful, both in grasses and cereals. Floral or reproductive tissues of *Carica papaya* and *Citrus* species have been proven to be an excellent source of embryogenenic material.

(iii) Culture Medium

Nitrogen Sources: Somatic embryos have been grown on wide range of media varying from low salt (White's medium) to high salt media (B_5, SH, MS). The form of nitrogen in the medium significantly affects *in vitro* embryogenesis. Both, reduced nitrogen (NH_4^+) and nitrate (NO_3^-) are beneficial in embryo initiation and maturation of embryos. White's medium, being particularly low in nitrate nitrogen and without reduced nitrogen, needs to be supplemented to support continuous somatic embryogenesis. For somatic embryogenesis, the optimal concentration and form of the nitrogen supply appears to be dependent upon the auxin concentration. Halperin and Wetherell (1965) reported that in the culture of wild carrot, raised from the petiolar segment, embryo development occurred only if the medium contained some amount of reduced nitrogen. The calli initiated on a medium with KNO_3 as the sole source of nitrogen failed to form embryos after removal of auxins. However, the addition of a small amount (5 mM) of nitrogen in the form of NH_4Cl in the presence of 55 mM KNO_3 allowed embryo development.

The embryos formed on amino acids (proline, serine, threonine) containing medium showed higher percentage of embryo conversion from callus culture. Of the other salts present in the culture media, potassium ion has been shown to be essential for somatic embryogenesis (Reinert, *et al.*, 1967). The presence of chelated iron in the form of iron EDTA in the medium is important for embryogenesis (Heberle-Bors, 1980).

(iv) Growth Regulators

(a) **Auxins:** Carrot is the classical example of somatic embryogenesis, where embryoids are produced in two-step process. In the first step synthetic auxins initiate callusing when present in the culture medium. When such callus tissue, in the second step, is transferred to auxin free medium or very low level of auxins, somatic embryos are formed. 2,4-D (0.5-1 mgl^{-1}) is the most commonly used auxin for induction of somatic embryogenesis. The presence of auxins in the callusing medium is essential for the differentiation of tissues or **proembryogenic masses** (**PEM**) into embryoids in second culture. The tissues maintained continuously in the auxin free medium would not form embryo. All the major species of cereals and grasses have been reported to regenerate plants *in vitro* through somatic embryogenesis (Vasil and Vasil, 1986).

The importance of auxins in embryogenesis is also suggested by the detailed work on callus tissues of *Citrus sinensis* (Kochba and Spiegel-Roy, 1977). Callus is raised from the nucellar tissues in a medium containing IAA and kinetin, which upon subculture developed embryoids. However, repeated subcultures showed gradual decline in embryogenesis, which was again improved by transferring the cultures in auxin-free medium. Addition of very low concentration of IAA inhibits embryogenesis. This observation suggested that after repeated subcultures, for a long time, the callus is habituated or phytohormone autonomous. This means that they are now able to grow on a standard medium which is devoid of growth hormones. The cells appear to have developed the capacity to synthesize adequate amount of both auxins and cytokinins which they require for the growth and somatic embryogenesis. However, after a few subculture, habituated callus of *C. sinensis* again shows decline in embryogenic potential. When the habituated callus tissue is exposed to irradiation, the process of somatic embryogenesis is again improved, as irradiation is known to breakdown auxins. This observation reveals that high level of endogenous auxins produced by habituated callus tissue inhibits the process of somatic embryogenesis. But when the tissue is irradiated, the high level of endogenous auxin is lowered to a level which restores their embryogenic potentiality.

However, there are reports where somatic embryoids could be developed, e.g., carrot, by manipulating sucrose concentration or pH value, the medium being without growth regulators (Kamada *et al.*, 1989; Smith and Krikorian, 1990).

(b) **Cytokinins:** Of the different cytokins, effect of BAP and zeatin has been studied in carrot tissue cultures. BAP has been found to be promoting callusing but it inhibits embryogenesis. This may be due to selective stimulation of non-embryogenic cell components of the culture. However, zeatin has been found to be beneficial for embryogenesis at a concentration of 0.1 µM. Steward *et al.* (1964) reported the importance of coconut milk (cotaining a source of cytokinin) in embryogenesis.

(c) **Gibberellins:** Gibberellins are rarely incorporated in primary culture media used for the maintenance of the line. However, they have proved useful in a number of of cases in fostering embryo maturation or in stimulating the rooting and subsequent growth of plantlets, e.g., *Santalum album, Panicum maximum, Citrus sinensis* and *Zea mays*.

(d) **Growth Inhibitors:** There is a little work on the role of inhibitors, especially ABA a naturally occurring growth regulator. When added to non- inhibitory level (0.1-1 M) to *Carum* cultures, ABA permitted embryo maturation to proceed but inhibited abnormal proliferation and repressed precocious germination. Working with anther cultures, Nitsch and Nitsch (1969) observed that ABA did not prevent embryo development from *Nicotiana tabacum* microspore but did inhibit their germination.

Growth promoters, even if not added to culture media, are found in embryogenic cultures (Carr and Reid, 1968). ABA, anti-auxins and other growth inhibitors may serve to promote somatic embryo maturation by countering the effects of growth promoters. Their addition to culture medium may help in organization and maturation of somatic embryos.

(e) **Polyamines:** There are certain evidences which suggest positive role of polyamines in embryogenesis. Increase in the endogenous level of polyamines and the enzymes for their biosynthesis concomitant with the induction of embryogenesis in carrot and suppression of somatic embryogenesis by the inhibitors suggest the involvement of polyamines in somatic embryogenesis. However, their exact role in embryogenesis is yet to be established.

(v) Activated Charcoal

The addition of activated charcoal to the medium has proven useful for somatic embryo development. Activated charcoal has been particular useful in anther culture. Analysis of its effects show media with charcoal had substantially lower levels phenylacetic and p-OH benzoic acid compounds that inhibited somatic embryogenesis. Activated charcoal has been shown to absorb hydroxymethylfurfural, an inhibitor formed by sucrose degradation during autoclaving as well as excessive amount of auxins and cytokinins.

(vi) Carbohydrates

Sucrose appears to be most effective reduced carbon source for somatic embryogenesis, although many other mono- and disaccharides can be successfully employed. Raising the sucrose concentration in the primary medium to 12% benefited the formation of embryogenic callus from the scutellum of immature embryos of *Zea mays*.

(vii) Physical Conditions

Light: Somatic embryogenesis has occurred under a variety of light/dark condition. In one case, *Nicotiana tabacum*, high light intensities are required. In other cases, *Daucus* and *Carum*, embryos maturation proceeds more normally in complete darkness.

Temperature: Usually, $25 \pm 3°C$ is ideal for embryogenesis to take place. Low temperature has been proven to be useful in androgenic embryos. In *Citrus*, nucellus derived cultures embryogenic potential was found to reduce as the temperature was lowered from 27°C to 12°C.

IMPORTANCE OF SOMATIC EMBRYOGENESIS

1. Somatic embryos in liquid culture offer a potential system for large-scale plant propagation.
2. Somatic embryogenesis has advantage over organogenesis as is provides an organized structure, from which plants could easily be obtained.

3. Plants developed from somatic embryos may in some cases is free of virus and other pathogens. *Citrus* plants propagated from embryogenic callus of nucellar origin are free of virus.
4. Somatic embryo could potentially be employed for synthetic seed production.

18 Haploid Production

The haploid refers to those plants which consist of gametophytic number of chromosomes. In angiosperms haploid cells are present only in anthers and ovules. While working with *Datura stramonium,* Blakeslee *et al.* (1922) reported for the first time the natural occurrence of haploid. However, through breeding technique the first haploid wheat was reported by Gaines and Aase (1926) in an intergeneric cross between *Triticum compactum* and *Aegilops cylindrica.* Since then its occurrence has been reported in all other members of this group. The haploid plants are of great value for breeding and for the study of the fundamental genetics of higher plants.

Haploid may occur spontaneously in nature (0.001-0.01%) or they may be induced experimentally. Spontaneous haploids occur as a result of apomixis or parthenogenesis. In such cases the unfertilized egg, male gametes or the synergids start to grow to form a haploid plant independent of any applied stimulus. Induced haploids can generally be obtained by the stimulation of the egg, synergids or sperms by a number of methods, including ionization irradiation and radioisotopes, thermal shocks, distant hybridization, delayed pollination, application of abortive pollens, spraying with various chemicals, chromosome elimination by culture of young embryos and *in vitro* culture of the excised anther, pollens, ovary and ovules.

Guha and Maheshwari (1964), working with *Datura innoxia* reported the first successful *in vitro* culture of anther. Subsequently, Bourgin and Nitsch (1967) obtained the first haploid plants from the anther of *Nicotiana.* So far anther or pollen culture has been reported from more than 134 species and hybrids, spreading within 25 families. Solanaceous members are very suitable for haploid culture; other families are Brassicaceae, Fabaceae, Liliaceae, Poaceae, and Ranunculaceae etc.

ANTHER CULTURE

1. Technique

Young plants grown under controlled condition of temperature, light and humidity are used as experimental material for anther or pollen culture. The knowledge of the life cycle of the donor plant is essential so that the buds could be collected at right stage (bud size is fixed as a visible marker).

The selected buds are surface sterilized and anthers dissected for inoculation. Utmost precaution is taken to ensure that the filament is removed as far as possible and the anthers are not injured while making dissection. Normally, 5-10 anthers are inoculated in a culture vessel.

The anther cultures are maintained in culture room where the conditions are highly controlled.

2. Induction of Androgenesis

Androgenesis refers to the *in vitro* regeneration of plants either from anthers or pollens. There are two modes of androgenesis (Fig. 18.1):

(i) **Direct Androgenesis:** In this type anthers or pollens undergo change through cell divisions to form organized non-zygotic embryos called somatic embryos or embryoids. These embryoids subsequently grow into complete plants.

(ii) **Indirect Androgenesis:** In contrast to the former type, here the anthers/pollens undergo repeated division to form unorganized mass of parenchymatous tissues called callus. These calli subsequently may either follow embryogenesis to form embryoid or organogenesis to form plant directly.

Normally pollen divides into two unequal cells: lager vegetative cell and smaller generative cell. The former grows into pollen tube, while latter divides to form two male gametes. However, in anther culture pollens form sporophyte through any one of the following pathways:

(i) **Pathway I:** The microspores divide by an equal division and two identical daughter cells contribute to embryoid formation e.g., *Brassica napus*.

(ii) **Pathway II:** The division of the microspore is unequal resulting in the formation of a vegetative and a generative cell. The sporophyte arise through further divisions in the vegetative cell, while the generative cell either does not divide or may divide once or twice before degenerating e.g., *Nicotiana tabacum, Hordeum vulgare, Triticum aestivum, Capsicum annum*.

(iii) **Pathway III:** The uninucleate microspore undergoes a normal unequal division but the pollen embryos are formed from the generative cell alone e.g., *Hyoscyamus niger*.

(iv) **Pathway IV:** The division of microspore is asymmetrical. Both the vegetative and generative cells divide and contribute in the development of the sporophyte, e.g., *Datura innoxia, D. metel, Atropa belladonna*.

Irrespective of the early pattern of microspore divisions, the embryogenic pollen grains ultimately become multicellular and burst open, gradually assuming the form of globular embryo. This is followed by post globular embryogeny until the development of a plant. There is an alternative pathway, where the multicellular mass liberating from the bursting pollen grains form callus which later differentiate into a complete plant (Fig. 18.2).

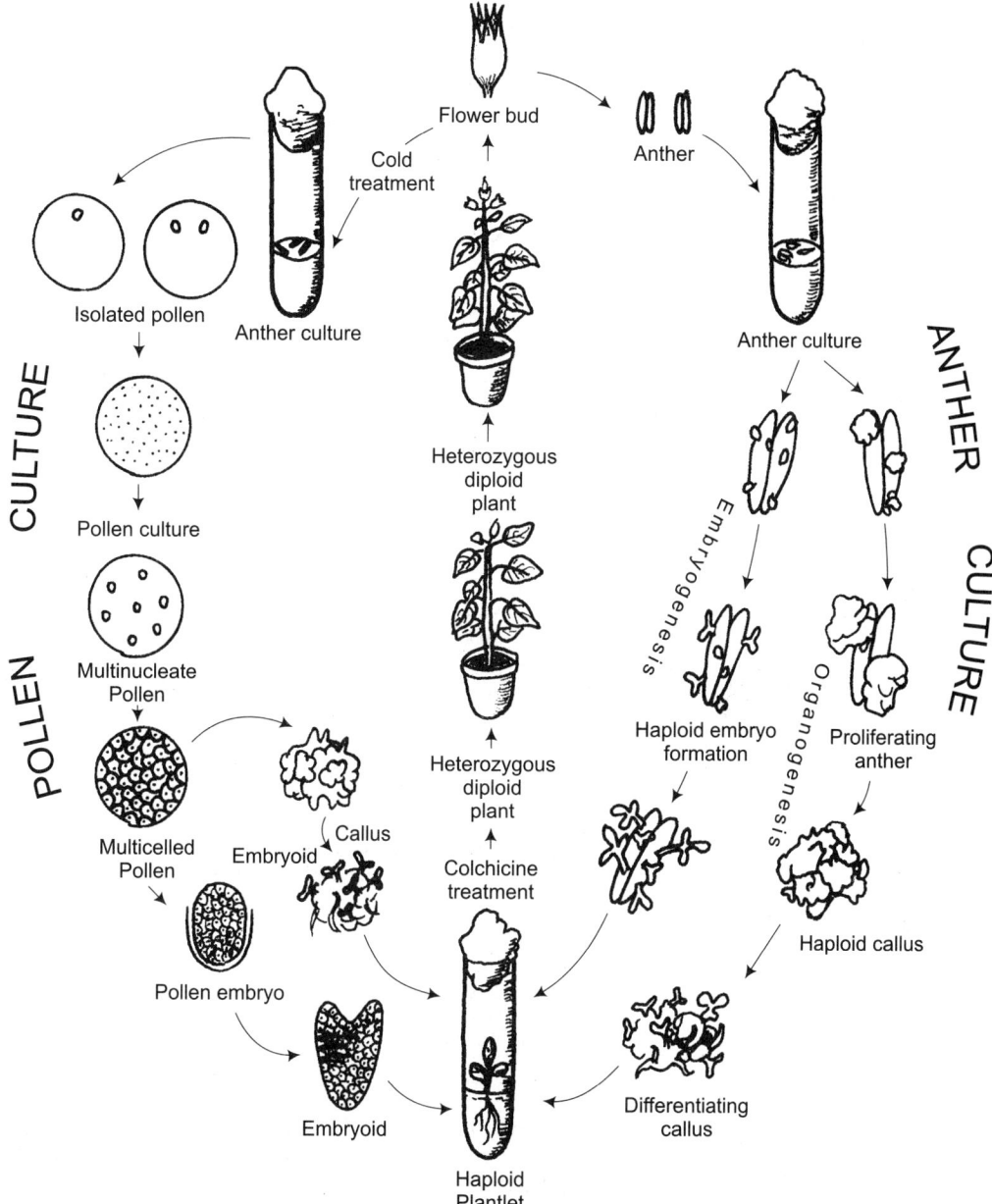

Fig. 18.1 Diagrammatic illustrations showing various modes of androgenesis and haploid formation by anther and isolated pollen culture. The homozygous plants are obtained by treating haploids with colchicine.

18.4 Plant Embryology: Classical and Experimental

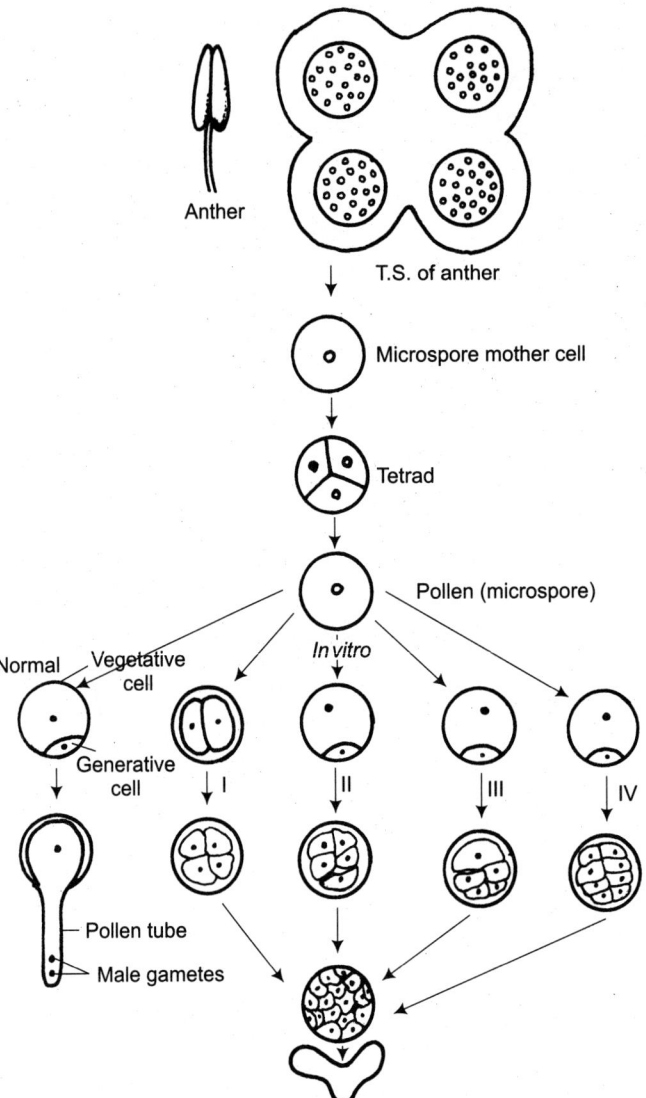

Fig. 18.2 Diagrammatic representation showing various modes of division of microspore *in vivo* and *in vitro*. A microspore may follow any one of the four pathway when cultured *in vitro* to form a multicellular pollen grains, the latter forms embryoid.

3. Factors Influencing Anther Culture

(i) Genotype of the Donor Plant

The genotype of the donor plant has a significant role in determining the androgenic response. It has been reported that various species and varieties exhibit different growth resposes in culture. Gresshoff and Doy (1972) working with 43 varieties of *Lycopersicon esculentum* and 18 lines of *Arabidopsis thaliana* could induce haploidy in 3 varities only in each case. The same trend is observed in rice (Guha-Mukherjee, 1973).

It may be worthwhile to select those genotypes which readily respond to culture medium because androgenic response of recalcitrant genotypes can hardly be improved by manipulation of culture medium.

(ii) Composition of Culture Medium

The composition of the medium is one of the most important factors determining not only the success of anther culture but also the mode of development. The basal media commonly used are Murashige and Skoog's medium (1962), N_6 medium (Chu, 1978) or Nitsch's medium (1972). Pollen embryogenesis can be induced on simple mineral-sucrose medium in plants like tobacco (Nitsch, 1969) and *Hyoscyamus niger* (Raghavan, 1975), yet for androgenesis to be completed, addition of certain growth regulators is required. For instance, cereal anthers require both auxins and cytokinins and their optimal growth depends on the endogenous level of growth regulators. However, to promote direct embryogenesis simple media with low level of auxins are preferred. Complex media enriched with auxins such as 2, 4-D encourage the formation of callus, which causes genetic instability and thus should be avoided.

Sugars are indispensable in the basal medium, as they are not only the source of carbon but are also involved in osmoregulatioin. Normally, 2-4% sucrose is routinely used; however, higher concentrations have yielded better results. All *Brassica* species require 12-13% sucrose for androgenesis in anther and pollen cultures. According to Dunwell and Thurling (1985), high sucrose concentration favours better survival of pollen grains, thus improving the frequency of androgenesis in *Brassica napus*. However, Last and Bretell (1990) have reported that replacement of sucrose with maltose produced better androgenic response. Iron is another important component whose presence in the medium is essential for embryoid formation. Incorporation of activated charcoal (0.5-2%) into the nutrient medium has been shown to stimulate androgenesis in anther culture of some of the plants either by removing the inhibitors or by regulating excessive of phytohormones, thus optimizing the hormonal combination.

Addition of the extracts of anthers and potato tubers has also been observed to stimulate androgenesis in tobacco and wheat respectively. Supplementing the media with serine and glutamine has also improved the culture of isolated pollen.

(iii) Stage of Microspore or Pollen Development

Selection of the appropriate stage of pollen grains is very critical in the induction of androgenesis. Generally, the pollens in mid- uninucleate stage are highly responsive. However, in tomato and tobacco androgenesis has been induced at pollen tetrad stage. The bicellular stage of pollen is the best for *Atropa belladonna* and *Nicotiana sylvestris* and absolutely necessary for *Nicotiana knightiana*. In the pollens of rice and most of the *Brassica* species embryogenesis is observed at late uninucleate stage. Custodio *et al.* (2005) in their observation with carob tree (*Ceratonia siliqua*) also confirmed the importance of pollen stage in anther culture.

The stage of pollen development at culture may also affect the ploidy level of the androgenic plants. In anther culture of *Datura innoxia* and *Petunia* sp. the uninucleate pollens produce haploids, while binucleate pollens form plants of higher ploidy.

(iv) Effect of Temperature and Light

Optimum temperature for anther culture is 25 ± 3°C; however, improved response is found when a panicle or bud is pre-treated to low temperature. Nitsch (1974) reported that in *Nicotiana tabacum* bud treated at 5°C for 72 h yielded 58% embryoid, while 21% result was observed from buds maintained at 22°C for the same period.

Occasionally, pretreatment of anthers at high temperature initially may also be stimulating in some of the species. A high temperature shock (30-35°C) for the initial 1-4 days of culture is essential to induce androgenesis in most of the *Brassica* species (Sharma and Bhojwani, 1985).

The frequency of haploids formation is better under light, although in some species light does not seem to be necessary for the induction of androgenesis. For pollen culture of *Datura innoxia*, *Nicotiana tabacum* and *Annona squamosa* an initial incubation of cultures in dark followed by diffused light was found to be suitable. Isolated pollens are more sensitive to light than anther cultures (Nitsch, 1977). In of the *Brassica juncea* and *Hordeum vulgare* species, light is harmful for anther culture.

(v) Physiological Status of Donor Plant

It has been observed that the age and physiological status of the donor plant considerably influence the frequency of androgenesis. Flowers from relatively young plants at the beginning of the flowering season are more suitable than the flower buds taken from the plants at the end of their growing season.

Temperature at which plants have been grown is important, since in *Datura* plants grown at 24°C gave a higher frequency of androgenesis (45%) than those grown at 17°C (8%) (Nitsch and Noreel, 1973). However, *Brassica napus* yielded better result with anthers excised from the plants grown at lower temperature. The removal of old flowers in *Datura* (Nitsch, 1975) and the apical portion of wheat inflorescence (Picard, 1973) caused an increase in the frequency of androgenesis.

Exposure of the donor plants to stresses, such as nutrient stress and water stress are also reported to promote androgenesis.

POLLEN CULTURE

Although anther culture proved to be quite efficient for the production of haploids, however, a few demerits are associated with anther culture are: (a) other parts of anther, in addition to pollen grains, contribute in androgenic response with the result that a population of plants with various ploidy level is obtained, (b) percentage of response is very low, may be because of toxic material released by older pollens suppress the androgenic response of younger pollens and (c) albinos are produced. These difficulties may be prevented by culture of isolated pollens. The culture of isolated pollens or microspores offer the following additional advantages:

  (i) Uncontrolled effects of the anther wall and other associated tissues are eliminated and various factors governing androgenesis are better regulated.
  (ii) The sequence of androgenesis can be observed starting from a single cell.

(iii) Pollens are ideal for uptake, transformation and mutagenic studies, as pollen may be evenly exposed to foreign genes, chemicals or mutagens.
(iv) Unlike callus, pollen is transformed directly into an embryo, thus would be most suitable for understanding the physiology and biochemistry of androgenesis.
(v) Pollen culture is much efficient method in term of embryoid formation.

In 1953, Tuleke was able to obtain callus from isolated pollen culture of *Ginkgo biloba*. Subsequently, pollens of other gymnospermous plants were also put under culture conditions to induce callus formation. The first report of the tissue formation in isolated pollens of an angiosperm was from Kameya and Hinata (1970). They used **Hanging Drop Culture** method to culture isolated pollens from *Brassica oleracea* and hybrid *B. alboglabra* x *B. oleracea* at low temperature. Their method involved placing a drop of medium containing 50-80 grains on the cover glass, which is then inverted over a cavity slide and sealed with paraffin. Before inverting the cover glass, a column of paraffin is raised in the centre, so that when inverted it touches the bottom of the cavity slide. This facilitates the aeration as well as the movement of the pollens when the slide is rotated. Cell clusters were formed from isolated pollens after 4 weeks in medium containing coconut water.

Later, Sharp *et al.* (1972) using "**Nurse Culture**" technique induced isolated pollens of *Lycopersicon esculentum* to form haploid callus (Fig. 18.3). In this method, the anthers are placed horizontally on top of the basal medium within a French square container. A filter paper disc is placed over the intact anther and about 10 pollen grains (from suspension) are

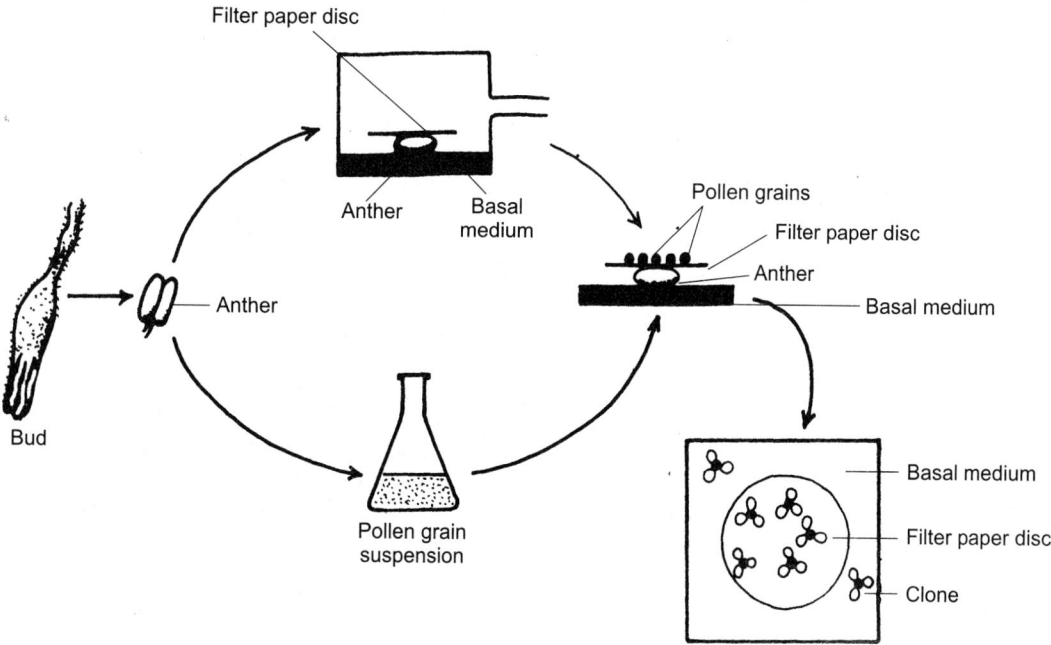

Fig. 18.3 Schematic representation for the culture of isolated pollen of tomato by nurse culture method.

then placed on the filter paper disc. Within 2 weeks green parenchymatous clones were formed on the filter paper disc, which were haploid.

For raising haploids from isolated pollens anthers from diverse species or callus could be effectively used as nurse tissue. Subsequently, it was shown that extract of cultured anthers has equal nursing effect on the growth and development of microspores.

Based on the study of anthers of diverse species, Nitsch (1974) developed purely a synthetic medium from isolated pollen grains. The ingredients of this medium are as follow:

Constituents	Amount (mg/l)
KNO_3	959
NH_4NO_3	725
$MgSO_4, 7 H_2O$	185
KH_2PO_4	68
$CaCl_2, 2 H_2O$	166
Fe.EDTA	5ml
Glutamine	730
Serine	105
Inositol	4505
Sucrose	2000
pH	5.8

The initial success of pollen culture was based on some kind of nurse culture. For most of the cereals pre-culture of anther for 2-7 days before isolation of pollens is found to give better results (Cho and Zapata, 1988; Pescitelli et al., 1989).

The crushing of the anthers in order to obtain pollens brings injury which affects the overall response of pollens in culture medium. Sunderland and Roberts (1977, 1979) developed "**Float Culture**" technique to obtain pollens without any injury. In this technique whole anthers, isolated from cold-treated buds are floated on liquid medium. In a few days the anthers burst open to release pollens, including pollens at various stage of development, directly into medium.

The efficiency of isolated pollen culture for haploid production may be increased by the "**Gradient Centrifugation**" technique (Fig. 18.4). In this technique, the advantage of pollen dimorphism is reaped. It has been observed in cereals that there are occurrences of two structurally different types of pollens within the same anther. Some of the pollens are large, starched- filled, and deeply stained with acetocarmine, other are comparatively smaller, highly vacuolated and faintly stained with acetocarmine. The latter types are embryogenic pollens, whereas the former types are non-embryogenic pollens.

Anthers at the proper stage of development are gently macerated to obtain a microscopic suspension. After removing the debris either by sieving or filtration, the suspension is layered on sucrose solution (30% or 50% percol and 4% sucrose solution) and centrifuged at 1000 rpm for 5 min. The embryogenic pollens being lighter in weight forms a band on the top of sucrose solution, while the heavier non-embryogenic pollens settles at the bottom. The upper band containing embryogenic pollens are pipetted out, suspended in the washing

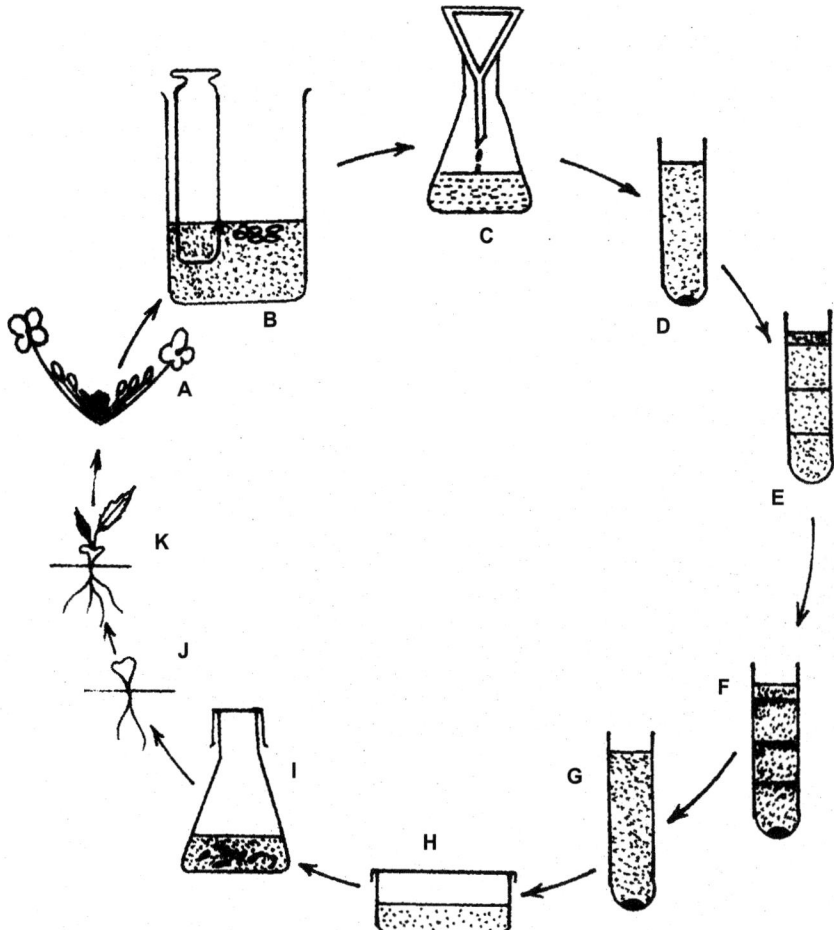

Fig. 18.4 Diagrammatic presentation of Gradient centrifugation technique for isolation and culture of microspores of *Brassica napus*. **A.** A parent plant used for isolation of buds and surface sterilization. **B.** Buds are crushed in glass homogenizer, containing B_5 medium-13% sucroser to release pollen grains. **C-D** Filtration of the mixture through 42μm nylon mesh to remove large debris and centrifugation at 1000 rpm for 3 min. **E.** After discarding the supernatant the pellet are loaded on the gradient solution 24/32/40% Percoll. **F.** The mixture is again centrifuged at 1000 rpm for 5 min. The upper two layers are pipetted out and mixed with B5- 13% sucrose solution. **G.** The mixture is centrifuged at 1000 rpm for 3 min and the supernatant is removed. **H.** The pollen grains are suspended in NLN medium adjusting the pollen density to $2\text{-}5 \times 10^{4.}$ The suspension is plated as thin layer in petri plate and incubated in dark at 32°C for 3-5 days. **I.** The regenerated tissues/embryos are transferred to 18 ml of hormone free NLM medium in contained in conical flask and maintained for a week on the shaking machine at 60 rpm, 32°C. **J, K**. Finally, regenerated embryoids are transferred to solid medium, B_5-2% sucrose to develop roots and shoots.

medium, centrifuged again and the supernatant removed. Finally, the pellets, thus obtained are cultured in the medium for enhanced percentage of haploids (Wenzel *et al.*, 1975; Rashid and Reinert, 1981).

Diplodization of Haploid Plants

Haploids plants obtained from anther, ovary or ovule culture may grow normally upto flowering stage, however, viable gametes are not formed due to absence of one set of homologous chromosomes. Spontaneous duplication of chromosomes from pollen derived haploid plants has, however, been observed but its frequency is fairly low. This can be substantially enhanced by the application of some chemicals.

A simple procedure designed to achieve diplodization involves immersion of very young haploids in filter-sterilized solution of cholchicine (0.4%) for 2-4 days followed by their transfer to a culture medium to allow their further growth.

INDUCTION OF GYNOGENESIS

Plants obtained through anther/pollen show low regeneration potential and large numbers of albinos are encountered. Moreover, the anther culture is not successful where anthers are recalcitrant to cultural conditions or plants are male sterile. Therefore, development of plants from unfertilized cells of female gametophyte (embryo sac) in ovary or ovule is only effective alternative to anther or pollen culture.

San Noeum in 1976 was the first to demonstrate that gynogenesis could be induced *in vitro* from the ovary culture of *Hordeum vulgare*. Subsequently, Zhu and Wu (1979) obtained haploid plants from cultured unpolllinated ovaries of *Triticum aestivum* and *Nicotiana tabacum*. This technique has also proved successful in rice. Haploids have also been obtained from unpollinated ovules of *Gerbera jamesonii*, *Lilium davidii*, *Zea mays* etc. (Yang and Zhou, 1990).

The stage of female gametophyte at culture is crucial. Mostly, embryo culture at nearly mature embryo sac stage gave best result. Rice is an exception, where inoculation of ovary at 1-4 nucleate stage of female gametophyte development turned out to be most responsive (Zhou *et al.*, 1986).

A pre-treatment of low temperature for 24-48 h has been reported to be beneficial. Culture media commonly used for gynogenesis are either N_6 (Yang and Zhou, 1982) or MS (Murashige and Skoog, 1962). Phytohormones incorporated in medium are NAA, 2,4-D, MCPA, BAP. Initial dark induction for a few days to culture is beneficial.

Production of haploids through gynogenesis is more tedious, therefore restricted to a few species as compared to androgenesis. However, evidences reveal that through gynogenesis percentage of haploids and production of green plants are higher than androgenic plants.

PRODUCTION OF HAPLOIDS BY BULBOSUM TECHNIQUE

Kasha and Kao (1970) developed haploid barley through the gradual and selective elimination of chromosomes during the process of hybrid embryo development. The method entails crossing *Hordeum vulgare* (2n = 14) with *Hordeum bulbosum* (2n = 14). In nature seeds produced from such a cross developed for about 10 days only and then begin to abort. However, if the immature embryos are dissected out 2 weeks after the pollination and cultured on B_5 medium without 2, 4-D, they continue to grow. Almost all the plants originating from such embryos are monoploids (n = 7).

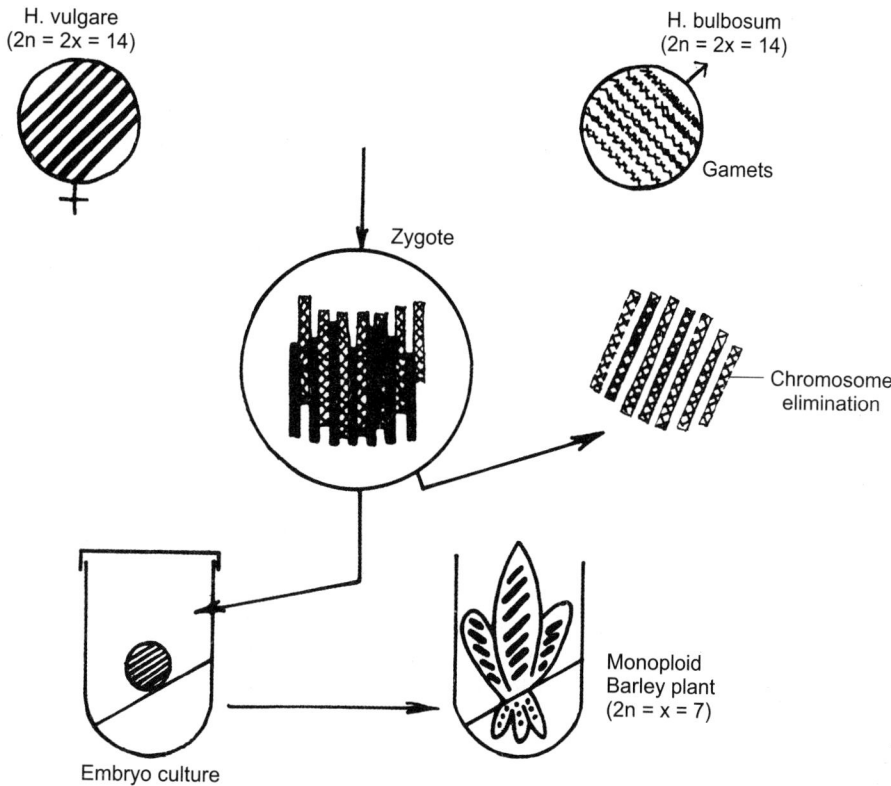

Fig. 18.5 Diagrammatic representation showing the formation of monoploid barley by crossing diploid *Hordeum vulgare* x *Hordeum bulbosum* followed by culture of excised embryo; one set of chromosome is eliminated.

Kasha and Kao (1970) presented evidence to show that these monoploids are not caused by parthenogenesis but by the elimination of the *H. bulbosum* chromosomes (Fig. 18.5). This elimination has been attributed to be under the genetic control.

Monoploid wheat has also been obtained by this technique when a cross is made between *Triticum aestivum* (2n = 6x = 42) with *H. bulbosum* (2n = 2x = 14). The chromosomes of the bulbosum are eliminated in the process and monoploid, *T. aestivum* (n = 3x = 21) is obtained.

The exact mechanism of chromosome elimination is not well understood but this technique has advantages, such as the haploid production is not depended upon the genotype; frequency of haploid production is very high; there are no chances of polyploids as callus is not formed; this technique can be extended to plants of economic importance.

APPLICATIONS OF HAPLOIDY

In vitro anther and pollen culture for the production of haploid and homozygous diploid have proved an important tool for the fundamental and applied plant biology.

1. Utility of Anther and Pollen Culture for Basic Research

Haploids derived from anther and pollen cultures are useful in cytogenetical studies, such as identification of recessive characters, study of genetic recombination, mode of differentiation (organogenesis or embryogenesis), production of aneuploids, understanding of meiotic behaviour etc.

2. Use of Anther and Pollen Culture for Mutational Studies

Normally *in vivo*, the majority of mutations are recessive, therefore, not expressed in diploid cell. Haploid cell lines have been utilized to study the effect of various mutagens. A number of cell lines have been successfully isolated that are resistant to environmental stresses or herbicides, diseases, drugs etc.

3. Use of Anther and Pollen Culture for Plant Breeding and Crop Improvement

Homozygous lines of the cross-pollinating species and hybrids are highly desirable to increase the efficiency and production of homozygous plants. The conventional method to produce homozygous plants is time consuming and labour-intensive, requiring 7-8 recurrent cycles of inbreeding. On the other hand, homozygous plants can be obtained in single generation by diplodization of the haploids.

The success of any crop improvements depends upon on the extent of genetic variability in base population. In this regard callus cultures are rich source of genetic variability and can be incorporated in plant breeding programme. In China and Japan, by anther culture new varieties of rice, wheat, maize etc., have been developed with improved agronomical characters.

4. Application of Haploid Culture for Horticultural Plants

In some plants of horticultural importance, haploid culture is highly desirable. *Freesia*, a horticulturally important plant, propagates vegetatively by corms. Normally, it requires 8-10 years to produce clones, which is long enough from commercial point of view. However, with anther culture clones can be produced quickly.

5. Anther Culture and Alkaloid Content

Amongst other uses, homozygous plants obtained through anther culture have also been employed for the selection of breeding lines of *Nicotiana tabacum* for high alkaloid contents. Likewise, *Hyoscyamus niger* having high alkaloid content could be obtained by anther culture.

6. Genetic Transformation

In order to overcome the poor regenerative ability of monocot and some dicot, the pollen embryos, which show good regeneration ability, can be used as a potential recipient cell for expression of alien gene. In *Brassica napus* recombinants have been regenerated successfully either through micro-injection of pollen embryo or co-cultivation of *Agrobacterium tumefaciens* with microspore and microspore derived embryos.

LIMITATION OF HAPLOID CULTURE

The practical application of haploids is still limited as desired success with anther and pollen culture has been largely confined to a few families, such as Brassicaceae, Poaceae and Solanaceae. Other problems associated with pollen or anther cultures are:

1. Low regeneration ability of anther/pollen.
2. Occurrence of large number of albinos.
3. Instability of genetic characters in pollen/ anther derived plants, which may result undesirable features.
4. Association of ploidy with haploids.

19 | *In Vitro* Pollination and Fertilization

Fertilization is the process that involves plasmogamy and karyogamy. The fusion of male and female gametes to form a zygote serves a very valuable source of new combination of genetic materials. Events preceding fertilization are pollination, germination of pollens on stigma, formation of pollen tube, growth of pollen tube within the style, penetration of pollen tube into the embryo sac and bursting of pollen tube to release two gametes. Out of the two gametes one fuses with egg to form zygote and another fuses with secondary nucleus to form primary endosperm nucleus. The zygote develops into embryo and primary endosperm nucleus to endosperm of a mature seed.

Barring the instances of apomictic seeds and parthenocarpic fruits, seeds and fruits which form more than 90% of the economic products of crop plants are the result of fertilization (Shivanna and Rangaswamy, 1998).

In nature, plants are equipped with pre- and post-fertilization devices to discard incompatible pollens and therefore, allow only the right mating or compatible types to germinate and to bring about fertilization. The pre-fertilization barrier devices are: (i) rejection of alien pollens by preventing their germination; (ii) slow growth of the pollen tube; (iii) excessive length of the style to be traveled by pollen tube; (iv) bursting of the pollen tube in the style; and (v) early abscission of the ovary. On the other hand Post-fertilization barrier is the failure of hybrid embryos to become mature due to absence of synchrony in the growth of embryo and endosperm.

In order to overcome pre-fertilization barriers, some of the important practices are: (i) bud pollination (ii) stub pollination (iii) heat treatment of the style and (iv) mixed pollination. In addition to the above-mentioned techniques, the *in vitro* fertilization is one of the potential methods to bypass the pre-fertilization barriers and to introduce new characteristics of genetic information from more distantly- related species, thus widening the gene pool.

IN VITRO POLLINATION

(i) Technique

The technique of *in vitro* pollination of ovule was for the first time developed by Maheshwari and his students at the University of Delhi with *Papaver somniferum* (Kanta et al. 1962). The *in vitro* pollination was originally referred to as ***in vitro* fertilization** or **test tube fertilization** but in the present time both the segments have their distinct identities. Therefore, in the present chapter both have been discussed separately. Before starting any experiment, study of the different aspects of floral biology of the experimental plant is essential. Such study involves: (i) anthesis, (ii) dehiscence of anthers, (iii) pollination, (iv) pollen germination, pollen tube growth and penetration (v) fertilization and (vi) seed maturation.

The sterilization of the materials is the pre-requisite for initiation and establishment of any *in vitro* study. Here two important parts to be used are pistil and anther. Flower buds to be used as female partner are emasculated before anthesis and bagged in order to prevent undesirable pollination. The buds are brought to the laboratory before the anthers burst. The pistil is obtained after removal of sepals and petals. The whole pistil of the ovary alone is sterilized with a suitable reagent and finally washed with the autoclaved distilled water. For ovule or placental pollination the ovary wall is carefully peeled to obtain either ovules or the ovules together with the placentae.

(ii) Types of Pollination

The *in vitro* pollination is of four types, depending on the design of experiment whether collected pollens are deposited on the cultured stigma, ovary, ovules or placentae (Fig. 19.1). Generally, the pollens deposited directly on the cultured part of the pistil give better result than that spread on the medium around the ovules.

(a) Stigmatic Pollination

In this technique the anthers are collected usually before anthesis and allowed to dehisce on sterile filter paper under aseptic condition. To carry out stigmatic pollination the pistil, from the emasculated or bagged flowers are collected a day before or on the day of anthesis. All the floral parts except the pistil are removed and the pistil is surface sterilized. The stigma is pollinated with the prepared sterile pollens and the treated pistil is immediately returned to culture medium for the seeds to be set.

Such *in vitro* stigmatic pollination, cross pollination in *Nicotiana rustica* and *Petunia violacea* (Rao and Rangashwamy, 1975; Shivanna, 1965) and self-pollination in *Antirrhinum majus* (Usha, 1965) have been described, where subsequently fertilization and seeds setting took place successfully.

(b) Ovarian Pollination

The intra-ovarian pollination was developed at Delhi University by Maheshwari and his associates (Kanta, 1960; Kanta and Maheshwari, 1963). The flowers used as female were emasculated and bagged 2-day before anthesis. On the day of anthesis the ovary was surface sterilized by wiping it through cotton soaked in ethanol and the pollen suspension injected into the ovary through a hole using hypodermic syringe. Another hole was made opposite the point of injection to allow the air to escape. The suspension was injected until the ovarian

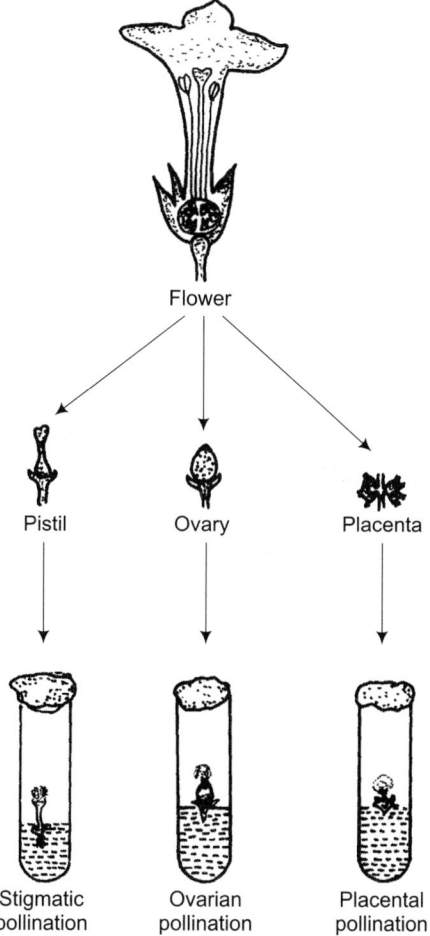

Fig. 19.1 Diagrams showing types of *in vitro* pollination.

cavity is filled and the liquid started oozing out of other hole. After introducing the pollen suspension the holes were sealed with petroleum jelly. This method led to successful production of seeds in many members of Papaveraceae (*Argemone mexicana, A. ochroleuca, Eschscholzia californica, Papaver rhoeas, P.somniferum*).

(c) Ovular Pollination

For carrying ovule pollination, the stigmatic, stylar and ovary wall tissues are completely removed to obtain naked ovules which are then dusted with pollens and cultured in a nutrient medium until seed maturity. Pollen germination, growth of pollen tube, entry of pollen tube, release of gametes and double fertilization are achieved sequentially and finally mature seeds are obtained. The first successful experiment was established with *Papaver somniferum* (Kanta et al., 1962). The success prompted the Delhi School to test the efficacy of this new technique with other plants species such as *Eschscholtzia californica, Argemone mexicana, Nicotiana rustica* and *N. tabacum* (Kanta and Maheshwari, 1963) and *Dicranostigma franchetianum* (Rangaswamy and Shivanna, 1969).

(d) Placental Pollination

As injury is caused during excision of ovule, therefore, an alternative method was developed. In this method the ovary wall is meticulously peeled, all the ovules within the ovary are retained intact on the placenta, pedicel is shortened, and pollen grains are dusted all over the intact ovules and such pistils are cultured on the culture medium. The refined technique, termed as **placental pollination,** not only prevented any injury to the ovules and also retained more of the maternal tissues which is known to be favourable for the development of seeds. In *Petunia axillaries* placental pollination was effective in overcoming the self-incompatibility (Rangaswamy and Shivanna, 1967).

(iii) Factors Affecting Seed-set after *In Vitro* Pollination

(a) Genotype

Genotype of the plant is playing very important role in establishing *in vitro* pollination. Higgins and Petlino (1988) showed the genotypic differences while experimenting with the maize. They showed full kernel development depends on ovule-to-cob tissue ratio. In case of 4:24 ovule–to-cob ratio (cob pieces with 24 ovaries were taken but only 4 ovaries per pieces were pollinated and the rest ovaries were removed) best results were obtained which vary from 19-36%, depending on the genotype.

(b) Explant

There are evidences which suggest that in different species success of *in vitro* pollination depends on the type of explants. In *Petunia hybrida* for seeds set placental pollination gives better result than pistil pollination. However, in the former case presence of style is essential. In maize large number of ovaries attached to cob tissues fully developed seeds than single ovary. Gengenbach (1977a) reported that cob pieces with one or two ovaries did not form any fully developed kernel, four ovary blocks developed only small kernel whereas ten ovary blocks had one or two fully developed kernel of large size.

The floral envelopes (lemma and palea) play an important role in seed-set. Ovaries excised soon after pollination of *Triticum aestivum* and *T. spelta* develop in culture only when floret envelopes remain intact.

(c) Physiological State of the Explant

The physiological state of the ovule or pistil influences the seed-set after *in vitro* pollination. The incidence of seed-set is higher when the ovules are excised 1-2 days after anthesis than on the day of anthesis. According to Gengenbach (1977b) the optimal stage of maize spike for *in vitro* pollination is 3-4 days after silking. Pollen germination and growth of the pollen tube stimulate the synthesis of protein which may block the entry of the pollen tube into the ovary. Therefore, it is necessary to find out the blockage site in the pistil so that their removal may improve the result of *in vitro* pollination.

(d) Culture Medium

The composition of the culture medium plays a significant role in establishing pollen germination, growth of the pollen tube resulting fertilization and seed sets. It is, therefore, information regarding the effect of various phytohormones and other supplements to the basal medium on seed development needs to be investigated as this practice improves the chance of success for *in vitro* pollination and fertilization. For *in vitro* culture of pollinated ovules in most of the species a modified Nitsch's medium is widely used (Table 19.1). However, the medium developed by Steward and Hsu (1978) has been recommended for raising intraspecific and interspecific hybrids from young fertilized ovules. The presence of 10 µg l^{-1} IAA or 0.1µg l^{-1} kinetin significantly improves the number of seeds per ovary. According to Gengenebach (1985), the nitrogen source does not affect fertilization frequency in excised *in vitro* pollinated maize ovaries but complete amino acid mixture is required for optimal kernel development and growth.

Osmolarity of the culture medium affects the development of the excised ovules. Generally, sucrose has been used at a concentration of 4-5%.

Table 19.1

Constituents	Amount (mg/l)
$CaNO_3$	500.00
KNO_3	125.00
KH_2PO_4	125.00
$MgSO_4, 7H_2O$	125.00
$CuSO_4, 5H_2O$	0.025
Na_2MoO_4	0.025
$ZnSO_4, 7H_2O$	0.50
$MnSO_4, 4H_2O$	3.00
H_3BO_3	0.50
$FeC_6O_5H_7, 5H_2O$	10.00
Glycine	7.5
Ca-pantothenate	0.25
Pyridoxine HCl	0.25
Thiamine HCl	0.25
Niacin	1.25
Sucrose	50,000.00
Agar	7,000.00

After Kanta and Maheshwari (1963)

IN VITRO FERTILIZATION

During 1960s when pollination of the cultured ovules was devised and used to overcome self-and interspecific-incompatibility, no further progress could be made towards *in vitro* fertilization because no technique was available to isolate sperms and eggs, also the protoplast technology was in its infancy. However, remarkable progress was made during 1970s and 1980s toward protoplast culture and somatic hybridization that set pathway for the experimental embryologist to isolate sperms and eggs and to fuse them (Fig. 19.2). With the establishment of recombinant DNA technology, the process of isolation and transformation was more simplified.

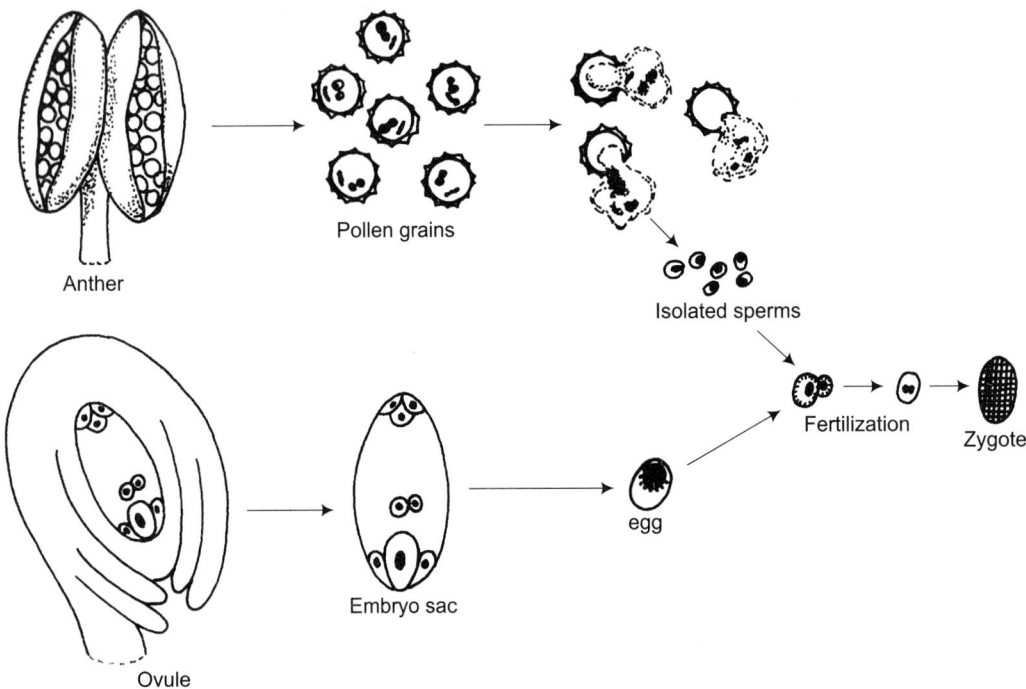

Fig. 19.2 Diagrammatic representation of isolation and fusion of male and female gametes to form zygote.

(i) Isolation of Sperms

Isolation of sperms involves two major steps: (i) release of pollens either from the anthers (3-celled pollen species) or from the pollen tubes (2-celled pollen species) and (ii) purification and maintenance of released sperms. The composition of medium used for the release and maintenance of sperms is variable from species to species (Cass, 1977; Kapil and Sokhi, 1977).

Isolation of sperms from the pollens which are shed at 3-celled stage, however, involves two methods: (i) physical method and (ii) osmotic shock method. In the physical methods the pollen grains are homogenized vigourously to bring about rupture of pollens whereas the osmotic shock method involves the incubation of pollens in a hypotonic solution to induce pollen rupture and release of sperms.

In 2-celled pollen species the sperms are formed within the pollen tube and the studies on such species are limited. In two species *viz.*, *Gladiolus* and *Rhonodendron* a semi- *in vivo* technique was applied to isolate sperms (Shivanna *et al.*, 1988). This technique involves combination of both *in vivo* and *in vitro* techniques. *In vivo* the pollens are allowed to germinate on stigma and grow into style to certain length. The styles are then cut and cultured on suitable medium where pollen tubes emerged into the medium through the cut end of style. The sperms are isolated from the pollen tubes either by enzymatic maceration or osmotic bursting of their tips.

When pollen grains are used for isolation of sperms, the medium into which sperms are released contains pollen debris. Exine fragments and larger particles are removed through filtration. The remaining contaminants, particularly starch grains, are removed through percoll or sucrose density gradient technique. The viability of the purified sperm is assessed mostly through fluorescein diacetate test (Shivanna and Rangaswamy, 1992). Isolated sperm cells have been shown to be metabolically active as they synthesize both RNA and proteins (Cass, 1977).

(ii) Isolation of Embryo Sac and Egg

Embryo sac and their components (egg cell, central cell and antipodals) are generally isolated by combining enzymatic maceration of ovules and then their micro-dissection under an inverted microscope (Theunis et al., 1991; Dumas and Mogensen, 1993). Generally cellulase, driselase, β-glucuronidase, hemicellulase, macerozyme, pectinase and pectolyase have been used in different concentrations and combinations which vary from species to species. In a few species embryo sacs have been successfully isolated from fixed and fresh ovules (Zhou, 1987; Yang and Zhou, 1992).

Isolation of egg cell is more difficult than that of sperms and time consuming. Moreover, the experience and skill of the investigator is much important. In maize, the Kranz group isolated only 40 eggs from ovules in 2-3 hours (Kranz et al., 1995; Kranz and Dressenhaus, 1996). The viability of the isolated egg is assessed through fluorecein diacetate test.

(iii) Fusion

The first success was achieved in maize in the Laboratory of Kranz at the University of Hamburg. The fusion of gametes was performed in a micro-droplet of 1-2μl fusion medium (540mos mol per kg^{-1} H_2O with mannitol solution); overlaid with mineral oil on a cover glass. The egg and the sperm protoplasts were individually picked up in micro-capillaries and transferred to the fusion medium by using hydraulic system and a computer-controlled dispensed-diluter. The egg and the sperm were aligned by dielectrophoresis and the fusion was achieved by application of one or a few pulse (0.9-1.0kVcm^{-1} for 50μs) of direct current (Kranz et al., 1991; Fig. 19.3).

Gametic fusion was generally accomplished in less than 1 sec. Cell wall formation around the freshly formed zygote initiated within 30s after fusion and karyogamy occurred in less than 1 hour from fusion (Kranz et al., 1995). The early initiation of cell wall prevents polyspermy. The composition and the osmolarity of the fusion medium are critical in the gametic fusion as in the fusion of protoplasts. Usually, fusion under the electric pulse may result in some membrane change; therefore, attempts have been made to bring about fusion without electric pulse, i.e. simply by manipulation of the fusion medium. The presence of calcium in the fusion medium has effectively replaced the use of electric pulse (Faure et al., 1993).

After achieving in vitro fusion of sperm and egg protoplasm, it took another two years to standardize the condition for embryogenesis and plantlet formation from in vitro formed zygote (Kranz and Lörz, 1993). This became possible only through the nurse culture technique. The in vitro formed zygotes of maize were cultured on a semipermeable membrane of Millicell-CM dish (12mm diameter) filled with 0.1ml of nutrient solution placed in the middle of a petri plate filled with 1.5 ml of nutrient medium containing fast growing

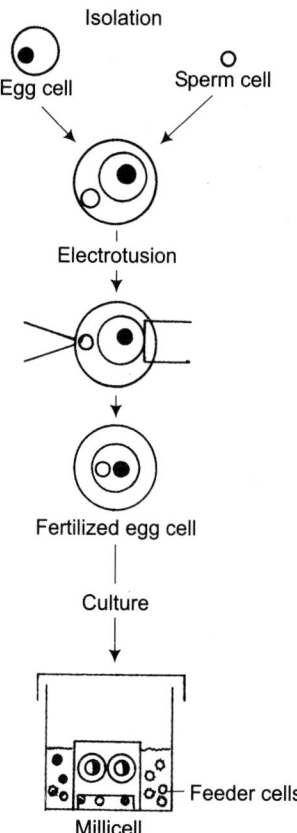

Fig. 19.3 Electrofusion of male and female gametes (After Kranz et al. 1991).

non-morphogenic cell suspension (feeder cells) culture derived from maize embryo or microspores. The culture was maintained under 16 hrs light and 8 hrs dark cycle with about 50µE m^{-2} s^{-1} light intensity. As high as 85% of zygote underwent divisions within 42-46 hours after *in vitro* fusion and formed embryoids which later developed into fertile plants (Kranz et al., 1995).

Interestingly the sperms have been found to fuse with central cells comparable to that of endosperm (Kranz, 1997). Further, attempt to fertilize the egg of maize with related taxa showed production of zygote which entered into a few divisions but the plantlets could not be obtained. Recently some success has been achieved towards *in vitro* fertilization and nurse culture of zygotes which formed microcalli but failed to develop embryoids or plantlets (Kovacs *et al.*, 1995).

(iv) Microinjection of Sperms into Egg and Central Cell

Dumas and his associates (Mathys-Rochon *et al.*, 1994, 1997) were successful in injecting the sperm nucleus into the egg or the central cell. The embryo sac was immobilized into a low melting point agarose as the latter did not affect the viability of the embryo sac. However, fertilization and cell division are yet to be achieved.

The potential use of cultured sperms, eggs and zygotes to achieve genetic transformation has been demonstrated in a recent paper from Dumas laboratory (Leduc *et al.*, 1996). They have reported success not only microinjecting DNA into cultured maize zygote but also transient expression of the micro-injected genes in the zygote.

APPLICATION OF *IN VITRO* POLLINATION AND FERTILIZATION

Potential applications are in the areas of plant breeding, genetics and understanding the fundamental processes of pollination and fertilization. A few important applications are described as under:

(i) Plant Breeding

In vitro pollination and fertilization have been extended to raise interspecific and intergeneric hybrids otherwise unknown in nature. *Melandrium album* and *M.rubrum* were used as ovule parents and 15 species belonging to different families were used as pollen parents. After pollination, cultured ovules underwent normal development into seeds containing viable embryos (Zenkteler *et al.*, 1992).

Another field where *in vitro* pollination and fertilization has found application is overcoming the self-incompatibility. *Petunia axillaries* and *P. hybrida* are self-incompatible species where barrier exists in the ovary and as a result pollens can not fertilize the ovule. *In vitro* placental pollination, however, can bring about normal development of embryo and endosperm formation.

(ii) Haploid Production Through Parthenogenesis

Another application of *in vitro* pollination is the production of haploids. Haploid sporophytes are invaluable in plant breeding because by a mere doubling of their chromosome number homozygous diploids can be obtained. Hess and Wagner (1974) reported that in their attempts to obtain haploids of *Mimulus luteus* through anther culture failed, however, placental pollination with the pollens of *Torenia fournieri* haploid plants were obtained which were successfully transplanted to soil. A fully developed plant revealed that it is not a hybrid rather haploids of ovule parent *Mumulus*. However, in the absence of detailed anatomical and cytological investigations, one of the possible explanations is the gradual elimination of chromosomes in *Torenia*. Such parthenogenetic development of haploids in cultures from *in vitro* pollinated but unfertilized ovules has also been reported in *Hordeum vulgare*, *Nicotiana tabacum* and *Triticum aestivum*.

(iii) Fundamental Study

Besides its practical applications in plant breeding and haploid production, *in vitro* pollination and fertilization has been employed to study physiology of pollen germination. For example, rather it is difficult to obtain *in vitro* germination of pollens of Brassicaceae.

However, study with the ovules of *Brassica oleracea* which were dipped in calcium chloride solution induced pollens to germinate. This suggests that secretion of certain chemical by ovule is essential for achieving chemotactic growth of pollens.

(iv) Production of Stress Tolerant Plants

Petolino *et al.* (1990) reported that *in vitro* pollination at higher temperature (38°C) resulted in production of heat tolerant plants of maize, which showed improved agronomical performances at higher temperature when compared to plants which were regenerated through *in vitro* pollination at normal temperature (28°C).

(v) Genetic Transformation

The isolation of gametes, fertilization and subsequent production of plants has opened the pathway for genetic manipulation for better qualities.

20
Embryo Culture

In angiosperms, embryo represents the beginning of the sporophyte. Normally, the zygote undergoes embryogenesis in the post-fertilization stage within the ovule and thus an embryo is formed inside the seed. The typical embryo is a bipolar structure consisting of a radicle and plumule which later develops into root and shoot systems, respectively.

Hanning (1904), for the first time successfully, obtained plantlets from embryo culture. The embryos were dissected out from seeds of *Raphanus sativa, R. landra, R. candatu, Cochleria danica* and cultured them in a simple medium containing mineral salts and sucrose. Subsequently, several workers have reported embryo cultures. Laibach (1925, 1929) reported embryo rescue from interspecific cross of *Linum perene* and *L. austrianum*, thus highlighted the practical application of embryo culture. Van Overbeek (1941, 1942) showed that mature embryos were self-nourishing and can grow in simple medium, while immature embryos failed to grow or grew feebly. However, enriched medium supplemented with coconut water showed successful results from pro-embryo.

The objective of embryo culture is to understand the growth pattern of proembryo or mature embryo and the various controlling parameters. The technique can also be successfully employed for developing hybrid embryos, which normally abort prematurely. There are two types of embryo culture:

TECHNIQUE

Embryos of the seed plants are generally enclosed within the ovules, which in turn, are covered with ovary. Thus, embryos already exist in sterile condition; hence the disinfection is not desirable. Instead mature seeds, entire ovules or fruits are surface sterilized and the embryo aseptically removed from the surrounding tissues.

(i) Culture of Immature Embryos or Proembryos

The term proembryo refers the early stage of the development of embryo that precedes cotyledon initiation. Globular or heart shaped stages of embryos are appropriately called as proembryo. The objective of such culture is to understand the control of differentiation and the nutritional requirements of immature embryos.

Practically, dissecting out such embryos from unripe or hybrid seeds is difficult and generally requires a complex nutrient medium to produce plants. However, the success per cent is very low.

(ii) Culture of Mature Embryo

Culture of mature embryo in large seed is relatively easier because mature embryo is visible. Seeds are placed in autoclaved distilled water for 24 hours after proper surface sterilization to make embryo prominent. A simple culture medium is required, which contains sugar, mineral salts and agar for growth and development. The success rate is very high.

For example in rice, mature seeds are dehusked and put into 70% alcohol for 1 minute followed by $HgCl_2$ (0.1%) treatment for 15-20 minutes for surface sterilization. Final washing is performed with autoclaved distilled water under Laminar Airflow and sterilized seeds are dipped in autoclaved distilled water for 24 hours for excision of mature embryos to be cultured (Fig. 20.1). In orchids, entire capsule is surface sterilized and seeds are dissected out and cultured intact for embryo culture. In Orchids, seeds are tiny and consists undifferentiated embryo completely filling the seed as functional endosperm is lacking and the seed coat is reduced to a thin membranous structure.

Capsule of *Capsella bursa-pastoris* is a good material for the culture of immature and mature embryos. Here the sterile capsule is dissected for the isolation of ovule at placental region. Ovule is splitted longitudinally and the tissue is teased to release immature embryo together with attached suspensor. For obtaining mature embryo, a small incision is made on the side lacking the embryo and a pressure is applied with blunt needle to set the embryo free which is culture on MS medium (Fig. 20.2).

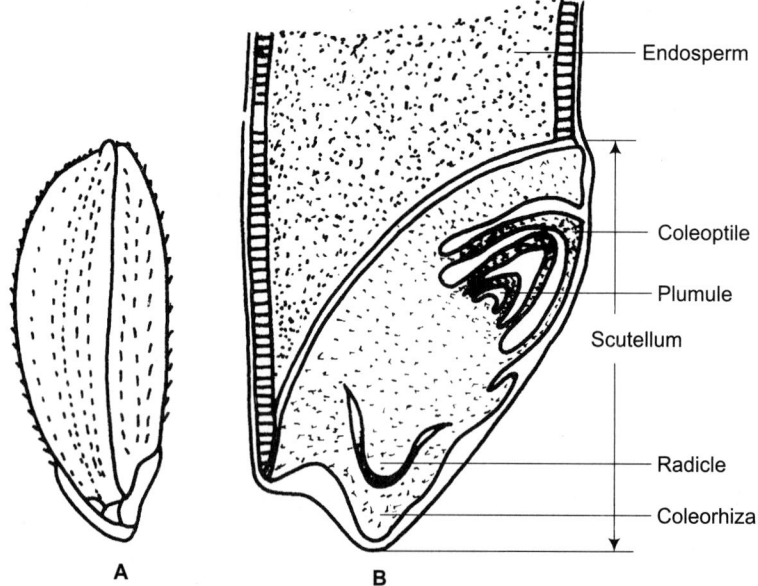

Fig. 20.1 A. Diagram of rice grain with glumes. **B.** L.S. of seed to show position of embryo.

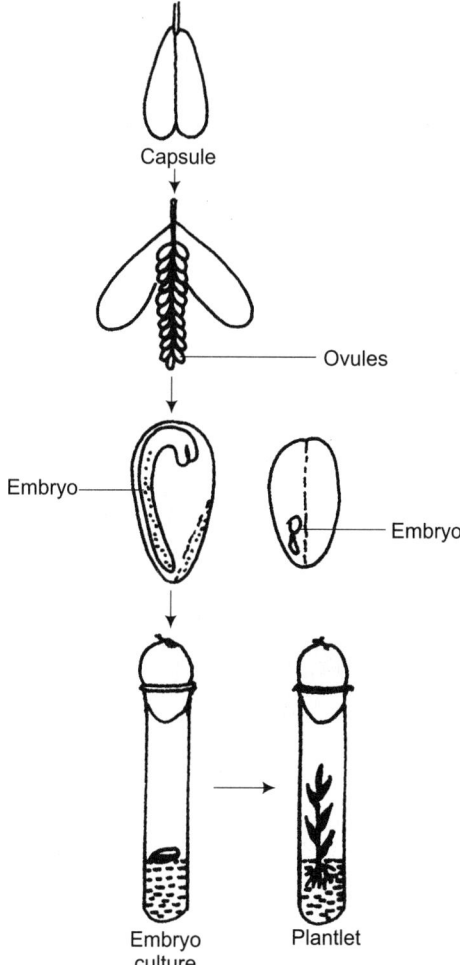

Fig. 20.2 Diagram showing isolation and culture of embryo of *Capsella bursa-pastoris*.

EMBRYO-ENDOSPERM TRANSPLANT

For culture of immature embryo, which aborts at very early stage of development, endosperm of the same species is used as nurse tissue to ensure the survival and development of immature embryos. This has been nicely described by Williams and De Lautour (1980) for hybrid embryos, developed from crosses of various species of *Trifolium*, *Lotus* and *Ornithus*. They implanted excised hybrid embryo into a cellular endosperm dissected from a normally growing ovules of any one of the parents. The nurse endosperm together with hybrid embryo were cultured on nutrient medium for obtaining interspecific or intergeneric hybrid plants which otherwise could not be developed (Fig. 20.3).

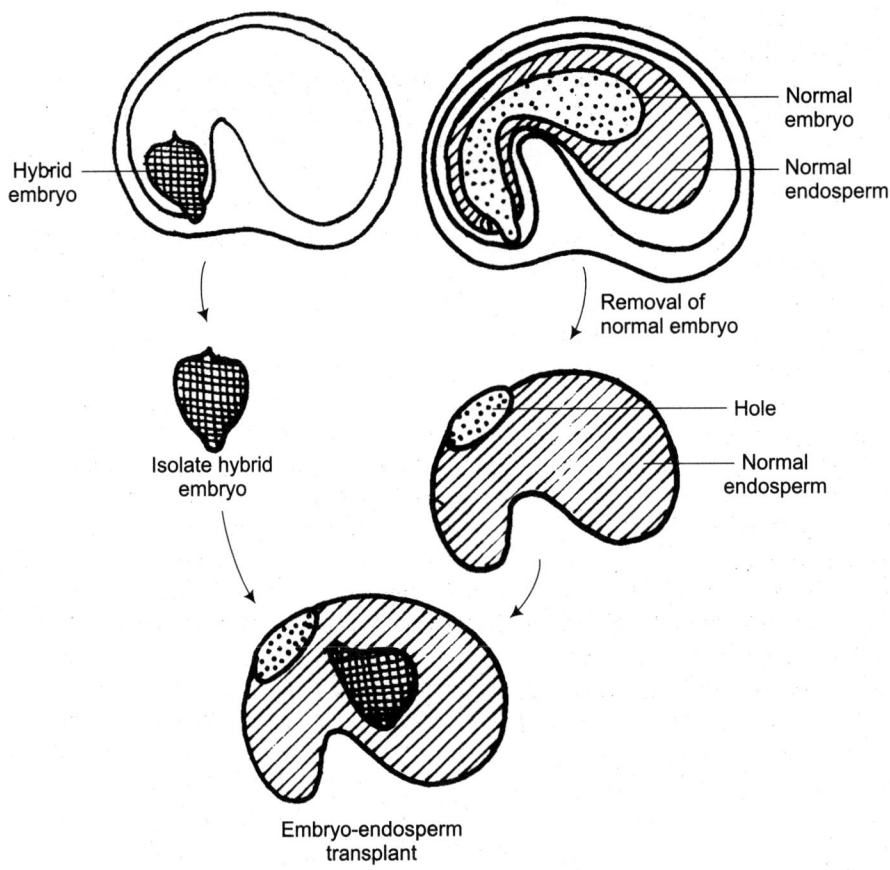

Fig. 20.3 Embryo-endosperm transplant.

CULTURE MEDIUM

The most important aspect of embryo culture work is the crucial selection of the medium to sustain continued growth of the embryos. Generally, mature embryos grow on a simple medium, whereas culture of immature embryo requires progressively complex nutritional requirement.

Inorganic nutrients of MS, B_5, and White's media with certain degree of modification are the most widely used basal media for embryo culture work. However, Monnier (1978) modified the MS medium and formulated new mineral solution, which ensured higher survival of cultured immature embryo of *Capsella*. This medium contained higher levels of potassium and calcium and reduced level of ammonium and FeEDTA and double the concentration of MS micronutrients.

Table 20.1: Composition of media for uninterrupted growth of *Capsella* globular embryo.

Constituents	Media (amount in mg/l)	
	M1 (external ring)	M2 (central zone)
Macronutrients		
KNO_3	1900	1900
NH_4NO_3	990	825
$CaCl_2 \cdot 2H_2O$	484	1,320
KH_2PO_4	187	170
$(NH_4)_2SO_4$	407	370
KCl	420	350
Micronutrients		
Na_2 EDTA	37.3	—
$FeSO_4 \cdot 7H_2O$	27.8	—
$MnSO_4 \cdot H_2O$	33.6	33.6
$ZnSO_4 \cdot 7H_2O$	21	21
H_3BO_3	12.4	12.4
KI	1.66	1.66
$Na_2MoO_4 \cdot 2H_2O$	0.5	0.5
$CuSO_4 \cdot 5H_2O$	0.05	0.05
$CoCl_2 \cdot 6H_2O$	0.05	0.05
Organic		
Glutamine	—	600
Sucrose	—	1,80,000

After Monnier (1976)

Media constituents for the *in vitro* growth of young or immature embryos are different. Therefore, it requires the transfer of embryos from one medium to another in order to obtain their full development. Monnier (1976) devised a culture method for the uninterrupted growth of globular embryo of *Capsella* to maturity without shifting them from their original position in the culture plate (Table 20.1). By this method, embryo can be grown in both solid and liquid medium at the same time. The first agar medium is liquefied by heating and then poured around the central glass container. After solidification of the medium, the container is removed. The medium will form external ring. A second medium of different composition is poured into the central ring. Embryos are cultured on the second medium in the central ring. As a rule of diffusion the embryos are subjected to the action of variable medium with time (Fig. 20.4). The composition of both media is different. Adopting this technique Ko *et al.* (1983) obtained uninterrupted growth of immature embryo of rice.

It has been demonstrated that osmotic concentration has a significant influence on the growth of excised embryo (Mauney, 1961; Maheshwari and Rangaswamy, 1965). Sucrose is most commonly used source of energy for embryo culture and also maintains osmolarity which is very important for immature embryos. For this function the optimum concentration varies with the stage of embryo development. Mature embryo grows fairly well with 2% sucrose but the younger embryos require higher level of carbohydrates. This is in harmony

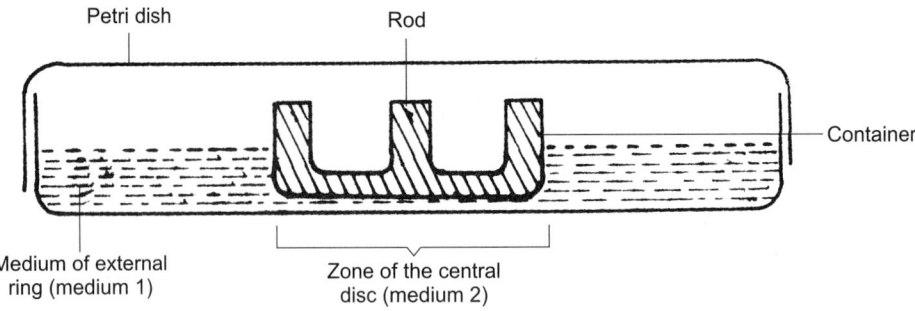

Fig. 20.4 Device of Monnier for embryo culture.

with the observation that *in situ* the proembryos are surrounded by fluid of high osmolarity. 8-12% of sucrose is generally, sufficient for the culture of proembryos. However, immature embryos of *Capsella* can grow on low sucrose concentration (2%) when IAA (0.1 mg l^{-1}), Kinetin (0.001mg l^{-1}) and adenine sulphate (0.001mg l^{-1}) are added to basal medium. Addition of glucose, maltose, raffinose, lactose or mannitol, however, improves the response. Incorporation of growth adjuvants, such as coconut water (CW) and casein hydrolysate (CH) have been found to give encouraging results.

The nitrogen, especially NH_4^+, has a significant role for growth and differentiation of embryos. Various amino acids have been tested for embryo culture. The glutamine has been found to be the superior source of nitrogen for embryos.

Van Overbeek *et al.* (1942) observed that in *Datura* young embryos either failed to germinate or develop feebly in culture medium containing minerals and sugars. They added non-autoclaved CM (coconut milk) in the above medium in an attempt to culture young embryos. On this modified medium normal seedlings developed from embryos as small as 150-200 µm (heart- shaped) and as young as 10 DAP. The growth promoting factors in the coconut milk referred to as **'embryo factor'**. In order to find a suitable substitute for CM for the culture of very young embryos wide range of natural extracts, such as CH (casein hydrolysate), dried brewer's yeast, skimmed milk and diffusate from the endosperm of *Ginkgo* and the seeds of several angiosperms have been tried. The alcohol diffusate from young seeds of *Luminus* and mature seeds of *Sechium* proved to be as effective as CM.

Kent and Brink (1947) reported that promotion of embryo growth and inhibition of precocious germination of immature barley embryos could be achieved by water extracts from dates, bananas, hydrolysate of wheat-gluten and tomato juice.

With a view to develop synthetic media, attempts have been made to replace the 'embryo factor' of CM with chemically defined substances. To promote the growth of barley proembryo, CM could be substituted by phosphate-enriched White's medium fortified with glutamine and alanine as major amino acids and five other amino acids as minor sources of N_2 at pH 4.5. The survival rate of the embryos was considerably increased when the concentration of KCl, KNO_3 and certain organic components was increased 5-10 times (Norstog, 1967, 1973). The modified Norstog's medium proved to be the best medium for the culture of 250 µm long embryos of five varieties of *Hordeum distichum*.

Use of auxins and cytokinins are generally not recommended because callusing is induced. A very low level of GA_3 (0.01mg/l) promotes embryogenesis. There are reports that

ABA, on the other hand, has the same results in barley and *Phaseolus* embryos. However, pre-bloom and bloom time spray of benzyladenine, a cytokinin, drastically improves embryo recovery, germination and plant development in Flame seedless grape variety (Bharthy et al., 2005).

The molecular studies of Siskoo *et al.* (2003) in interspecific crosses of Cucurbita sp. showed that the nuclear DNA contents of hybrids was intermediate and differed significantly from those of the parental species.

Excised embryos grow well in a medium with pH 5 to 7.5. This is within range of the pH of ovular sap.

For the best growth of the embryos the temperature is within the range of 25-30°C. The optimum temperature of *Datura tatula* is reported to be 35°C. Light is not critical for embryo culture. On the contrarily, in barley light suppresses the precocious germination of immature embryo. However, there are some examples where initial dark incubation is important for the proper growth of embryo.

There is a significant role of suspensor in embryo development. It has been shown that in cultures the presence of a suspensor is critical, particularly for the survival of embryos. Yeung and Sussex (1979) observed that attachment of the suspensor with young embryos of *Phaseolus coccineus*, or its placement in close proximity in the culture medium strongly stimulated the development of the embryo to maturity when compared to those cultured without the suspensor. The requirement of the suspensor may be substituted by the addition of GAs or ABA in the culture medium.

APPLIED ASPECTS OF EMBRYO CULTURE

(i) Overcoming Seed Dormancy and Shortening of the Breeding Cycle

Seed dormancy is a natural phenomenon, where there is temporary arrest of metabolic functions. This temporary suspension of growth is definitely either due to internal factors itself or due to some external factors. Prolonged seed dormancy presents special problem in cultivation of horticultural and crop plants. Conventional methods of breaking the seed dormancy has its own limitation, therefore, embryo culture has been successfully used to break seed dormancy. Orchid growers are commercially utilizing this technique for profuse production of orchid plantlets, where normal seed germination takes place in the soil in association with mycorrhizal fungi.

In breeding practice prolonged dormancy and slow growth of the seedling is not desirable. Here, the embryo culture is one of the effective alternatives because this method can reduce the breeding cycle. For example, using embryo culture Randolph and Cox (1943) could shorten the life cycle of *Iris* from 2-3 years to less than 1 year. *Rosa* normally takes whole year to come into flowering. However, through embryo culture it has been possible to produce two generation in a year.

Seed maturation in soybean and sunflower, takes 50-60% of the life cycle duration (120-150) days. The life cycle of sunflower could be reduced to half by *in vitro* culture of 10-day-old immature embryos (Fig. 20.5).

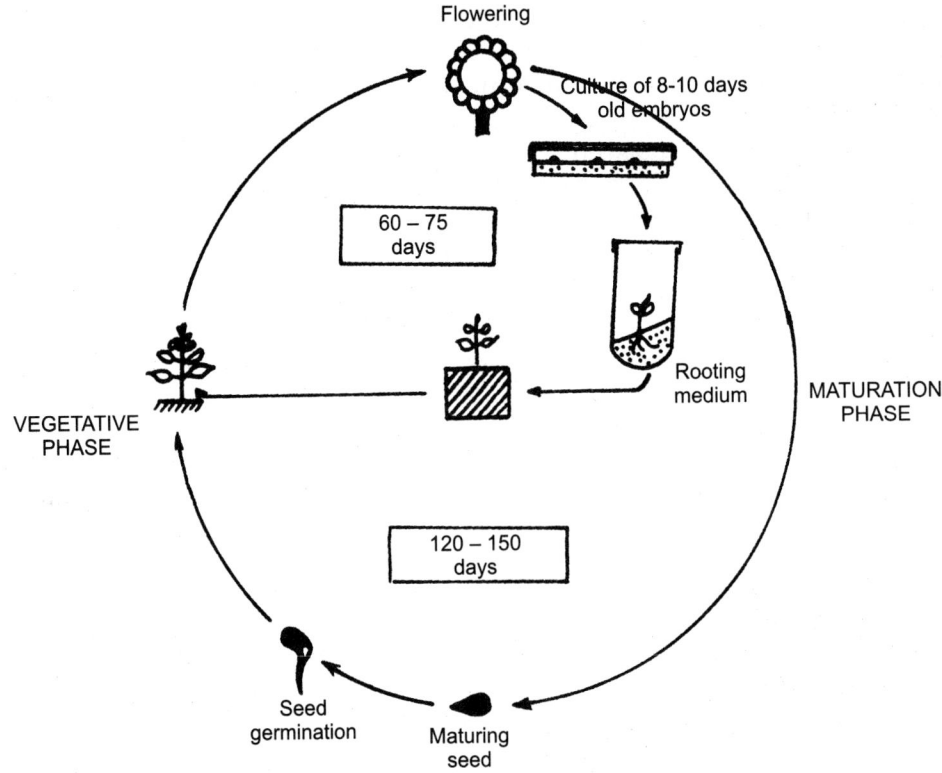

Fig. 20.5 Diagrammatic representation of the application of embryo culture in shortening the life cycle duration (After Serieys, 1992).

(ii) Overcoming Seeds Sterility

There are large numbers of plants known for their seeds sterility as they fail to germinate for e.g., peach, apricot, banana, kachhu etc. This natural sterility barrier in the seeds could be overcome by embryo culture.

'Makapuno' coconuts are very expensive and most relished for their characteristic soft fatty endosperm in place of a liquid endosperm. Under normal conditions the coconut seeds fail to germinate. In an attempt to achieve seed germination and complete transmission of makapuno trait, De Guzman *et al.* (1971) obtained 85% success in raising field-grown trees with the aid of embryo culture.

(iii) Haploid Production

A novel application of the embryo culture technique has been in the production of haploid through systematic elimination of chromosomes through distant hybridization. In the cross between *Hordeum vulgare* as female and *H. bulbosum* as male a viable fertilization is followed by zygote formation. But the chromosomes of *H. bulbosum* are rapidly eliminated from the cells of developing embryos. As a result monoploid embryo is formed where both cell division and the developmental processes are slow (Fig. 18.5).

(iv) Culture of Abortive Hybrid Embryo or Embryo Rescue

The most useful application of embryo culture is to overcome the barrier of interspecific or intergeneric cross. Embryo culture has been successfully employed to produce agriculturally and horticultural useful hybrids.

For example, *Lycopersicon esculentum* is very susceptible for virus, molds and nematodes, while *L. peruvianum*, a wild species, is relatively resistant. In a cross between the two species, the fruit develops normally but the seeds containing underdeveloped embryos do not germinate. However, a viable hybrid has been raised from such seeds by embryo culture.

Other example is *Melilotus officinalis* (sweet clover) which is a forage, but harmful to cattle because of the high content of coumarin. Therefore, attempts have been made to hybridize this species with other low coumarin species, *M. alba* and to culture the hybrid embryo on nutrient medium to raise plant. Employing the similar technique hybrid plants have been raised from *Phaseolus vulgaris* x *P. acutifolius* and *Lathyrus clymenum* x *L. articulata*.

A significant improvement in the frequencies of hybrid production from the intergeneric crosses between *Hordeum* x *Secale*, *Hordeum* x *Triticum*, *Hordeum* x *Agropyrum*, *Triticum* x *Aegilops* and *Triticum* x *Secale* was achieved with the aid of embryo culture.

Leng *et al.* (2006) have showed that even the immature embryos from the fruits of parthenocarpic cultivars could be cultured for the development of plantlets. Tian *et al.* (2008) successfully cultured hybrid embryos obtained by cross between wild Chinese *Vitis* species, which show high resistance to fungal diseases and disease prone seedless grape cultivars (*Vitis vinifera*), widely growing in Europe, America and Asia in order to obtain disease resistant varieties.

(v) Seed Testing

Embryo culture has provided a means for rapid testing of the viability of a particular batch of seeds. In seed testing practices, it is very useful and reliable method to predict the viability of seeds.

(vi) To understand the *in vivo* Growth of Embryo

Embryo culture is a useful technique to understand the fundamental aspects of embryogenesis, such as the effect of nutrients, phytohormones and other chemicals and physical factors on embryonal growth and differentiation.

21 | Culture of Different Parts of Pistil and Seed (Organ Microculture)

In order to unravel the developmental morphology, a variety of plant parts or organs could be cultured *in vitro*. They are cultured in isolation as far as possible to avoid correlative influences from other members/tissues with which they are in continuity.

A female gametophyte consists embryo sac surrounded by spoprophytic parental tissues, nucellus and integument (Fig. 21.1). Pollination and fertilization results seed-set, thus seed is a mature ovule which consist embryo (fertilized zygote), endosperm (fertilized secondary nucleus) and seed coat (arising from integuments) with parts of nucellus or may be completely consumed.

We have already studied the *in vitro* pollination and fertilization and their applications (Chapter-18). *In vitro* pollination and fertilization of the excised ovules have led to the successful seed-set in *Papaver*, *Eschscholtzia*, *Nicotiana*, *Argemone* and others which indicated possibilities of obtaining viable hybrid seedlings in interspecific and intergeneric crosses and in case of self-incompatibility. The ovules culture technique has enabled us to gain information on the physiology of lint fibre growth from cotton ovules (Beasley *et al.*, 1970s).

OVARY CULTURE

Ovary is ovule-bearing region of a pistil. After fertilization ovary matures into fruit and ovules into seeds. In order to study the precise requirements for fruit development and fruit physiology, Laibach and Kribben (1949) devised a method by which young flowers or fruits of *Cucumis sativus* could be excised and grown under aseptic condition.

La Rue (1942) obtained rooting of the pedicel and limited growth of the ovaries. Following the first attempt considerable success has been achieved in culturing the pollinated ovaries of several species. The technique was further developed by Nitsch (1951), who successfully reared ovaries of *Cucumis* and *Lycopersicon* excised from pollinated flowers on synthetic medium to develop into mature fruits (Fig. 21.2). These fruits developed viable seeds but the size of the fruits were smaller than those formed *in vivo*. In cultures the ovaries failed to grow into full-sized fruit (in restricted space of culture vial). Maheshwari and Lal

21.2 Plant Embryology: Classical and Experimental

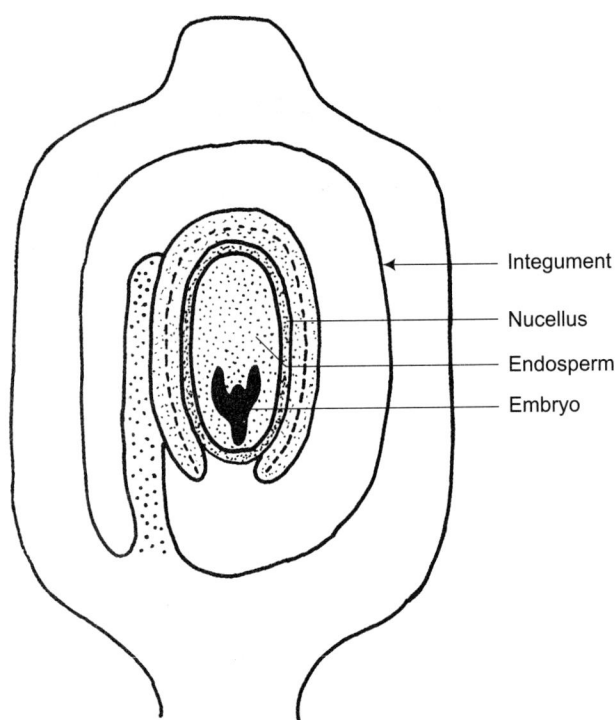

Fig. 21.1 Diagram showing L.S. of an ovary.

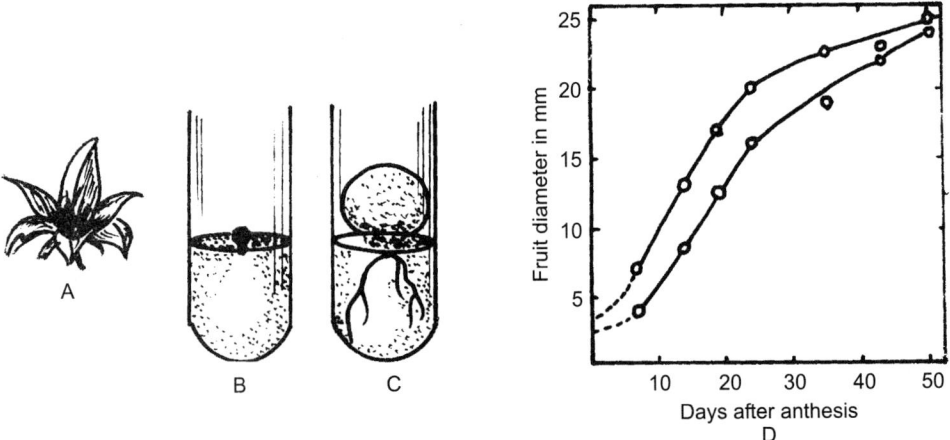

Fig. 21.2 The *in vitro* culture of detached tomato ovary. **A.** Tomato flower, two days after pollination. **B.** Ovary planted aseptically in a culture tube after the removal of petals, stamens, style and parts of sepals. **C.** Formation of a small tomato in the culture tube. **D.** Growth curves of tomato ovaries *in vitro* (After Nitsch, 1951).

(1958, 1961), however, obtained nearly normal sized fruits. Subsequently, attempts have been made to culture ovary for a better understanding of the factors controlling fruit development but with varying degree of success.

Pollinated or unpollinated flowers are collected from the healthy plants and after thorough washing with tap water, flower is immersed in sterilizing reagent (sodium hypochloride) for 5-7 minutes, final washing is done with autoclaved distilled water. Under the laminar airflow cabinet the flower is dissected for obtaining ovary, which is placed on agar solidified nutrient medium. The culture is incubated in 25 ± 3°C under 1500-3000 lux light intensity.

Excised ovary of *Iberis amara*, *Anethum graveolus* and *Hyoscyamus niger* from pollinated flowers could be grown on simple nutrient medium containing mineral salts and sucrose. However, addition of IAA or coconut water helped in larger fruit formation when compared with *in vivo* grown fruits.

In view of the smaller size of the fruit due to restricted growth in culture vials, Ito (1961, 1966) developed a partial sterile culture technique, to overcome this problem, in which only the long flower stalk is inserted into the aseptic nutrient medium through an opening into the stopper, thus leaving the ovary free to grow outside the culture vial.

In vitro culture of ovaries has revealed that the floral organs, in addition to their protective functions, have significant role in the development of embryo and fruit formation. In monocots "**Hull Factor**" is essential for the proper development of the embryo (La Croix *et al.*, 1962). In dicotyledonous plants removing the calyx also results in poor development of fruits (Guha and Johri, 1966; Richards and Rupart, 1980).

Supernumerary embryos (polyembryony) development has been reported in the members of Umbelliferae (Johri and Sehegal, 1963, 1966). Budding and shoot formation was promoted in a basal medium supplemented by CH, IAA or YE.

IMPORTANCE OF OVARY CULTURE

(i) Study of ovary culture helps in understanding the development of embryo and fruit formation.

(ii) The effect of phytohormones on parthenocarpic fruits development can be studied from the culture of unpollinated ovary.

(iii) Role of floral organs can be studied from the *in vitro* culture of ovary because floral organs play a significant role in fruit development.

(iv) Ovary culture has been successfully employed to overcome various impediments such as the failure of pollen germination on stigma, or slow or insufficient growth of pollen tube as well as precocious abscission of flower. Inomata (1977, 1979) raised interspecific hybrids between sexually incompatible parents *Brassica campestris* x *B. oleracea* through ovary culture.

(v) Through ovary culture polyembryony could also be induced successfully in various species.

(vi) Seed dormancy could be overcome from ovary culture.

OVULE CULTURE

An ovule or the megasporangium is embryo sac containing structure, which is surrounded by integument. Embryo sac consists egg apparatus, secondary nucleus and antipodals. Ovules are attached by means of their stalk or funicles to the placentae. Fertilization results in the formation of embryo and endosperm within the ovule, which soon matures into seed. The embryo consisting root and shoot primordial within the seed develop into a seedling at the cost of nutritive materials contained in the endosperm.

In vitro ovule culture helps to understand the factors that regulate the development of zygote. The first report of ovule culture dates back to 1932, when White cultured the ovules of *Antirrhinum* and obtained haploid callus.

Sterile ovules are obtained from pollinated or unpollinated flowers as per the experimental design. The dissected ovules are inoculated on defined medium supplemented with auxins and cytokinins (Fig. 21.3). The cultures are maintained at 25 ± 3°C in light (16 hrs; 3,000 lux).

Nirmala Maheshwari and Lal (1958) for the first time obtained mature seeds through the ovule culture of *Papaver somniferum* containing proembryo at the two celled-stage surrounded by a layer of endosperm nuclei in Nitsch medium (1951). She observed that kinetin triggered the growth and differentiation of embryo in the ovule. However, this initial rate of growth was not maintained for long and the length of embryo in cultured ovule was less than that of embryos in nature. The additive role of placental tissues in the growth and the maturation of seeds have been emphasized by Chopra and Shabharwal (1963) in pollinated ovules of *Gynandropsis gynandra*.

The growth of the embryo is related to the age of the ovule at culture. Eid *et al.* (1963) concluded that the difference in the rate of growth is associated with the developmental stage of endosperm present in the ovule at culture.

IMPORTANCE OF OVULE CULTURE

(i) Test Tube Pollination and Fertilization

A significant development of the ovule culture has been test tube pollination and fertilization. By this technique the pollens could be germinated together with the ovule culture and fertilization can be induced.

(ii) Application of Ovule Culture in Plant Breeding

In many interspecific or intergeneric crosses, ovule cultures have been successfully employed to develop hybrid seedlings. Viable hybrids have been obtained through ovule cultures from the interspecific crosses in *Abelmoschus* species and also in and *Brassica* species. A hybrid has been successfully developed by culturing the fertilized ovule obtained from the cross between *Lolium perenne* and *Festuca rubra*.

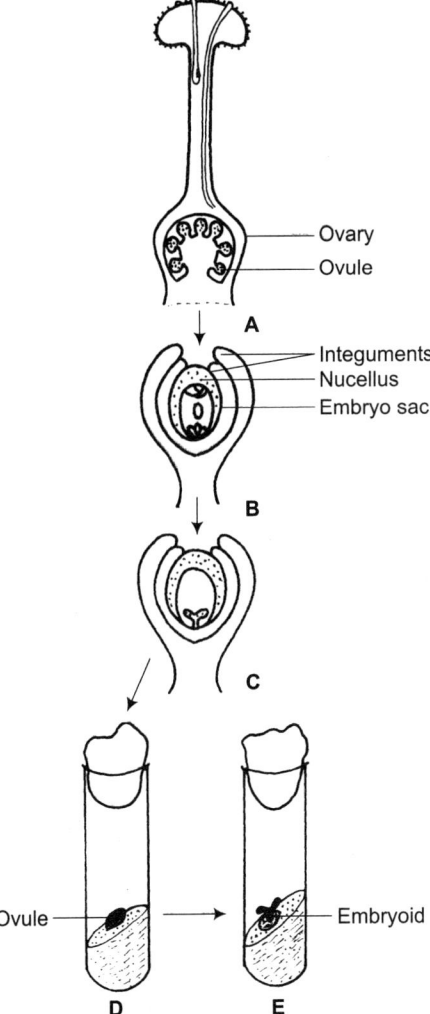

Fig. 21.3 A. Diagram showing pistil with pollen germination on the stigma. **B.** Diagram of an ovule showing different parts. **C.** Ovule with mature embryo. **D.** Culture of ovule. **E.** Regeneration of embryoid.

(iii) Production of Haploid Callus and Plants

In 1971, Uchimiya obtained haploid callus by culturing unfertilized ovules of *Solanum melongena*. However, plants have been successfully regenerated from the ovule culture of *Gerbera jamesonii*, *Helianthus annus* and many more species.

(iv) Ovule Culture and Angiospermic Parasite

The work on culture of ovule and seeds of angiospermic parasite enables the unraveling of the intricacies of host- parasite relationship and development of embryos in seeds that are shed at immature stage (Rangan and Rangaswamy, 1968).

Studies on ovule cultures of *Orobanche aegyptica* and *Cistanche tubulosa*, root parasites, have demonstrated that the formation of shoots *in vitro* can be induced even in the absence of any stimulus from the host.

(v) Ovule Culture of Orchids

In vivo germination of orchid seeds requires fungal association and the maturation of the seed capsule of many orchids demands long time. In order to overcome such barriers fertilized ovules have been successfully cultured to regenerate complete plants.

NUCELLUS CULTURE

Nucellus represents a homogenous non-vascularized tissue responsible for experimentation of different kinds in the studies of cyto-differentiation (xylogenesis), induction of adventive embryony, derivation of callus and regeneration, plantlets regeneration and other plant attributes.

Besides zygotic embryos, the adventitious embryos are also of much value since they are genetically uniform and reproduce the characters of maternal plant/parent without inheriting the variations brought about by the gametic fusion. In addition, embryos originating from the nucellar tissue in *citrus* are not only of help in obtaining virus–free plants but also of great advantage in citrus culture for propagating desirable varieties.

Through tissue culture technique, factors affecting the formation of adventive embryos have been studied. Nucellus polyembryony has been reported from the natural polyembryonic species of citrus (*Citrus microcarpa*) by Rangaswamy (1961). He observed that in *Citrus microcarpa* nucellar tissue, when freed from the integuments and cultured in suitable medium, there were continuous production of nucellar embryos. Bitters *et al.* (1972) extended these studies to several other mono- and polyembryonic, as well as seedless varieties of *Citrus*.

Through a series of experiments on nucellar embryony it has been concluded that the lack of stimulus of pollination or fertilization, can no longer be considered as a limiting factor for nucellar embryogenesis (Button and Bourman, 1971; Mitra and Chaturvedi, 1972).

Most of the studies focus on nucellar culture of both mono- and polyembryonic types of *Citrus* varieties. Attempts to induce nucellar polyembryony in *Luffa cylindrica* and *Trichosanthes anguina* (Rangashwamy and Shivanna, 1976) have been unsuccessful. Nucellar polyembryony has been reported in *Cymanchum vincetoxicum* (Haccius and Hausner, 1976) and *Vitis vinifera* (Mullins and Srinivasan, 1976). However, these taxa do not show natural polyembryony.

ENDOSPERM CULTURE

Endosperm is the most common nutritive tissue, which is ultimately digested during seed germination and utilized for the initial growth of the seedling. Unlike Gymnosperm, the development of endosperm in angiosperm is postponed until after fertilization. Endosperm is a post-fertilized product developing as a result of fusion of secondary nucleus and one of the male gametes. The endosperm is predominantly a triploid tissue (**3n**).

In many plants the endosperm is consumed by the developing embryos, as seen in the members of Papilionaceae. The seeds of such plants are called **non-endospermic or exalbuminous**, where the cotyledons become much thickened due to storage of food materials and gives nutrition to growing points during germination. In other category of plants, e.g. *Ricinus, Triticum, Zea* etc., the seeds are called **endospermic or albuminous**, where endosperm provides nutrition to growing points. However, there are some angiospermic families, which do not form endosperm at all e.g., Orchidaceae, Podostemaceae and Trapaceae.

The endosperm tissue offers an excellent system for experimental morphogenic studies. Besides being triploid, it is homogenous mass of parenchymatous tissue, lacking the differentiation of vascular elements. All the other tissues have been reported to form embryos in nature; however, there is no report of the endosperm doing so (Bhojwani and Bhatnagar, 1990).

In nature the endosperm cells store large quantity of materials in the form of starch, proteins, and lipids. It is therefore, an ideal system to study the metabolism of these natural products (Chu and Shanon, 1975). Role of endosperm in the growth and development of embryo is well understood because improper development of endosperm may lead to the abortion of embryo as it has a direct and dynamic influence on the differentiation of embryos. The technique of endosperm culture may be profitably exploited as an alternative to conventional plant breeding technique (cross of tetraploid and diploid for raising triploids) in plant improvement programme (Bhojwani, 1984).

Morphogenic differentiation of the endosperms of large number of species has been well established. The first report of the culture of endosperm of maize is by La Rue (1947), where only callus was raised. The culture of mature endosperm is more responsive than the immature embryo. Organ differentiation in endosperm cultures has been demonstrated in parasitic as well as autotrophic species: *Exocarpus cupressiformis* (Johri and Bhojwani, 1965), *Jatropha* (Srivastava, 1971), *Taxillus vestitus* (Nag and Johri, 1971) *Oryza* (Nakano *et al.*, 1975) and other taxa.

Technique

Members of Euphorbiaceae, Santalaceae and Loranthaceae have proved most amenable to endosperm culture. It is worthwhile to mention here that the endosperm could respond only when harvested at proper stage. Usually, in cereals endosperms undergo certain changes after 12 days after pollination (DAP), rendering tissue unable to respond.

Seeds having massive endosperm are decoated; surface sterilized with suitable disinfectant and after two to three washes in sterile distilled water, inoculated on the nutrient medium. Association of embryo with mature endosperm has a significant role for the proliferation of endosperm, which suggests concept of **'embryo factor'**. However, in some species embryo factor may be replaced by pretreatment of endosperm with GA_3 (Srivastava, 1970). Sometimes soaking the seeds in water for different periods also eliminates the need for association of the embryo during initial stage of endosperm culture. The number of proliferating endosperm pieces increased if the soaked seeds were allowed to germinate as there was a direct relationship between the number of days after seed germination and the number of cultures showing endosperm proliferation.

In *Taxillus*, the embryo had an adverse effect on bud differentiation from endosperm. The number of buds per culture and the number of cultures forming buds was comparatively higher when endosperm halves were cultured compared to that when endosperm with the embryo intact was cultured. However, subsequent development of the buds was better in the presence of embryo. The position of the endosperm on culture medium had a significant effect on the differentiation and number of buds (Fig. 21.4). Irrespective of the position of endosperm on the medium, the buds invariably first appeared along the injury. When the half-split endosperm of *T. vestitus* was planted with the cut surface in contact with the medium (White's medium with 5 mg l^{-1} kinetin) 100% cultures formed 12-18 buds, whereas if the cut surface was kept away from the medium only 1-3 buds developed in 30% cultures. Thus, the site of bud formation in these segments was almost predictable.

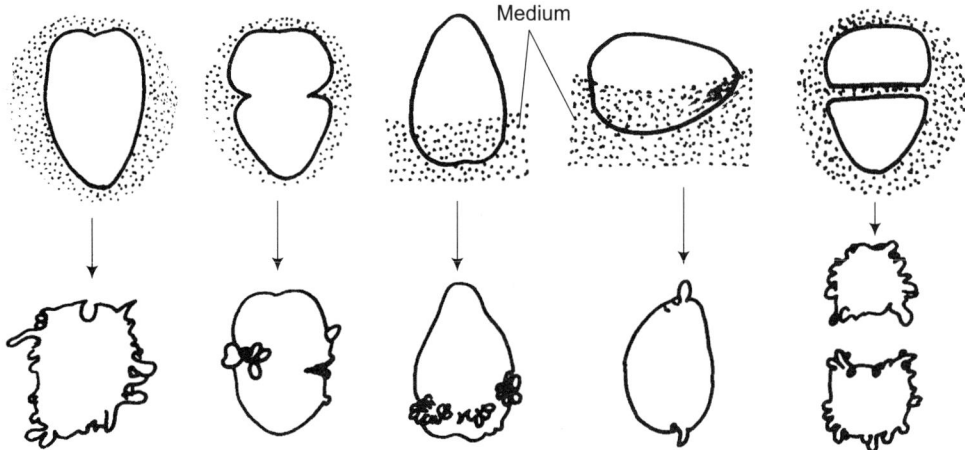

Fig. 21.4 Effect of injury and the position of endosperm in medium for bud differentiation in *Taxillus vestitus* (After Nag and Johri).

Factors

To induce callus the endosperm cultures of maize are maintained in dark. However, to induce differentiation the cultures need to be transferred to light condition. *Ricinus* cultures grow better in continuous light. The optimum temperature reported to be 25 ± 3°C and the pH varies (5.6-5.8) from species to species.

The basal medium of White or MS is generally used in endosperm culture. A combination of auxin, cytokinin and source of organic nitrogen such as yeast extract (YE) or casein hydrolysate (CH) was necessary for the better growth of endosperm tissues. Sucrose (2-4%) has proved to be the best source of carbohydrate. *Asimina triloba* is an exception where growth of endosperm is supported by the medium containing starch as a source of carbon.

The endosperm may form buds directly or it may first proliferate into a callus mass followed by organogenesis/embryogenesis.

Application of Triploid Culture

Triploid production by conventional method involves crossing of tetraploid with diploid. This method requires hard working and is not successful in most cases because of the high sterility of tetraploids. Therefore, endosperm culture is an alternative effective measure to obtain triploids. Triploid plants are self-sterile and usually seedless. This feature is highly desirable for the commercially important fruits such as apple, banana, mulberry, grapes, and watermelon. Triploid plants having timber value are more useful than their diploid or tetraploid partners. For example, the triploid of *Populus tremuloides* have better quality pulp wood. The Trisomics from triploids may be useful in gene mapping.

22 Synthetic Seeds

Synthetic or artificial seeds are the seed-like structures, which are made by encapsulating the *in vitro* raised embryoids with hydrogel which can grow into seedling under suitable conditions (Fig. 22.1).

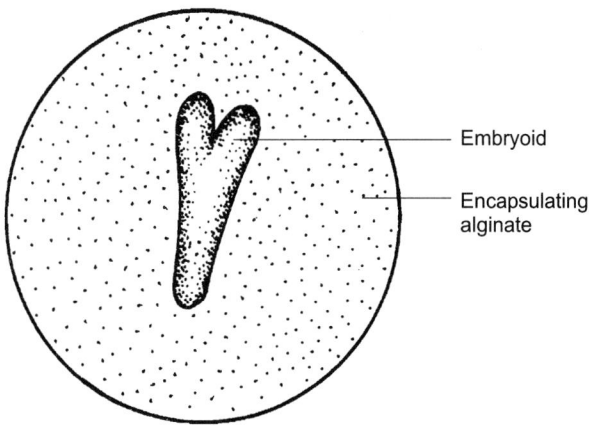

Fig. 22.1 Diagram showing artificial seed.

We know that there are two passages for the regeneration of a plantlet through tissue culture; it is either through the organogenesis or embryogenesis. In the latter case, embryoids are formed, which are non-zygotic embryos having polarized ends and can grow into a complete plant under aseptic conditions. Since they are not enclosed within any protective covering, therefore, they get contaminated or undergo desiccation if transferred to soil directly. So, it is visualized that a technique should be evolved to encapsulate the *in vitro* raised embryoid in order to protect them from contamination and desiccation.

The concept of **'synthetic seed, synseed or artificial Seed'** was first introduced by Toshio Murashige in 1977. Kitto and Janick (1982) for the first time coated clumps of carrot embryoids with polyoxyethylene to develop artificial seeds. Subsequently, Redenbaugh *et al.* (1986) used more effective hydrogel, such as sodium alginate for encapsulation of somatic embryos of alfalfa and celery, which developed into plantlets. Synthetic seed technology offers many useful advantages on a commercial scale. The resultant plant production from the synthetic seeds will be uniform and the direct delivery of somatic embryos will save many sub-cultures to obtain plants from regenerated embryos. The encapsulated embryoid could also be packed with pesticides, fertilizers, nitrogen fixing bacteria and even microscopic worms (Redenbaugh *et al.*, 1986; Ohishi *et al.*, 1995), which ensure the successful establishment of plants in the soil.

In addition to somatic embryos, axillary buds, adventitious buds and shoot tips have also been used in the preparation of synthetic seeds (Redenbaugh, 1993; Bapat and Rao, 1988, 1990; Ganpati *et al.*, 1992, 1994). The main advantage of using these vegetative propagules is that they can be useful delivery systems for tissue cultured plants by eliminating rooting and hardening phases by direct sowing the encapsulated beads in the soil. This is an effective alternative method, especially in those plant species where the explants fail to establish embryogenesis (Table 22.1).

Table 22.1 Crop plants in which artificial seed production and plant conversion has been demonstrated

In vitro propagules for incapsulation	Crop	Reference
Somatic embryos	Alfalfa	Rendenbaugh *et al.*, 1984
	Celery	Rendenbaugh *et al.*, 1986
	Carrot	Kitoo & Janick, 1985
	Brassica	Rendenbaugh *et al.*, 1987
	Lettuce	Rendenbaugh *et al.*, 1987
	Sandalwood	Bapat & Rao, 1988
	Brinjal	Rao & Singh, 1992
	Rice	Suprassana *et al.*, 1994
	Horse radish	Shigeta & Sato, 1994
Axillary buds/ adventitious buds	Mulberry	Bapat *et al.*, 1987
		Machi, 1992
	Eucalyptus	Huang & Zhu, 1990
	Vitis	Huang et al., 1990
Shoot tip	*Carum carvi*	Furamanova.1991
	Banana	Ganpati *et al.*, 1992
	Cardamon	Ganpati *et al.*, 1994

TECHNIQUE

Redenbaugh *et al.*, (1988) identified various stages for successful production of somatic embryos and their utilization as artificial seeds:

1. Selection of desired crop based on both technological and commercial potential.
2. Optimization of somatic embryogenesis system from cultured cells.
3. Optimization of embryo maturation.

4. Automation of embryo production.
5. Production of mature synchronized embryos.
6. Encapsulation of embryo with necessary adjuvants.
7. Coating of encapsulated embryos.
8. Optimization of green house and field conditions for the conversion of embryos to plant.
9. Delivery system for artificial seed in term of increasing productivity.

For the encapsulation, it is a general practice to coat the propagule with matrix, which serves as synthetic endosperm. The coat should be non-toxic, provide protection against mechanical injury during handling and allow the development and conversion of propagule into seedlings without any undesirable change (Redenbaugh and Ruzin, 1989). Several chemical agents, such as sodium alginate, potassium alginate, guar gum, agar, sodium pectate, carboxylmethyl cellulose etc. could be used for encapsulation. Of these, sodium alginate has been widely used.

Establishment of callus culture and induction of embryogenesis is pre-requisite for developing synthetic seeds. Maturation of somatic embryo means the completion of embryo development stages i.e. passing through globular-shaped, heart-shaped till torpedo-shaped stage is attained, when it changes into a bipolar structure with root and shoot meristem on two opposite extremes connected by a vascular system. For example, in mature seed of rice, callus is established on MS (Murashige and Skoog, 1962) medium supplemented with 2,4-D. Thereafter, callus showed development of the somatic embryos on MS medium incorporated with IAA and KN. Mature embryoids (2-3 mm) are then used for encapsulation.

One of the standard methods of encapsulation is Gel Complexion via a dropping procedure. Isolated somatic embryos are mixed with 5% (w/v) sodium alginate prepared either in MS or White's medium and dropped from the tip of the funnel into solution of calcium chloride to form calcium-alginate beads. The encapsulated embryos complexed in calcium salt for 20 min; then rinsed with autoclaved distilled water and stored in an airtight container to prevent from drying out. The encapsulated embryos develop into plantlets on nutrient media or filter paper moistened with $1/4^{th}$ MS basal medium. The latter gives higher frequency of germination (Fig. 22.2). Other method is an automated encapsulation process, which is a quick method of artificial seed production. Wen-Guo Wang *et al.* (2007) have shown that in *Pogonatherum paniceum* addition of activated charcoal enhanced the conversion percentage and healthy plantlets.

For the development of synthetic seeds from axillary buds and shoot tips, the propagules are carefully dissected from *in vitro* cultures and after drying out with the help of filter paper propagules are immersed in 3% sodium alginate solution prepared in MS medium. The propagules are then transferred into a 2.5% solution of calcium nitrate to obtain encapsulated beads.

There are two types of synthetic seeds; **hydrated** and **desiccated**. Hydrated synthetic seeds consist of somatic embryos individually encapsulated in hydrogel, such as calcium alginate. Desiccated synthetic seeds are produced by coating a mixture of somatic embryos in polyoxyethylene glycol. In this method, coated seeds are allowed to desiccate for several hours on a Teflon surface on sterile hood. The dried mixture is then hydrated and placed on a medium for ensuring embryo survival. The desiccated synthetic seeds are closer to true

22.4 Plant Embryology: Classical and Experimental

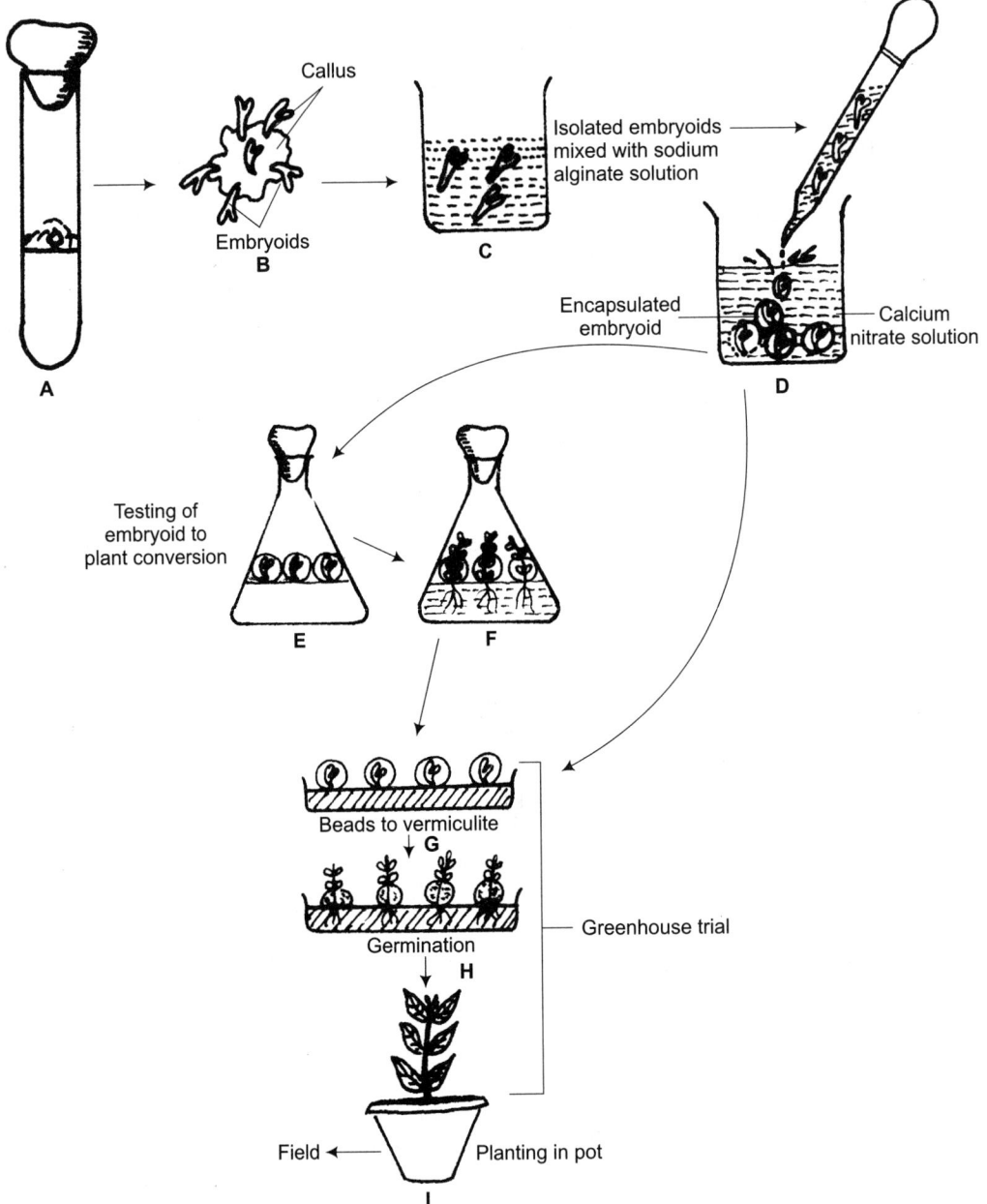

Fig. 22.2 Diagrammatic representation of different steps of synthetic seed production.

seeds. Hydrated capsules are difficult to store because of active respiration. Also, the capsules are difficult to handle because they are very wet and tend to stick together. In addition, Ca-alginate capsules lose water rapidly and dry down to hard pellets within a few hours when exposed to atmosphere. However, this barrier can be overcome by coating the capsule with watertight film composed of polyvinylcloride (PVC), polyvinylacetate (PVA) and bentone as thickener to form pharmaceutical type capsule (Dupuis *et al.*, 1994).

Enhancement of Conversion

Onishi *et al.* (1994) and Sakamoto *et al.* (1995) have described a protocol for enhancing conversion percentage for carrot and celery embryos produced in biorectors. It includes the following three sequential treatments:

(i) *Promotion of embryo development,* by culturing the embryos for 7 days in medium of high osmolarity (with 10% mannitol) under light (16 hours photoperiod with 300 lux of illumination). It increased the size of embryos from 1-3 mm to 8 mm and chlorophyll contents.

(ii) *Dehydration of embryos* to reduce the water content from 95-99% to 80-90% by keeping them for 7 days on 2-7 layers of filter paper under 16 hours photoperiod and 14 $\mu E\ m^{-2}\ s^{-1}$ irradiance.

(iii) *Post dehydration culture* on SH medium containing 2% sorbitol, 0.01 mg l^{-1} and 0.01 mg l^{-1} GA_3, in air enriched with 2% CO_2 under 16 hours photoperiod at 20°C for 14 days to acquire autotrophic nature and reserve food.

These authors also modified the bead quality by impregnating the gel beads with 3% sucrose microcapsules coated with a mixture of 8% Elvax 4260 and bees-wax and 0.1% Topsin M as the fungicides. The microcapsule releases the sucrose over a period 3-21 days at 20°C. To facilitate the emergence of embryos during germination they made the gel capsules self-breaking under humid conditions. Such synthetic seeds showed 50% conversion 2 weeks after sowing.

Limitation

(i) Synthetic seeds cannot be stored for long time because of the development of anaerobic ambience within the capsule as the locked tissue is under continuous respiration. To overcome this limitation, two possible solutions are either to have a smaller ratio of capsule volume to embryo volume so that gas diffusion can readily take place or to induce the growth of resting embryo using phytohormones in the encapsulating medium.

(ii) Conversion from encapsulated embryos is very low.

(iii) Sensitive to infections by microorganisms.

Significance of Synthetic Seeds

(i) Artificial seeds help to study the role of endosperm and seed coat formation.

(ii) The encapsulation of somatic embryos (hydrated or desiccated) provides a potential method to combine the advantage of clonal propagation with the low-cost, high- volume capabilities of seed propagation.

(iii) Artificial seeds can be useful in handling transportation and delivery of embryos in a safer and easier way.

(iv) The encapsulated embryos can be packed with nitrogen fixing bacteria or pesticides to ensure survival and setting of embryoid successfully into soil.

(v) Dormancy is the common feature of natural seeds but by means of artificial seeds the dormancy period can be reduced to a great extent, thereby shortening the life cycle of a plant.

(vi) Encapsulated somatic embryos can withstand the stress conditions.

(vii) The resultant plant population from the synthetic seeds will be uniform and the direct delivery of somatic embryos will save time bypassing many sub-cultures to obtain plantlets from regenerated embryos.

(viii) Encapsulated embryos can be stored for the preservation of valuable germplasm and could be made available throughout the year.

Appendix
Experiments in Embryology

EXPERIMENT-1

COLLECTION OF POLLEN GRAINS

The importance of pollen morphology as an expression of geographical distribution, genomal constitution, taxonomy, phylogeny and evolution of plants has been amply demonstrated in the recent years.

Therefore, the collection of the pollen grains in viable condition is the primary requirement for any study on pollen. Pollen responses vary considerably among individuals of a species as well as among different samples from the same individual. Responses of pollen collected during early, middle and late period of the flowering season and at different times of the day also vary. Environmental changes, genotypic differences, vigor and the physiological status of the plants contribute to such variability. Generally, pollens collected soon after anther dehiscence gives optimal response. For uniform results it is desirable to use pollens collected from plants of known genotype, same age and grown under similar conditions.

In many species it is convenient to excise flower/inflorescence/flowering twigs the previous evening and keep them overnight in the laboratory with their cut ends dipped in water. By the following day the anthers would dehisce and a gentle tap on such flower held over a watch glass, sheds the pollen grains.

Collection of Pollens from Brassicaceae

There are some of the plants which are easily available and therefore, could be conveniently used for pollen collection and study. One of such species is *Brassica* and other crucifers. Here anther dehiscence occurs soon after the anthesis. As there is high frequency of insect visit following anthesis, most of the pollen is lost if the flowers are left on the plant. In these systems, it is necessary to collect and excise the flowering buds in the morning just before the anthesis for pollen collection. After removing the sepals and petals the flowering buds are maintained under a table lamp or sun light (in a Petri dish) for 1-2 hours until anthesis. Pollens, thus shed can easily be collected.

Collection of Pollens from Solanaceae and Convolvulaceae

Petunia and *Nicotiana* are good Solanaceous members, while in Convolvulaceae family *Ipomoea* and *Argyreia* could be used for pollen collection. The anthers shed considerable amount of pollens into corolla tube. Flowers are inverted over a Petri dish and gently tapped to dislodge the pollens from the corolla tube and from the dehisced anthers.

Collection of Pollens from Papilionaceae

One of the excellent materials for pollen collection is *Crotolaria* spp., a Papilionaceous member. In *Crotolaria* and many Papilionaceae members anthers generally dehisce 12-24h before anthesis and release pollens inside the keel. On the day of anthesis, however, very little pollens are left in the keel, as most of it has been used for pollination or by foraging insects. Also, pollens collected from open flower have very poor germination ability. In such systems, therefore, pollens are collected from flower buds excised before anthesis but only after anther dehiscence. Pollens can be collected by removing the standard and wing petals and splitting open the two keel petals along the line of fusion over watch glass. In this way large quantity of pollens could be collected easily.

For Long-term Storage

For long term and many biochemical experiments, the pollens collected must be free from any contamination. Pollens lose their germination ability when treated directly with surface sterilizing agent. Therefore, adopting indirect method, excised mature undehisced anthers are washed with Cl_2- water, or 10% sodium hypochloride, or 0.25% mercuric chloride, or 70% ethanol. Excessive reagent is removed by blotting anthers in sterile filter paper and are paced in dry sterile Petri dish until they dehisce.

EXPERIMENT-2

IN VITRO GERMINATION OF POLLENS

In vitro method is the most commonly used technique in pollen physiology. This technique provides a simple experimental method to study the physiology and biochemistry of pollen germination and pollen tube growth as well as the responses of the pollen system to physio-chemical factor. Generally for laboratory study such pollens are collected which could geminate quickly in culture medium.

The culture medium for germination of pollens mainly contains sucrose, boric acid and calcium nitrate. The following are the some of the pollen culture media formulated by different scientists:

Constituent	Amount in mg/l		
	1	2	3
Boric acid	100	10	100
Calcium chloride	—	362	—
Calcium nitrate	300	—	400
Magnesium sulphate	200	—	200
Potassium nitrate	100	100	100
Tris	60-130	—	—
TAPS	—	—	4.86g/l
Sucrose	10%	20%	20%

1: Brewbaker and Kwack's medium (1963)
2: Robert's medium (Roberts et al., 1983)
3: Hodgkin and Lyon's medium (1986)

However, for those pollens where germination of pollens require longer period, the culture medium needs proper sterilization. As the scoring of the pollens takes considerable time and all the cultures can not be scored simultaneously, the cultures are, therefore fixed by applying following fixatives:

(i) FAA (formaldehyde: glacial acetic acid: 70% ethanol, 5:5:9 v/v/v).
(ii) Acetic alcohol (glacial acetic acid: ethanol, 1:3 v/v).
(iii) Acetocarmine (1-2%).
(iv) Formaldehyde (2-4%).
(v) Ethanol (10%)

Add the fixative to the culture in the ratio of 1:1, mix thoroughly and lower a coverglass. The culture is ready for scoring and data be fitted in the following table:

# no.	Microscope individual Field no.	No. of pollen in field	No. of pollens germinated	% germination	Length (units in ocular micrometer) of straight pollen tubes in each field										
					pollen tube no.										
					1	2	3	4	5	6	7	8	9	10
1.														
2.														
3.														
4.														
5.														
6.														
7.														
8.														
9.														
10.														
Total	N	n			L									

Per cent pollen germination: n/N × 100
Total no. of pollen tubes measured from all fields: F
Total length of all F pollen tubes (in units of ocular micrometer): L
Mean tube length: F/L (convert this value into μm)

EXPERIMENT-3

TESTS FOR POLLEN VIABILITY

Pollen viability is critical for any studies on pollen biology. It is necessary to know the extent of pollen viability of the pollen sample to be used for experimentation or pollination. With stored pollen also, its viability needs to be monitored under different storage conditions. A few important tests for pollen viability have been described:

Tetrazolium Test

The tetrazolium test is based on the reduction of a colorless soluble tetrazolium salts to a reddish insoluble substance called formazan, in the presence of dehydrogenases. Nitroblue tetrazolium and 2, 3, 5-triphenyl tetrazolium chloride are the most commonly usd tetrazolium salts. With many species tetrazolium test has proved satisfactory in assessing pollen viability.

Fluorochromatic Reaction (FCR) Test

The FCR test to assess pollen viability was introduced by Heslop-Harrison and Heslop-Harisson (1970). When pollen grains are mounted in fluorescein diacetate (FDA) solution, the nonpolar nonfluorescent FDA readily enters the pollen cytoplasm. Cytoplasmic esterase hydrolyses FDA and release fluorescein, which is polar and fluorescent. Unlike FDA Fluorescein passes sparingly through an intact membrane and therefore accumulates in the cytoplasm of viable pollen grains and gives a bright green or yellowish green fluorescene under the fluorescene microscope. If the plasma membrane of the vegetative cell is not intact, fluorescein readily comes out into the mounting medium in the preparation and so the pollen grains fail to fluoresce brightly. Also, pollen grains do not fluoresce if they lack esterases which hydrolyse FDA. Thus, the FCR test assesses two properties of a pollen grain: the integrity of the plasma membrane of pollen and the activity of estrase capable of hydrolyzing the fluorescein ester.

EXPERIMENT-4

ANILINE BLUE FLUORESCENCE METHOD TO STUDY POLLEN GERMINATION AND POLLEN TUBE GROWTH IN THE PISTIL

In investigation on reproductive biology, studies on the structural aspects of pistil and the details of the pollen- pistil interaction are essential.

From time to time several methods have been standardized to localize pollen tubes in the pistil. Most of these methods are based on clearing the pistil and/or dissecting the transmitting tissues followed by staining. The most satisfactory technique presently used by several investigators is the fluorescent method. The pollen tubes invariably develop callose deposition, the tubes can be readily detected under the fluoresce microscope following treatment with 0.005% aqueous aniline blue solution or sirofluor.

EXPERIMENT-5

ISOLATION OF EMBRYO FROM DICOT SEED

Ladies finger (*Lycopersicon esculentum*) is one of the good materials for the isolation of embryo of different developmental stages. In order to dissect out the embryos the fruits of different ages are collected and with the help of needle and forceps incision is made at the point of placenta to tear out the developing seeds.

Following three stages of the embryo can be obtained and can be examined under dissecting microscope:

(1) **Globular embryo:** It consists of a suspensor and globular mass consisting of 16-celled embryo. The uppermost cell of the suspensor is slightly swollen and is known as vesicular cell which is haustorial in nature. The lowermost cell of the suspensor is hypophysis.

(2) **Heart-shaped embryo:** Slightly older seed also shows suspensor and embryo. The suspensor has vesicular cell and hypophysis. The embryonal mass becomes heart-shaped and shows distinction into dermatogen, periblem and plerome.

(3) **Horseshoe shaped embryo:** It consists of small suspensor with swollen terminal cell. The cotyledon is large and bends to become horse-shoe shaped. The cotyledon enclosed a small plumule (shoot) which is terminal in position. The portion of the embryonal axis which is above the level of cotyledon is called hypocotyl.

EXPERIMENT-6

EMBRYO CULTURE OF TOMATO

Embryo culture is very important in tissue culture study as it provides the sterile plant immediately bypassing the dormancy period.

Tomato seeds are surface sterilized with mercuric chloride (0.01%) and soaked for 24h in autoclaved distilled water. With the help of sterile scalpel incision was made in the sharper end of tomato seed to isolate embryo on sterile brown paper or petri dish. The isolated embryos are inoculated in MS (Murashige and Skoog's medium, 1962) under laminar flow and placed in culture room where cultures are maintained under 16h light and 8h dark period and controlled temperature, light and humidity.

After 4 days of incubation, tiny plantlets are seen arising.

EXPERIMENT-7

CALLUS INDUCTION AND REGENERATION IN RICE SEED

The seeds of rice varieties are obtained from field in order to get callusing and regeneration.

About 250 cc of Murashige and Skoog's (MS) medium is generally prepared for callus induction and regeneration supplemented with 2,4-D (2mg/l). The pH is adjusted to 5.8 and dispensed into 24 tubes, each containing about 10 cc, which are plugged with cotton bungs. The medium containing tubes are autoclaved at 121°C (1.06Kg/cm^2) for 15-20 min and allowed to cool down gradually. The culture tubes are put in culture rack and placed slightly in slant position. After 2-3 hours the medium is ready for the inoculation of seeds.

The rice seeds collected from the field are surface sterilized before to be implanted in culture tubes as they contain plenty of contaminants. The seeds are at first dehusked to obtain rice grains, which are treated with 70% alcohol for 1 min followed by 0.01% HgCl2 for 15-20 min. The treated seeds are finally rinsed, under the laminar airflow, with double distilled autoclaved water for 5-6 times to make the seeds free of any trace of sterilizing reagents in addition to the so-called contaminants. Now the processed seeds are ready for inoculation.

2-3 sterile seeds are usually implanted into each culture tubes under laminar airflow. The inoculated culture tubes are maintained in culture room under controlled conditions; temperature (25 ± 30C), light period (16 h light and 8 h dark), light intensity (1500-3000 lux) and relative humidity (60%).

Within a week or so hypotrophy is clearly observed at the embryonal, which soon develops into callus. Together with the growth of callus, short coleoptile is also seen emerging. However, in 4-6 week time profuse callus is obtained, which is compact, nodular and cream colour suggesting embryogenic nature. The developed callus may either be sub-cultured or transferred into differentiating medium for regeneration of plantlet.

In order to obtain plantlets, seed-derived callus is excised precisely from the seeds under laminar airflow to avoid the transfer of parental tissues. Initially the callus is cut into small pieces, measuring about 4-6 mm in diameter, before to be transferred into the said culture tubes adopting all necessary precautions.

MS+KN (4mg/l) + IAA (2mg/l) or MS basal medium is employed as regenerating medium. The healthy, compact and actively growing embryogenic callus is highly preferred for regeneration into complete plant.

It is from the 3rd day onwards the embryoids show its appearance, the shoot buds is evident initially with the green spots, which soon differentiate into shoots. Following shoot differentiation roots seem arising downward from the callus. Within 3-4 weeks a complete plantlet measuring 10-12 cm in height is regenerated.

However, for field transfer a very small plant measuring about 2-3 cm is selected for hardening condition before transfer into field condition.

Selected References

Bapat, V.A., Hatre, M. and Rao, P.S. 1987. Propagation of *Morus indica* L. (mulberry) by encapsulated shoot buds. *Plant Cell Rep.* **6**: 393-395.

Barathy, P.V., Karibasappa, G.S., Patil, D.C. and Agrawal, D.C. 2005. *In ovulo* rescue of hybrid embryos in Flame seedless grapes-influence of pre-bloom spray of benzyladenine. *Scientia Horticulturae* **106 (3):** 353-359.

Bhatnagar, S.P. and Sawhney, V. 1981. Endosperm-its morphology, ultrastructure and histochemistry. *Int. Rev. Cytol.* **73**: 55-102.

Bedinger, P. 1992. The remarkable biology of pollen. *Plant Cell.* **4**: 879-887

Bhojwani, S.S. and Bhatnagar, S.P. 1992. The Embryology of Angiosperms. Vikas Pub. House Ltd. New Delhi.

Bhojwani, S.S. and Razdan, M.K. 1996. Plant Tissue Culture: Theory and Practice Vol. 5 Elsevier Scientific Publications, Amsterdam.

Bhojwani, S.S., 1966. Morphogenetic behaviour of mature endosperm of *Croton bonplandianum* in culture. *Phytomorphology* **16**: 349-353.

Bozkhov, P.V., Filonova, L.H., Suarez, M.F. Helmersson,A., Smertenko, A.P., Zhivotovsky, B. and Arnold, S.V. 2000. VEIDase is a principle capase-like activity involved in plant programmed cell death and essential for embryonic pattern formation. *J. Exp. Bot.* **51**: 971-983.

Button, J. and Bornman, C.H. 1971. Development of plants from pollinated and unfertilized ovules of Washington Navel orange *in vitro. J. S. Afr. Bot.* **37**: 127-134.

Carol, A., Paula, F. and Rudall, J. 1999. Microsporonenesis in monocotyledons. *Ann. of Bot.* **84 (4):** 475-499.

Chu, C. 1978. The N6 medium and its application to anther culture of cereal crops. *In* : Proceedings of Symposium on Plant Tissue Culture Science Press, Peking, pp. 43-50.

Corner, E.J.H., 1976. The Seeds of Dicotyledons. Cambridge Univ. Press, U.K.

Coulter, J.M. and Chamberlain, C.J. 1903. Morphology of Angiosperms. D. Appelton and Co., New York.

Custodio, I., Filomena., C. and Romano, A. 2005 Microsporogenesis and anther culture in carob tree (*Ceratonia siliqua* L.). *Scientia Hoticulturae* **104 (1):** 65-77.

Davis, G.L. 1966. Systematic Embryology of Angiosperms. John Wiley, New York.

Eames, E.J., 1961. Morphology of the Angiosperms. McGraw Hill Book Co. Inc., New York.

Edlund, A.F., Swanson, R. and Preuss, D. 2004. Pollen and stigma structure function: the role of diversity in pollination. *The Plant Cell*: **16**: S84-S97.

Eid, A.A., Delanghe, E. and Waterkyn, L. 1973. *In vitro* culture of fertilized cotton ovules: The growth of cotton embryo. *Cellule* **69**: 369-371.

Evans, D.A., Sharp, W.R., Ammirato, P.V. and Yamada, Y. 1983. Hand Book of Plant Cell Culture Vol. 1. Macmillan Pub. Co., New York.

Faegri, K. and Vander, P. 1979. The Principle of Pollination Ecology. Pergamon Press, Toronto.

Fahn, A., 1997. Plant Anatomy. Aditya Books (P) Ltd., New Delhi.

Filanova, L.H., Arnold, S.V. Daniel, G. and Bozhkov, P.V. 2002. Programmed cell death eliminates all but one embryo in a polyembryonic plant seed. *Cell Death and Differentiation*. **9**: 1057-1062

Gamborg, O.L., Miller, R.A. and Ojima, K., 1968. Nutrient requirement of suspension culture of soybean root cells. *Exp. Cell Res.* **50**: 151-158.

Gautheret, R.J. 1939. Sur la possibilite de realiserla culture indefinite des tissue de tubercules of carrote. *C.R. Acad. Sci. Paris* **208**: 118-120.

Gifford, E.M. and Foster, A.S., 1989. Morphology and Evolution of Vascular Plants.W.H. Freeman, New York.

Guha Sipra and Maheshwari, S.C., 1964. *In vitro* production of embryo from anthers of *Datura*. *Nature* **204**: 497.

Gupta, P., Shivanna, K.R, and Mohan Ram, H.Y. 1996. Apomixis and polyembryony in guggul plant, *Commiphora wightii*. *Annals of Botany*. **78**: 67-72.

Haberlandt, G. 1902. Cultureversuche mit isolierteinp flanzenzellen.*Sitzungsberg. Math. Naturwiss. KI Kais Acad Wiss.*, Wien. **111**: 69-92.

Hanai, H., Matsuno, T., Yamamoto, M., Matasubayashi, Y., Kobayashi, T., Kamada, H. and Sakagami, Y. 2004. *Nature* **428** (6978): 81-84.

Heslop-Harrison, J. (ed.). 1971. Pollen Development and Physiology. Butterwort, London.

Heslop-Harrison, J. 1972. Sexuality in angiosperms. *In*: Steward, F.C. (ed.) *Plant Physiology* Vol. 6, Academic Press, New York.

Heslop-Harrison, J. 1978. Genetics and physiology of angiosperm incompatibility systems. *Proc R Soc Lond. Ser. B*. **202**: 73-92.

Heslop-Harrison, J., 1987. Pollen germination and pollen tube growth. *Int. Rev. Cyto*. **107**: 1-78

Heslop-Harrison, J. Shivanna, K.R. 1982. The receptive surface of the angiosperm stigma. *Annals of botany*. **41**: 1233-1258.

Horsch, R.B., Fraley, R., Rogers, S., Sanders, P., Lloyd, A. and Hoffman W. 1984. Inheritance of functional foreign gene in plants. *Science*, **223**: 496-498.

Huan T.V.L., Takamura, T. and Tanaka, M. 2003. Callus formation and plant regeneration from callus through somatic embryo structures in *Cymbidium* orchid. *J. Jap. Bot*. **78(3)**: 145-151.

Jensen, W.A. 1972. The embryo sac and fertilization in angiosperms. *Univ. Hawaii Harold L. Lyon Arboretum Lecture*, No. 3.

Jensen, W.A. 1973. Fertilization in flowering plants. *Bioscience*, **23**: 21-27.

Johri, B.M. 1963. Embryology and taxonomy. *In*: Masheshwari, P. (ed.) Recent Advances in the Embryology of Angiosperms. *Int. Soc. Plant Morphol.* Delhi, pp. 395- 444.

Johri, B.M. (ed.). 1982. Experimental Embryology of Vascular Plants. Springer-Verlag, Berlin, Hiedelberg, New York.

Johri, B.M. and Bhojwani, S.S. 1965. Growth response of mature endosperm in culture. *Nature* **208**: 1345-1347.

Johri, B.M. and Vasil, I.K. 1961. Physiology of pollen. *Bot Rev* **27**: 325:331

Kanta, K., Rangaswamy, N.S. and Maheshwari, P. 1962. Test tube fertilization in a flowering plant. *Nature* **194**: 1214-1217.

Kapil, R.N., 1962. Some recent examples of the value of embryology in relation to taxonomy. *Bull. Bot. Surv. India.* **4**: 57-66.

Kapil, R.N. and Bhatnagar, A.K. 1981. Ultrastructure and biology of female gametophyte in flowering plants. *Int. Rev. Cytol.* **70**: 353-403.

Kapil, R.N. and Bhatnagar, A.K. 1991. Embryological evidences in angiosperm classification and phylogeny. *Bot. Jahrb. Syst.*, **113**: 309-338.

Kapil, R.N. and Vasil, I.K. 1978. The integumentary tapetum. *Bot. Rev.* **44**: 457-490.

Kitto, S. and Janick, J. 1985. Production of synthetic seeds by encapsulating asexual embryos of carrot. *J. Amer. Soc. Hort. Sci.* **110**: 277.

Knox, R.B and Singh, M.B. 1987. New perspective in pollen biology and fertilization. *Ann Bot Suppl* **4**: 15:37

Konar, R.N. and Nataraja, K., 1969. Morphogenesis of isolated floral buds of *Ranunculus sceleratus* L. In vitro *Acta. Bot. Neerl.*, **18**: 680-699.

Koultunow, A.M., Bicknell, R.A. and Choudhury, A.M. 1995. Apomixis: Molecular strategies for the generation of genetically identical seeds without fertilization. *Plant Physiol.* **108**: 1345-1352.

Kranz, E. 1998. *In vitro* fertilization with isolated single gamete. In Hall, R (Ed.) Methods in Molecular Biology. Vol. 7. Plant Cell Culture Protocols. Humana Press, N.J.

Kranz, E and Dresshaus, T. 1966. *In vitro* fertilization with isolated higher plant gametes. *Trends Plant Sci.* **1**.82-89

Kranz, E and Lorz, H. 1993, T. 1993. *In vitro* fertilization with isolated, single gamete results in zygotic embryogenesis and fertile maize plant. *Plant Cell* **5**: 739-746.

Le-cheng,L., Jia-shu, C., Xiao- lin, Y., Xun, X. and Yong-jun, F. 2006. Expression of antisense BcMF3 affects microsporogenesis and pollen tube growth in *Arabidopsis*. *Agr. Sc. in China* **5(5)**: 339-345.

La Rue, C.D. 1942. The rooting of flowers in culture. *Bull. Torrey bot. Club.* **69**: 332-341.

Maheshwari, P. (ed.). 1963. Recent Advances in the Embryology of Angiosperms. International Society of Plant Morphologists, University of Delhi.

Maheshwari, P. 1971. An Introduction to the Embryology of Angiosperms. Tata Mc Graw Hill Publ. Co. Ltd. New Delhi.

Marshall, D.S. 2004. Flowering Plant Embryology. Nels R. Lersten, Blackwell Publishing.

Mascarenhas, J.P. 1975. The biochemistry of angiosperm pollen development. *Bot. Rev.* **41**: 259-314.

Mascarenhas, J.P. 1993. Molecular mechanism of pollen tube growth and differentiation. *Plant Cell.* **5**: 1303-1314.

Mathys-Rochon, E. 1992. *In vitro* fertilization in flowering plants. **In**: Y. Dattee *et al.* (Eds.) Reproduction Biology and Plant breeding. Springer. Berlin, pp. 197-224

Mauney, J.R., 1961. The *in vitro* culture of immature cotton embryos. *Bot. Gaz.* **122**: 204-209.

Meeuse, B.J.D., 1961. The Story of Pollination. Ronald Press Co., New York.

Mitra, G.C. and Chaturvedi, H.C. 1972. Embryoids and complete plants from unpollinated ovaries and ovules of *in vivo* grown emasculated flower buds of *Citrus spp. Bull. Torrey. bot. Club* **99**: 184-189.

Mongensen, H.L. 1992. The male germ unit: concept, composition and significance. *Intl. Rev. Cytol.* **140**: 129-147.

Murashige, T. and Skoog, F. 1962. A revised medium for rapid growth and bioassay with tobacco tissue culture. *Physiol. Plantarum* **15**: 473-479.

Nag, K.K. and Johri, B.M.. 1971. Morphogenetic study on endosperm of some parasitic angiosperms. *Phytomorphology* **21**:202-208.

Narayanswamy, S. 1994. Plant Cell and Tissue Culture. Tata McGraw Hill Pub. Co. Ltd. New Delhi.

Niimi, Y. 1982. Studies on the self incompatibility of *Petunia hybrida* in excised style culture. *Euphytica.* **31**: 787-793.

Nitsch, J.P. 1951. Growth and development *in vitro* of excised ovaries. *Am. J. Bot.* **38**: 556-576.

Nitsch, J.P., 1969. Experimental androgenesis in *Nicotiana. Phytomorphology* **19**: 389-404.

Nobecourt, P. 1937. Culture in serie de tissue vegetaux sur milieu artificiel. *C.R. Seanc. Soc. Bid.* **205**: 521-523.

Pandey, K.K., 1968. Compatibility relationship in flowering plants: role of the S-gene complex. *Am. Nat.* **26**: 163-216.

Raghavan, V. 1966. Nutrition, growth and morphogenesis in plant embryo. *Biol. Rev.* **41**: 1-58.

Raghavan, V. 1976. Experimental embryology of vascular plants. Academic press, London.

Raghavan, V. 1986. Embryogenesis in Angiosperms. A Developmental an Experimental Study. Cambridge University Press, U.K.

Rangan, T.S. 1982. Ovary, ovule and nucellus culture. *In*: Johri BM (ed.): Experimental Embryology of Vascular Plants: 105-130. Springer-Verlag, Berlin, Hiedelberg, New York.

Rangaswamy, N.S. 1963. Control of fertilization and embryo development. *In*: Maheshwari, P. (ed.) Recent Advances in the Embryology of Angiosperms. *Int Soc Plant Morphol*, University Delhi, pp. 327-353.

Rangaswamy, N.S. 1977. Application of *in vitro* pollination and *in vitro* fertilization. *In*: Reiner, J and Bajaj, Y.P.S. (eds) Applied and Fundamental Aspects of Plant Cell Tissue and Organ Culture. Springer, Berlin Heidelberg, New York, pp. 412-425.

Rangaswamy, N.S. 1982. Nucellus as an experimental system in basic and applied tissue culture research. *In*: Rao, A.N. (ed.) Proc. COSTED Symp. on Tissue Culture of Economically Important Plants. Univ. Singapore.

Rao, P.S., Suprassana, P.,Ganapathi, T.R and Bapat, V.A. 1998. Synthetic seed: concept, method and application. *In*: Srivastava, P.S. (ed.) Plant Tissue Culture and Molecular Biology. pp. 607-619. Narosa Pub. House, New Delhi.

Razdan, M.K. 1995. An Introduction to Plant Tissue Culture. Oxford and IBH Pub. Co. Pvt. Ltd., New Delhi.

Redenbauch, K., Nocol, J., Kossler, M.E. and Paasch, B.D. 1984. Encapsulation of somatic embryos for artificial seed production. *In vitro* **20**: 256-257.

Reinert, J. 1958. Morphogenese und ihre kontrolle an gewebekulturen aus karotten. Naturwissenchaften, **45**: 344-345.

Reinert, J. and Bajaj, Y.P.S. 1977. Applied and Fundamental Aspects of Plant Cell, Tissue and Organ Culture. Springer-Verlag, Berlin, Hiedelberg, New York.

Russell, S.D. and Cass, D.D. 1981. Ultrasructure of the sperms of *Plumbago zeylanica*. 1. Cytology and association with the vegetative nucleus. *Protoplasma.* **107**: 85-107.

Russell, S.D. 1982. Fertilization in *Plumbago zeylanica*: Entry and discharge of the pollen tube in embryo sac. *Can. J. Bot.* **60**: 2219-2230.

Sanchez-romero, C., Peran-quesada, R., Marquez- martin, B., barcelo-munoz, F. and Pliego-Alfaro, F. 2007. *In vitro* rescue of immature avocado (*Persea americana* Mill.) embryos. Scientia Horticulturae. **111(4)**: 365-370.

Schenk, R.U. and Hilderbrandt, A.U. 1972. Medium and technique for induction and growth for monocotyledonous and dicotyledonous plant cell culture. *Can. J. Bot.*, **50**: 199-204.

Shivana, K.R., 1979. Recognition and rejection phenomenon during pollen- pistil interaction. *Proc Indian Acad Sci* **88B**: 115-141.

Shivana, K.R. 1982. Pollen pistil interaction and control of fertilization. *In*: Johri, B.M. (ed.) Experimental Embryology of Vascular Plants. pp.131-174. Springer-Verlag, berlin.

Shivanna, K.R. and Johri, B.M. 1985. The Angiospermic Pollen: Stucture and Function. Willey Eastern Ltd.

Shivana, K.R. and Rangaswamy, N.S. 1998. *In vitro* fertilization-then and now. *In*: Srivastava, P.S. (ed.) Plant Tissue Culture and Molecular Biology. pp.1-17. Narosa Pub. House, New Delhi

Shivanna, K.R. and Sawhney, V.K. 1997. Pollen Biotechnology for Crop Production and Improvement. Cambridge University Press, U.K.

San-Noeum, L.H. 1976. Haploids d' *Hordeum vulgare* L. par culture *in vitro* d'ovairies non fecondes. *Ann. Amelior Plant.*, **26**: 751-754.

Singh, B. 1964. Development and structure of angiosperm seed-1. Review of Indian Work. *Bull. Natn. Bot. Gdns.* **89**: 1-115.

Stevens, V.A.M. and Murray, B.G. 1982. Studies on the heteromorphic incompatibility system: Physiological aspects of the incompatibility system of *Primula obconica*. *Theor. Appl. Genet.* **51**: 245-256.

Suwan, T. and Owens, J.N. 1997.Floral biology, pollination, pistil receptivity and tube growth of teak (*Tectona grandis* Linn). **79**: 227-241

Steward, F.C. 1958. Growth and development of cultivated cells III. Interpretation of the growth free cells to carrot plant. *Am. J. Bot.*, **45**: 709-713.

Teng, N., Huang, Z., Mu, X., Jin, B., Hu, Y. and Lin, J. 2005. Microsporogenesis and pollen development in *Leymus chinensis* with emphasis on dynamic changes in callose deposition. Flora-Morphology, Distribution, Functional Ecology of Plants. **200**: 265-263.

Tian, L., Wang, Y., Niu, L. and Tang, D. 2008. Breeding of disease- resistant seedless grapes using Chinese Wild *Vitis* spp.: I. *In vitro* embryo rescue and plant development. *Scienticia Hoprticulturae*. 117(2): 136-141.

Thorpe, T.A. (ed.) 1995. *In vitro* Embryogenesis in Plant.Kluwer, Dortrecht.

Van Der Pluijm, J.E. 1964. An electron microscopic investigation of the filiform apparatus in embryo sac of *Torenia fournieri*. *In*: Linskens, H.F. (ed.) Pollen Physiology and Fertilization. North- Holland Publishing Co., Amsterdam. pp. 8-16

Van Went, J.L. 1970a. The ultrastructure of the synergids of *Petunia*. *Acta. Bot. Neerl.* **19**: 313-322

Vasil, V. and Vasil, I.K. 1986. Regeneration in cereals and other grass species. *In*: Vasil, I.K. (ed.) Cell Culture and Somatic Cell Genetics of Plants, Vol. 3. Plant Regeneration and Genetic Variability. Academic Press, New York, pp. 121-150.

Wen-Guo, W., Sheng-Hua, W., Xiao-Ai, W., Xing-Yu, J. and fang, C. 2007. High frequency plantlet regeneration from callus and artificial seed production of rock plant *Pogonatherum paniceum* (Lam.) Hack. (Poaceae). *Scientia Horticulturae* 113 (2): 196-201.

White, P.R., 1943. A Handbook of Plant Tissue Culture. Jacques Cattell Press, England.

White, P.R. 1963. The cultivation of Animal and Plant Cells (2nd ed.). The Ronald Press, New York. pp. 228.

Xiao-peng, F., Jin-yi, H., Hui-rong, H. and Man-zhu, B. 2008. Cytological observation microsporogenesis in male-sterile lines of Chines pink (*Dianthus chinensis* L.). Agr. Sc. in China **7(5)**: 547-553.

Xiao- quan, Z., Xue-de, W., Yun-guo, Z., Wei, Z. and Pei-dong, J. 2007. Breeding for male-sterile line with barbadense nuclear background and cytological observation of its microsporogenesis. Agr. Sc. in China **6(5)**: 547-553.

Zhu, Z. and Wu, H. 1979. *In vitro* production of haploid plantlets from the unpollinated ovaries of *Triticum aestivum* and *Nicotiana tabacum*. *Acta Genet. Sin.* **6**: 181-183.

Zimmermann, U and Viken, S. 1982. Electric field induced cell-to-cell fusion. *J. Membranae Biol.* **67**: 165-182.

Subject Index

2,4-dichlorophenoxy acetic acid (2,4-D) 16.5
6-benzoylaminopurine (BAP) 16.5
6-dichlorobenzoic acid (dicamba) 17.2
6-furfurylaminopurine (kinetin) 16.5
6-isopetenylaminopurine (IPA) 16.5
6-methylpurine 9.10

A

Abrus 14.3
A. fructiculosus 14.3
A. laevigatus 14.3
A. ochroleuca 9.16, 19.3
A. odorum 13.4
A. nutans 13.4
Abelmoschus 21.4
Abiotic agents 7.4
Abscisic acid (ABA) 14.9, 17.5, 17.6, 20.7
Acacia 3.11
Acalypha indica 5.8, 6.6
Acanthaceae 8.2, 14.3
Acer 14.6
Acetic alcohol A.3
Acetocarmine A.3, 18.8
Acetone 14.8
Achlamydeous 2.2
Achyranthes 14.8
Acmopyle 15.7
Aconitum napellus 6.11
Acorus 10.11
Actinomycin D 9.10

Activated charcoal 17.6, 18.5
Adansonia digitata 7.7
Adelphous 2.4
Adhesive body 3.11
Adoxa 4.2, 6.3, 10.3
    moschaetillina 6.7
    type 6.5
Adventive
    embryony 13.1, 13.5, 13.6, 13.7, 15.2
    embryos 12.11, 21.6
    polyembryony 12.1, 12.5
Aegilops cylindrica 18.1
Aeginetia indica 8.10
Aegle marmelos 8.10
Aethionema 2.7
After-ripening 14.9
Agamospermy 13.3
Agar-agar 16.5
Agarose 19.8
Agave 3.10
    americana 13.2
Agrobacterium
    tumefaciens 18.13
Alathaca rosea 8.10
Albinos 18.6, 18.13
Albuminous 2.5, 10.1, 21.7
Alcohol 14.8, 16.5
Alectra thomsoni 3.4
Aleurone layer 10.13
Aleurone grains 10.13

Alginate 22.3
Alismataceae 15.9
Allium 6.3, 6.12, 15.9
    carinatum 13.8
    cernuum 4.3
    fistulosum 6.4
    odorum 12.7
    type 6.4, 6.12, 13.4, 15.9
Almus 5.3
Alnus rugosa 13.8
Alstonia 14.7
Allogamy 7.1, 7.3
Amaranthaceae 8.2
Amaranthus retroflexus 14.8
Amaryllidaceae 10.13, 11.8
Amaryllis 8.7
Amentiferae 8.10
Amoeboidal 3.4
Amorphophallus 7.6
Amphicarpy 2.7
Amphitropous 5.3
Amphycarpaea 2.7
Amylase 4.10
Anacardiaceae 12.5, 14.8
Anacardium occidentalis 7.7
Anaphase 3.4, 4.3
    I 3.6
Anatropous 5.2, 15.9
Androecium 2.1, 7.7
Androgenesis 18.2, 18.3, 18.5, 18.6, 18.7
Andrographis 14.3
Androgynophore 2.2
Androphore 2.2
Andropogon 14.9
Androsaemum officinale 11.6
Anemochory 14.6
Anemone 5.5
    nemeros 14.8
Anemophilous 4.1
Anemophily 7.4, 8.1
Anethum graveolus 21.3
Aniline Blue A.4
Annona 7.9, 9.8, 10.11
    squamosa 18.6
Annonanceae 10.4
Anona 3.3, 3.11

Antennaria 13.8
Anther 4.7, 4.8, 7.10, 15.3, 15.6, 18.1, 19.2, A.2
    culture 18.1, 19.9
    tapetum 15.2
    wall 3.2, 15.3
Anthericum 4.2
Anthesis 8.9, 19.2
Anthophore 2.2
Antipodal 6.5, 6.7, 6.11, 6.12, 8.16, 12.4, 12.7, 12.8, 15.4, 19.7, 21.4
    apparatus 6.1, 6.2
    cells 15.2, 15.6
Antipodalembryo 12.5
Antirrhinum 9.18, 21.4
    majus 8.12, 19.2
Apiaceae 8.2
Apical cell or terminal cell 11.1
Apocarpous 2.5
Apogamy 15.2
Apomictic embryo sac 13.7, 13.8
    species 13.8
Apomixis 13.1
Aposporic 12.9
    embryo sac 13.5
Arabidopsis thaliana 18.5
Arabinose 8.3
Araceae 15.7
Arachis hypogea 7.1
Arceuthobium minutissium 3.1
Archesporial cell 5.6, 15.7
Archesporium 3.2, 5.9, 15.6
Arceuthobium 3.1
Archichlamydeae 5.3
Argemone 9.16, 21.1
    mexicana 9.16, 14.7, 19.3
Arginine 16.4
Argyreia A.2
Aril 5.3, 14.4
Aristida 14.8
Aristolochia 14.7
    bracteata 12.4
Aristolochiaceae 8.7, 10.4
Aristotelia 5.10
Arrhostoxylus 9.18
Artabotrys 8.7
Artemisia 13.5
Artificial Seed 22.1

Arum 7.6
    italicum 3.6
    masculatum 10.14
Asclepiadaceae 3.11, 4.3
Asclepias 4.2, 4.4, 14.7
Ash 4.9
Asimina triloba 21.8
Asclepiadaceae 7.9
ASP-Allele specific proteins 9.11
Asparagines 9.10
Aspargine 16.4
Asperula 11.12
Asphodelus 5.3, 14.4
Aspidistra 6.12
Asteraceae 2.7, 7.6, 8.2, 8.16, 9.4, 9.15, 13.1
Asterad type 11.3
Atropa belladonna 18.2, 18.5
Autochory 14.6
Autotetraploidy 9.17
Auxins 11.13, 18.5, 21.4, 21.8
Avena fatua 14.9
Avicenia 14.11

B

β-glucuronidase 19.7
B_5 medium 17.4, 18.9, 18.10
B_6 (pyridoxine) 16.4
Bacillus amyloliquefaciens 4.14
Baculum 4.5. 4.6, 4.10
Balanophora 5.10
Balanophoraceae 5.3
Balanophoriaceae 11.15
Balsamita vulgaris 6.7
BAP 17.5, 18.10
Barleria cristata 10.7, 10.8
Barnase 4.14
Barstar 4.14
Bauhinia vahlii 14.6
Begoniaceae 15.7
Benzylaminopurine 9.18
Bergenia delavayi 12.4
Beta 4.4
    vulgaris 8.10, 11.7
Biotic agents 7.4
Bisporic Embryo Sac 6.1, 6.4
Bitegmic 5.3
Blumenbachia hieronymi 10.7, 10.8

Bombax ceiba 7.7
Boraginaceae 8.2, 10.4, 15.7, 15.8
Bougainvillaea 7.6
Brachiaria setigera 17.1
Brachysiphon 6.5
Bracteate 2.7
Brassica 9.4, 9.15, 9.18, 18.5, A.1
    alboglabra 18.7
    campestris 4.4, 9.10, 21.3
    insigne 7.7
    juncea 18.6
    napus 18.2, 18.5, 18.6, 18.9, 18.12
    oleracea 9.17, 19.7, 19.10
Brassicaceae 2.7, 8.2, 9.4, 11.3, 18.1, 18.13, A.1
Brewbaker and Kwack's medium A.3
Bromelliaceae 8.2
Bud pollination 9.15, 19.1
Bulbils 13.2
Bulbosum Technique 18.10
Bulbs 13.2, 14.8
Butomaceae 15.9
Butomus 15.9

C

C. reticulata 12.5
Ca-alginate 22.4
Ca-ions 8.12
Ca-pantothenate 19.5
Ca-salts 8.6
Cacataceae 8.7, 12.5
Caducous 2.3
Caecum 6.12
Caffeic acid 14.9
Cajanus 8.7
Calcium 20.4
    nitrate 22.3
Callase 4.8, 4.13
Callose 3.8, 4.5, 4.6, 4.7, 4.8, 9.9
    plug 9.9
Callus 12.10, 17.2, 17.6
    culture 16.5
    induction A.6
Calotropis 2.5, 7.7, 7.9, 14.7
Calyx 2.1, 2.3
Campanulaceae 8.10, 15.7, 15.8
Campsis radicans 7.7
Campylotropous 5.3, 15.9

SI.4 Plant Embryology: Classical and Experimental

Cannaceae 4.10, 11.14
Capitulum 7.6
Capparidaceae 5.3, 11.14
Capsella 11.9, 20.4, 20.5, 20.6
    bursa-pastoris 11.3, 20.2
Capsicum annum 18.2
Carbohydrate body 10.13
Carbohydrates 4.1, 4.9, 8.6, 8.7, 9.10, 14.2, 17.6, 20.6
Carboxylmethyl cellulose 22.3
Cardamine 2.7
Cardiospermum 8.13, 14.4
Careya arborea 7.7
Carica papaya 8.15, 17.2, 17.4
Carotenoids 4.9, 4.10
    pigments 4.5
Carpel 2.1, 2.5
Carthamus tinctorius 12.6
Carum 17.6
Caruncle 5.3, 5.4, 14.2, 14.3
Carya illinoensis 8.15
Caryophyllaceae 10.2
Caryophyllad Type 11.3, 11.6
Casein hydrolysate (CH) 16.5, 20.6, 21.8
Casuarina 5.10, 8.13, 15.7
Caudicle 3.11
Cell suspension (feeder cells) 19.7
Cellular endosperm 10.1, 10.3, 15.7
Cellulase 9.17, 19.7
Cellulase 4.9, 8.7
Cellulosic intine 4.7
Censer mechanism 14.7
Central Cell 4.12, 6.1, 6.2, 6.12, 8.15, 8.16, 19.7, 19.8
Centranthera hispida 10.7, 10.9
Centrosperales 6.12
Centrospermales 5.3
Ceratonia siliqua 18.5
Ceratophyllum 7.4
Ceriops 14.11
Cestrum 7.6
CH 21.3
Chaeoichris 6.4
Chalazal end 6.3-6.5, 6.12
    haustoria 15.7
    haustorium 10.7, 10.8, 10.9, 11.11
Chalazogamy 8.13
Chalazosperm 11.14, 11.15

Chasmogamous 7.2, 8.11
    flower 7.3
Cheiropterochory 14.6, 14.8
Cheiropterophily 7.7
Chelated iron 17.4
Chemotropic factor 8.12
    gradient 8.12
Chenopodiaceae 5.3
Chenopodiad Type 11.3, 11.7, 11.8
Chenopodifolia bonus-henricus 11.7
China rose 2.5
Chiropterophily 7.4
Chlorophyll 5.3
Cholchicine 18.10
Chrysanthemum cinerariaefolium 6.3, 6.7
Cicer 11.10, 11.11
Cimicifuga 11.15
Circinotropous 5.2, 5.3
Cistanche tubulosa 21.6
Citrus 12.1, 12.5, 12.11, 17.1, 17.4, 17.6, 17.7, 21.6
    microcarpa 21.6
    sinensis 12.5, 17.5
    trifoliate 13.7
Cl_2-water A.2, 16.5
Cleavage polyembryony 12.1, 12.2
Cleistogamous 3.3, 8.11
Cleistogamy 7.2
Clerodendron thomsonae 7.10
CO_2 Level 9.18
Coccoloba 10.11
Cochleria danica 20.1
Coconut milk 16.5, 17.2
    water (CW) 18.7, 20.1, 20.6
Cocos nucifera 14.7
Coenocytic 15.5
Coenomegaspore 6.5
Coffea arabica 14.12
Colchicine 4.14, 18.3
Colchicum autumnale 11.1
Coleorhiza 11.1
Columella 4.5, 4.6
Colutea 10.2
Commelina 7.3
    benghalensis 7.2
Complete 2.1
Composite Endosperm 10.10

Compound Pollen Grains 3.11
Con A 8.3
Connective 2.4
Convolvulaceae 9.4, A.2
Corms 14.8
Corolla 2.1, 2.3
Corona 2.4
Corpusculum 7.9
Corydales cava 11.15
Costalia 5.8
Costus 3.3, 5.8
Cotyledons 11.8, 11.9, 14.2, 14.5, 14.11, 21.7
Coumarin 14.9, 20.9
Crassinucellate 5.4-5.6, 5.9
Crassulaceae 15.5
Crepis 13.5, 13.8
    capillaris 8.16, 11.1
    foetida 9.5
Crinum 10.13
Crocus 8.2, 8.7
Crosomataceae 5.7
Cross-pollination 7.1, 7.2, 7.4, 7.6, 7.9, 7.11, 9.10, 9.15
Crotaderia jubata 6.9
Crotalaria 8.7, A.2
Crucifer type 11.3
Cruciferae 9.15
Crucifers A.1
Cryptic self-fertilization 8.11
Cucumber 2.5
Cucumis 21.1
    sativus 21.1
Cucurbita 3.4, 8.13, 8.16
    pepo 3.7
Cucurbitaceae 5.5, 8.10, 10.5, 10.7, 15.7
Culcitium reflexum 5.10
Culture 21.7
    Medium 17.4
Cuscuta reflexa 3.10
Cuticle 8.2, 9.9
Cyanastrum 11.15
Cymanchum vincetoxicum 21.6
Cyperaceae 4.12, 15.2, 15.3, 15.4
Cypripedium parviflorum 8.15
Cystaceae 5.2
Cysteine 16.4
Cystine 9.10
Cyto-differentiation (xylogenesis) 21.6

Cytokinesis 3.9, 4.4, 15.3
Cytokinins 8.12, 11.13, 14.9, 17.5, 18.5, 21.4, 21.8
Cytoplasm 4.6
Cytoplasmic Male Sterility (CMS) 4.13
Cytoplasmic bridge 3.8

D
D. metel 18.2
Dactylis glomerata 17.2
Dalapon 4.14
DAP 21.7
Datura 20.6
    innoxia 17.1, 18.1, 18.2, 18.5, 18.6
    stramonium 18.1
Datura tatula 20.7
Daucus 17.6
    carota 12.10, 17.1, 17.2, 17.4
Deciduous 2.3
Decussate 3.10
Delonix regia 2.3
Dendrophthoe 5.9
    neelgherrensis 12.7
Deschampsia caespitosa 13.9
Deschcampsia 13.3
Desiccated 22.3
Diadelphous 2.4
Dianthus 2.5
Diastase 5.7
Dichogamy 7.8
Diclamydeous 2.2
Diclinous 2.2
Dicliny or Unisexuality 7.7
Dicotyledonous Embryo 11.9
Dicraea 11.14, 15.5
Dicranostigma franchetianum 19.3
Dictyosomes 4.7, 8.6, 11.11, 11.13
Didynamous 2.4
Dielectrophoresis 19.7
Digera arvensis 6.12
Dimorphic 3.3
    flowers 9.1
    heterostyly 7.9
Dioecious 2.2, 7.8
Diplodization 18.10, 18.12
Diploid 12.3, 14.1, 21.7, 21.9
    parthenogenesis 13.3
    triploid 12.8, 12.9

SI.6 Plant Embryology: Classical and Experimental

Diplospory 13.3, 13.4
Direct Androgenesis 18.2
Dithecous 3.1
DNA 4.3, 4.13, 8.16, 9.10, 11.13, 19.9
Double fertilization 8.1
Drimys 3.11
Driselase 19.7
Drosera 3.11
    rotundifolia 11.7
Drusa 6.3
    oppositifolia 6.6
    Type 6.5, 6.6
Dry Stigma 8.2
Dyad cell 5.10

E

E. cupressiformis 15.8
E. eurhynchum 12.9
E. pedunculosum 12.9
E. sparteus 15.7
E. stricta 7.7
E. strictus 15.7
Ecballium elaterum 14.6
Echynocystis lobata 10.5
Egg 6.9, 8.1, 8.8, 8.13, 8.15, 8.16, 8.17, 10.15, 11.1, 12.1, 12.3, 12.7, 13.8, 14.1, 15.2, 18.1, 19.1, 19.7, 19.9
    apparatus 6.1, 6.2, 6.6, 6.7, 4.12, 15.4, 21.4
    cell 6.5, 6.6, 12.7, 19.7
Eidolon helvum 7.7
Ektexine 4.5, 4.6
Elaeagnaceae 5.7
Elaeocarpaceae 5.10
Elaiosome 14.4
Elatine 6.4
    hydropiper 6.4
Elatostema 12.8, 13.8
    acuminatum 12.9, 13.7
    euryhnchun 13.7
Elodea 8.16
    canadensis 13.2
Elytraria 10.11, 14.3
Embryo 6.3, 6.11, 8.15, 10.1, 10.8, 10.9, 10.12, 10.15, 11.8, 11.10, 11.11, 11.14, 12.1, 12.4, 12.5, 12.7, 13.3, 13.5, 13.8, 14.1, 14.2, 14.4, 14.5, 14.9, 15.2, 17.1, 17.2, 18.10, 18.13, 19.7, 19.9, 20.1, 20.4, 20.5, 20.6, 20.7, 20.9, 21.3, 21.7, 21.8, A.5
    Culture 20.1, 20.4, 20.7, 20.9
    factor 20.6, 21.7
    Rescue 20.9
    sac 4.12, 5.4, 5.7, 5.10, 6.1, 6.2, 6.4, 6.6, 6.7, 6.12, 6.13, 8.13, 8.16, 9.6, 10.2, 10.4, 10.7, 11.1, 12.1, 12.3, 12.4, 12.5, 12.6, 12.9, 13.3, 13.4, 13.5, 13.7, 15.2, 15.4, 15.8, 17.1, 18.10, 19.7, 19.8, 21.4
    sacs 5.1, 10.10, 12.7, 12.8, 19.7
Embryogenesis 13.8, 11.6, 16.7, 17.1, 17.5, 17.6, 20.1, 20.6, 20.9, 21.8, 22.1
Embryogenic pollen grains 18.2, 18.8
Embryogeny 11.2, 15.3, 15.7
Embryoid 12.10, 16.6, 17.1, 17.2, 18.2, 18.4, 18.6, 18.7, 19.8, 22.1
Embryony 15.2
Encapsulation 22.2, 22.3, 22.5
Endexine 4.5-4.7
Endoanaphase 3.4
Endometaphase 3.4
Endomitosis 3.4
Endoplasmic reticulum 4.8, 4.10, 8.5, 11.1
Endoprophase 3.4
Endosperm 6.11, 9.7, 10.1, 10.6, 10.7, 10.9, 10.15, 12.4, 12.6, 12.9, 13.7, 14.1, 14.2, 14.4, 14.9, 14.11, 15.2, 15.3, 15.7, 19.1, 19.8, 19.9, 20.3, 21.1, 21.4, 21.6-21.8
    culture 21.6, 21.9
    haustorium 10.5
    mother cell 6.12
Endospermic 2.5, 10.1, 21.7
    or albuminous 14.2
Endostome 5.3, 5.5
Endotelophase 3.4
Endothecium 3.2
Endothelium 5.7, 15.8
    or Integumentary Tapetum 5.7
Endozoochory 14.6, 14.7
Endymion 6.3
    Type 6.4, 6.12
Entomophilous 7.6
Entomophily 7.4, 7.3, 7.6, 8.1
Epacridaceae 15.3, 15.4
Ephydrophily 7.4
Epicalyx 2.3
Epicotyl 11.8, 14.3
Epidermis 3.2, 5.5, 8.2
Epigeal Germination 14.11
Epigynous 2.5
Epipetalous 2.4
Epistase 5.8, 5.9

Epizoochory 14.6, 14.8
ER 8.2, 8.4, 8.6, 8.7, 11.11, 11.13
Ericaceae 8.7
Eriospermum carpense 14.2
Erythrina variegata 7.7
Erythronium americanum 12.2
Eschscholtzia 21.1
    californica 9.16, 19.3
Esterase 4.10, 8.6, 8.7
Ethanol A.3
Ethephon 4.14
Ethyl methane sulphonate (EMS) 4.14
Eugamous 13.8
Eugamy 13.8
Eugenia jambos 13.6
Eulophia epidendraea 12.2, 12.3
Eupatorium 13.4, 13.8
    glandulosum 13.4
    Type 13.4
Euphorbia terracina 4.4
Euphorbiaceae 5.3, 5.5, 6.6, 14.3, 21.7
Ex-ovule 17.1
Exalbuminous 2.5, 10.1, 21.7
Exine 3.6, 4.5-4.7, 4.10, 4.11, 9.10, 15.6
Exocarpaceae 15.7
Exocarpus 12.2, 15.7
    cupressiformis 15.7, 21.7
    sparteus 10.7, 15.8
Exocytosis 8.4
Exostome 5.3, 5.5
Explants 16.3, 16.6, 17.2, 17.4, 19.4
Exudates 8.8, 8.9, 9.4

F
Fabaceae 2.7, 14.3, 18.1
Fagopyrum 8.13, 9.3
False polyembryony 12.1, 12.7
Fats 4.9, 14.2
Fe.EDTA 18.8, 20.4
Female 15.6
    gametes 8.1, 11.1
    gametophyte 5.1, 10.1, 10.15, 21.1
    megasporophyte 6.1
    potency 4.12
Fertile pollens 4.13
Fertilization 4.10, 5.8, 6.12, 8.1, 8.10, 8.11, 8.15, 8.16, 9.1, 9.5, 9.16, 10.10, 10.12, 10.15, 11.1, 12.3, 12.4, 12.5, 12.9, 13.2, 13.6, 13.8, 14.1, 15.2, 19.1, 19.2, 19.5, 19.8, 19.9, 21.1, 21.4, 21.6
Festuca 13.3
    rubra 21.4
Ficus benghalensis 14.7
    religiosa 14.7
Field transfer 16.7
Filament 2.4
Filiform appararus 6.9, 8.12
Flavanoids 4.9
Float Culture 18.8
Flower 2.1, 2.2, 7.3, 7.4, 7.6, 7.7, 13.3, 18.6, 19.2, 19.3, 21.2, A.2
    diacetate A.4
Fluorescein diacetate test 19.7
Fluorescence Method A.4, A.5
Fluorochromatic Reaction (FCR) Test A.4
Foot layer 4.5, 4.7, 4.8
Formaldehyde A.3
Fragaria grandiflora 12.7
Fraxinus 14.9
Freesia 18.12
Fritillaria 6.3, 6.6, 6.7, 8.10, 8.16, 10.1
    imperialis 13.3
    type 6.5, 6.6
Fruit 2.6 9.7, 10.12, 14.6, 15.7, 19.1, 21.1, 21.3
Funiculus 5.1, 5.2, 10.9, 14.4

G
GA 11.13
GA3 9.18, 20.6, 21.7, 22.5
Gagea 6.7
Galactose 8.3
Gallium 6.12
Gametes 4.1
Gametogenesis 6.13
Gametophyte 4.11
Gametophytic Self-incompatibility (GSI) 9.4
Gamma rays 4.14
Gamosepalous 2.3
Gamopetalous 2.3, 2.4
Gamophyllous 2.3
Garrya elliptica 8.10
GAs 20.7
Gasteria 9.8
Geitonogamy 7.1, 7.3
Generative Apospory 13.3
    cell 4.1, 4.3, 4.4, 4.13, 15.3
    nucleus 4.8

SI.8 Plant Embryology: Classical and Experimental

Genetic Transformation 18.12
Genetics of Gametophytic Self-Incompatibility 9.5
Genista monisperma 11.10, 11.11
Genisteae 11.10
Genotype 17.4, 18.4, 19.4, A.1
Gentianaceae 10.4, 11.15
Geodorum 6.4
Geranium phaeum 11.13
Gerbera jamesonii 18.10, 21.5
Germ pore 4.5, 4.10
Gesneriaceae 15.7
Gibberellic acid 14.9, 4.14
Gibberellins 17.5
Ginkgo 20.6
    biloba 18.7
Gladiolus 8.3, 8.7, 19.6
Globba 13.2
Globoids 10.14
Globular embryo A.5
Globularia vulgaris 10.9
Gloriosa 3.3, 5.10, 7.10
Glucose 8.3, 9.12, 20.6
Glucuronic acid 8.3
Glutamine 9.10, 18.5, 18.8, 20.6, 20.5
Glycine 9.10, 16.4, 19.5
Glycolipids 8.3
Glycoproteins 8.3, 8.6, 9.10
Gnetalean Theory 6.13
Golgi apparatus 8.4
Gonophore 2.2
Gossipium 8.16, 14.3, 14.7, 14.12
Gradient Centrifugation 18.8
Gravillea robusta 10.5
Greenhouse 16.6
Growth Inhibitors 17.5
GSI system 9.5-9.7, 9.10
Guar gum 22.3
Guava 2.5
Gunnera 6.5
Gymnarrhena 2.7
Gymnosperm 6.13, 10.1, 12.1, 14.1, 15.7, 21.6
Gynandropsis gynandra 21.4
Gynandrous 2.4
Gynoecium 2.1, 2.5
Gynogenesis 18.10
Gynophore 2.2
Gynostegium 7.7

H

Hieracium aurantiacum 13.5
    excellens 13.5
    flagellare 13.5, 13.6
Half-Closed Type 8.7
Haloptelia 3.7
Hanging Drop Culture 18.7
Haplanthus 14.3
Haploid 4.5, 6.1, 6.2, 6.7, 10.1, 12.3, 18.3, 18.5, 18.9, 18.10, 18.13, 19.9
    apogamy 13.1
    parthenogenesis 13.1
    Plants 18.10
    Production 20.8
    diploid 12.4, 12.8
    triploid 12.8
Haloptelia 3.7
Haplotelia integrifolia 4.4
Harnandia peltata 14.5
Haustorial pollen tube 8.17
Haustorium 6.9, 6.12, 11.4, 15.8
Heart-shaped embryo A.5
Hedychiyum garsnerianum 6.4
Helianthus 7.1, 7.9
    annus 21.5
Helobial endosperm 10.1, 10.4
    type 15.2
Helobiales 15.7
Helosis 10.3
Hemerocallis 9.8
Hemianatropous 5.3
Hemicellulase 8.6, 19.7
Heritiera 14.11
Herkogamy 7.9
Hermaphrodite 2.2
Heterocarpy 2.7
Heterofertilization 8.17
Heteromorphic Incompatibility 9.1
Heteropogon 14.9
Heterostyly 7.9
Heterozygotes 4.14
Heterozygous 9.2
Hibiscus 3.1, 7.7, 7.9
Hieracium 13.5
    excellens 13.6
    Type 13.5
Hilum 5.1, 14.2-14.4

Histidine 4.13
Hodeum distichum 20.6
Hodgkin and Lyon's medium (1986) A.3
Hollow style 9.10
    styled 9.8
Holocrine secretion 8.4
Holoptelia 14.6
Homogamy (Bisexuality) 7.1
Homorphic incompatibility 9.4
Homozygous 4.13, 18.3, 19.9
Hordeum 4.4
    bulbosum 18.10, 18.11
    distichon 8.10
    vulgare 18.2, 18.6, 18.10, 18.11, 19.9, 20.8
Horizontal gyratory shaker 17.3
Horseshoe shaped embryo A.5
Hot Water and High Temperature Treatment 9.17
Hull Factor 21.3
Hyacinthus orientalis 4.12, 4.13
Hybrid Embryo 20.9
Hydrated 22.3
    capsules 22.3
Hydrilla 5.9
Hydrobryum 11.14
Hydrocharitaceae 3.3, 7.4, 11.8, 15.5
Hydrochory 14.6, 14.7
Hydrogel 22.2
Hydrophily 7.4, 8.1
Hydrophyllaceae 15.7
Hydrostachyaceae 15.5
Hydroxymethylfurfural 17.6
Hyoscyamus niger 18.2, 18.5, 18.12, 21.3
Hypericum 13.5
Hyphaene 3.10
Hyphidrophily 7.4
Hypocotyl 11.1, 11.4, 11.9, 12.10, 14.2, 14.3, 17.2
Hypogeal Germination 14.11
Hypogynous 2.5
Hypophysis 11.7, 11.11
Hypostase 5.7, 5.8, 6.12

I

Ixeris dentata 13.4
Iberis amara 21.3
Immature Embryo 14.9, 20.1
Impatiens 8.16, 10.6
    parviflora 14.6
In vitro Fertilization 9.16, 19.2
    pollination 19.2
Incompatibility 9.17
Incomplete 2.1
Indirect Androgenesis 18.2
Indole-3-acetic acid (IAA) 8.12, 16.5, 17.5, 20.6, 21.3, 22.3
    3-butyric acid (IBA) 9.18, 16.5
Indotristicha 11.14
Induced embryonic determined cells (IEDCs) 17.2
Inferior 2.5
Inhibitors 14.9
Inositol 18.8
Integument 5.1, 5.3, 5.7, 8.13, 12.5, 12.7, 14.3, 15.2, 21.1, 21.4
Intercellular matrix 8.3, 8.6
Interspecific incompatibility 9.1
Intine 4.5, 4.8, 4.10, 4.11, 9.10
Intra-Ovarian Pollination 9.16
Intraspecific incompatibility 9.1
Iodina rhombifolia 10.7
Ipomoea 3.3
    tricocarpa 9.16
Iridaceae 4.10, 11.8
Iris 20.7
Iron EDTA 17.4
Irradiation 9.16
Isobilateral 3.10
Isoleucine 9.10
Isotoma longiflora 12.2
Ixeris Type 13.4

J

Jacaranda 14.6
Jatropha 21.7
Juglans 8.16
Juncaceae 11.8, 15.3
Juncus 4.4

K

Karyogamy 8.1, 19.1, 19.7
Kinetin 17.5, 20.6, 21.4
Kingiodendron pinnatum 14.5
KN 22.3
Korthalsella 3.7

L

L. articulata 20.9
L. austrianum 20.1
L. peruvianum 20.9
Labelia trigona 5.7
Labyrinth Seeds 14.5
Lactose 9.12, 20.6
Lactuca sativa 14.10
Lagerstroemia 14.6
Lamiaceae 2.5
Laminar Air Flow 16.5, 21.3
Langsdorffia 5.10
Lateral haustoria 10.10
Lathyrus clymenum 20.9
Lauraceae 4.10
Laurus 3.10, 9.12, 9.18
Legitimate pollination 7.9
Leguminosae 5.3
Lemna pausicosta 14.4
Lemnaceae 15.7
Leontodon 13.8
    hispidus 13.8
Lepeostegeres gemmiflorus 12.7
Leptotene 3.6
Lichi 14.4
Liliaceae 7.10, 8.2, 9.4, 18.1
Lilium 3.3, 6.6, 8.7, 8.9, 8.10, 8.12, 8.16, 9.4, 9.8, 9.17, 13.3
    bulbifera 6.6
    bulbiferum 13.3
    davidii 18.10
    henryi 9.14
    longiflorum 8.12, 9.17
Limnanthes douglasii 6.7
Limonium 9.3
Linear 3.10
Linum 2.5, 9.3
    grandiflorum 9.4
    perene 20.1
    usitatissimum 12.9
Lipids 4.1, 4.10, 4.11, 8.3, 8.7, 14.4
    RNA 6.9
Lobelia amoena 10.7, 10.8
    syphilitica 12.2, 12.3
Lodoicea 14.7
    maldivica 14.2
Lolium 9.18
    perenne 21.4
Lomantia 10.6, 14.4
Long-term Storage A.2
Loranthaceae 5.3, 8.13, 10.10, 11.2, 11.10, 12.7, 14.3, 15.9, 21.7
Loranthoideae 15.9
Losaceae 10.7
Lotus 11.10
    corniculatus 11.11
Luffa 14.3
    cylindrica 21.6
Luminus 20.6
Lupinus 11.10
    arboreus 14.8
    pilosus 11.11
Luzula forsteri 11.8
Lycopersicon 9.18, 10.11, 18.4, 21.1
    esculentum 18.7, 20.9, A.5
    peruvianum 9.14, 9.17
Lythrum salicaria 9.2

M

M. alba 20.9
M. rubrum 19.9
Macadamia ternifolia 10.5
Macerozyme 9.17, 19.7
Machaerocarpus californicus 6.4
Macronutrients 20.5
Macrosolen cochichinensis 11.15
Macuna gigantean 7.7
Magnolia 7.9
Maize 4.10
Malacophily 7.4
Male 4.11
    gamete 8.1, 8.15, 12.9, 14.1, 18.1
    gametophyte 9.4, 9.5
    Germ Unit (MGU) 4.4
    potency 4.12
    Sterility 7.8, 4.14, 18.10
Maleic hydrazide 4.14
Malpighiaceae 6.6, 8.11
Malt extracts 16.5
Maltose 20.6
Malus 9.17, 10.2, 13.5, 13.8
    hupehensis 12.7
    seiboldi 13.8
Malva 7.8
    neglecta 8.10
Malvacaeae 3.7, 8.10

N

N. langsdorfii 9.17
N. tabacum 9.16, 19.3
N/10 HCl 16.5
N/10 NaOH 16.5
N6 medium 18.5, 18.10
NAA 9.18, 18.10
Najas 7.4
NaOCl 16.5
Naphthalene-1-acetic acid (NAA) 16.5
Neck cells 8.11
Necrohormone 4.12
    Theory 12.10
Nectary 2.4
Nelumbo nucifera 14.7
Nemec-Phenomenon 4.12
Nemophila 10.7
Neotieae 3.11
Nexine 4.5, 4.6
Niacin 19.5
Nicolaia 5.8
Nicotiana 3.3, 9.4, 9.15, 11.6, 11.7, 17.6, 18.1, 21.1, A.2
    glauca 9.17
    knightiana 18.5
    rustica 9.16, 19.2, 19.3
    sanderae 9.5
    sylvestris 18.5
    tabacum 3.2, 9.17, 12.9, 17.1, 17.6, 18.2, 18.6, 18.10, 18.12, 19.9
Nigella 14.7
Nigritella nigra 13.6
Nitsch's medium 18.5
Nodule 10.3
Non-embryogenic pollens 18.8
Norstog's medium 20.6
Nucellar beak 5.6
    embryogenesis 21.6
    epidermis 5.8
    polyembryony 12.6, 21.6
Nucellus 5.1, 5.4, 5.5, 5.9, 12.5, 15.2, 21.1
    Culture 21.6
Nuclear endosperm 10.1
Nurse 19.8
    Culture 18.7, 19.8
    endosperm 20.3

Nutritional media 16.3
Nyctaginaceae 5.5
Nyctanthes 7.6

O

Obturator 5.8, 8.13
Ochna serrulata 13.7
Oenothera 6.3, 8.7, 8.12, 8.16, 9.17, 10.14, 12.10
    acaulis 15.4
    lamarkiana 6.4
    organensis 9.8, 9.10
    type 6.1, 6.4, 6.12, 15.4
    organensis 9.16
Oenothra type 15.2
Omithochory 14.6
Onagraceae 8.17, 10.1, 15.2, 15.4, 15.5
Ononis 11.10
    fructicosa 11.11
Open Style 8.7
Operculum 14.4
Ophiopogon 3.7
Ophrydeae 3.11
Opposition S-allele 9.5
Opuntia aurantiaca 13.7
Orchidaceae 3.11, 4.3, 7.9, 10.1, 12.2, 12.5, 14.9, 21.7
Orchis machulatus 8.15
Organogenesis 16.7, 18.2, 21.8, 22.1
Orhotropous (Atropous) 5.2
Ornithophily 7.3, 7.7
Ornithus 20.3
Orobanche aegyptica 21.6
Orobancheae 14.9
Oroxylum indicum 7.7
Orygia decumbens 14.2
Oryza 21.7
    sativa 11.1, 12.9
Ostreya 5.10
Ovary 5.1, 8.12, 9.8, 9.10, 9.16, 10.10, 14.1, 18.1, 19.3, 21.2
    culture 18.10, 21.1, 21.3
    wall 5.8
Overcoming Incompatibility 9.14
Ovarian pollination 19.2, 19.3
Ovular Pollination 19.3
Ovule 5.2, 5.3, 5.5, 5.7, 5.8, 6.6, 6.12, 8.1, 8.12, 12.1, 12.7, 14.3, 14.4, 15.2, 15.3, 18.1, 18.10, 19.2, 19.5, 20.2, 21.1, 21.4
    Culture 21.4, 21.6

Mammaliochory 14.6
Mangifera 12.1, 12.6, 12.11
Mangifera indica 12.5, 12.11
Mannitol 20.6, 22.5
    solution 19.7
Mannose 8.3
Massulae 3.11, 4.3
Mature Embryo 11.9, 20.2
Maximovicii 6.7
MCPA 18.10
Mechanism of Inhibition 9.4
Medicago sativa 12.9
Megagametogenesis 6.3, 6.12
Megasporangium 21.4
Megaspore 5.9, 6.1, 6.2, 12.7, 13.3, 13.5, 15.2, 15.7
    mother cell 5.5, 5.6, 5.9, 12.7, 13.4, 13.5, 15.3, 15.4
Megasporium 5.1
Megasporogenesis 5.6, 5.9, 6.1-6.3, 15.2
Megasporophyll 2.1, 2.5, 14.1
Meiosis 3.8, 4.5, 4.13, 9.6, 9.7, 13.1, 15.3, 15.4, 6.12
Meiotic Prophase 3.8
Melampodium divaricatum 12.6
Melandrium album 19.9
Melanopodium divaricatum 12.5
Meliaceae 14.8
Melilotus 14.8, 6.12
    officinalis 20.9
Mendok 4.14
Mentor Pollens 9.14
Mercuric chloride A.2
Mesogamy 8.13
Mesophyll cells 17.2
Metaphase 4.3
Metaxenia 10.12
Methionine 9.10
MGU 4.5
Microcalli 19.8
Microfibrils 8.7
Microinjection 19.8
Micronutrients 20.5
Micropylar haustorium 10.9
Micropores 3.10, 4.1, 4.5
Microsporogenesis 4.5
Micropteris pusillus 7.7
Micropylar 15.8
    end 6.3
    haustorium 10.6, 10.8

Micropyle 5.1-5.3, 5.5, 5.8, 8.12, 11.1, 14.3, 15.2
Microsporangium 2.4, 2.5, 3.1, 8.1
Microspore 2.4, 3.7, 3.8, 3.10, 4.1, 4.3, 4.5, 4.6, 4.7, 4.10, 4.12, 15.3, 15.4, 17.6, 18.5, 18.6, 18.13, 19.7
    mother cell 15.2
    tetrads 15.3
    tapetum 4.11
Microsporogenesis 3.9, 4.5
Microsporophyll 2.1, 2.4, 3.1
Microtubules 3.6, 4.7
Middle layer 3.3
Millicell-CM dish 19.7
Mimosaceae 3.11
Mimulus luteus 19.9
Mirabilis 7.9
Mitochondria 3.6, 3.7, 4.4, 8.2, 8.6, 8.15, 11.11, 11.13
Monadelphous 2.4
Monochoria karasakowii 10.10
Monoclamydeous 2.2, 2.4
Monocot 2.5, 4.5, 21.3
    Embryo 11.10
Monocotyledons 5.2, 5.3, 8.7, 11.1, 11.8, 14.1, 14.2, 15.2
Monoecious 2.2, 7.8
Monoploid 18.10, 18.11
    embryo 20.8
Monosaccharides 8.3
Monosporic embryo sac 6.1
Monothecous 2.4, 3.1
Moringa 14.6
Mosaic Endosperm 10.11
Mucilaginous layer 8.2
Muda 6.4
Mumulus 19.9
Murashige A.6
Murashige and Skoog's medium 18.5, 18.10, 20.2, 20.4, 22.3, A.5
Musa 3.3, 3.10, 5.10
Mussaenda 7.6
Myosotis hispida 11.7
Myriophyllum 11.11
    alterniflorum 11.12
    intermedium 5.6
Myristica 10.11
    fragrans 14.4
Myrmecocohory 14.6, 14.8
Myrmecophily 7.4
Myrtaceae 12.5

Oxalidaceae 8.2
Oxalis 7.2, 7.9, 8.13

P

p-OH benzoic acid 17.6
Phaseolus acutifolius 20.9
Phaseolus coccineus 11.13
Phaseolus hysterinus 11.13
Papaver rhoeas 9.16
Phaseolus vulgaris 11.13
Plumbago zeylanica 4.4
Phaseolus multiflorus 11.13
Papaver somniferum 19.3
Pachytene 3.6
Paeonia 15.5, 15.7
    lactiflora 15.6
Paeoniaceae 15.7
Pandanus 6.4
Panicoidae 13.5
Panicum maximum 17.5
Papaveaceae 8.7
Papaver 14.7, 21.1
    rhoeas 19.3
    somniferum 9.16, 19.2, 19.3, 21.4
Papaveraceae 19.3
Paphiopedilum maudae 8.15
Papilionaceae 10.1, 21.7, A.2
Papilla 9.9
Papillae 8.2, 8.10
Pappus 2.3
Paraffin 18.7
Paramular bodies 8.3, 8.6, 8.7
Parasexual Hybridization 9.17
Parietal cell 5.6, 6.2
Parthenium 13.5, 13.8
    argentatum 9.5, 13.8
Parthenocarpic 2.6
Parthenogenesis 15.2, 18.1, 19.9
Parthenogenetic 13.8
Paspalum 8.12
Passiflora 7.8, 10.11
Pectic 8.2, 8.6
Pectinase 9.17, 19.7
Pecto-cellulosic 4.6, 8.2
Pectolyase 19.7
Pedicellate 2.2
Peduncle 17.2

Pellicle 8.2, 8.3, 8.10
Pelliphylum peltatum 12.4
Penaea 6.3, 6.5
    type 6.5, 6.6, 6.13
Penaeaceae 6.5
Pentaphragma 15.7
    horsfieldii 15.7
Peperomia 6.3, 10.1, 10.14
    hispidula 6.5
    pellucida 6.5
    type 6.5
Percoll 19.7
Perfect 2.1
Perianth 2.4
Pericarp 14.9, 15.7
Perigynous 2.5
Periplasmodium 3.4
Perisperm 11.14
Peroxidases 8.6
Persera americanum 8.10
Petiole 17.2
Petunia 3.7, 9.4, 9.15, 9.17, 10.11, A.2
    axillaries 7.8, 19.4, 19.9
    hybrida 8.6, 8.11, 9.11, 9.12, 9.14, 9.18, 19.4, 19.9
    violacea 19.2
Pfr 14.10
pH 16.5, 18.8, A.6
Phagocytotic 8.4
Phaseolus 11.10, 20.7
    aureus 14.8
    coccineus 11.12, 11.13, 20.7
    vulgaris 20.9
Phenolic compounds 8.3
Phenylalanine 9.10
Philydraceae 11.8
Phoenix dactylifera 10.12
Phosphatases 4.10, 4.11, 8.6, 8.7
Photoblastic 14.10
Phseolus coccineus 11.13
    multiflorus 11.11
Phylogenetic embryology 15.1
Phylogeny A.1
Phytochrome pigments 14.10
Pilate exine 4.5
Pin flowers 9.1
Pin 9.4
Pinus 7.4, 14.6

Piperaceae 5.2, 11.2, 11.14
Pistil 2.5, 8.12, 9.1, 9.5, 9.7, 9.11, 19.2, 19.3, A.4, A.5
    factor 9.15
Pistillate 2.2
Pistillode 2.5
Pisum sativum 14.8
Placentae 5.1, 5.5, 10.9, 19.2-19.4, 21.4
Placental haustorium 11.11
    pollination 19.2-19.4, 19.9
Plasma membrane 8.15
Plasmalemma 4.6-4.8, 4.10
Plasmlemma 8.5
Plasmodesmata 4.4, 8.6, 8.15, 10.13, 11.11
Plasmodesmatal connection 3.8
Plasmogamy 8.1, 19.1
Plastids 3.6, 3.7, 4.3, 4.4, 4.10, 6.9, 8.6, 8.15, 11.11
Plumbagella 6.3
    micrantha 6.7
    Type 6.5, 6.7
Plumbaginales 5.3
Plumbago 4.4, 6.3, 8.16
    capensis 6.6, 6.9
    Type 6.5, 6.6
Plumule 11.15, 14.3, 14.11, 20.1
Poa 5.10, 13.3, 13.5, 13.8
Poaceae 8.2, 9.4, 13.1, 14.12, 18.1, 18.13
Podocarpus 15.7
Podostemaceae 5.3, 6.12, 10.1, 15.5, 21.7
Pogonatherum paniceum 22.3
Polar nuclei 8.14, 6.12, 15.4
Pollen 4.6, 4.11, 6.12, 7.7, 7.8, 7.10, 8.8-8.10, 8.14, 9.2, 9.4, 9.5, 9.12, 10.11, 15.3, 15.4, 15.6, 18.1, 18.7, 19.1, 19.9, A.1, A.2
    Adhesion 8.9
    chamber 15.7
    Culture 18.6, 18.7
    embryo sac 4.13
    embryos 18.2
    exine 9.8
    factors 9.15
    germination 8.10
    grains 3.8, 3.11, 4.1, 4.4, 4.7-4.10, 7.1, 7.6, 7.10, 8.1, 8.10, 8.12, 9.2, 10.12, 15.2, 18.4, A.4
    hydration 8.9
    sterility 4.12
    tetrad 4.3
    tube 4.4, 5.8, 8.1, 8.3, 8.6, 8.7, 8.10-8.13, 8.16, 9.9, 9.16, 9.18, 12.3, 12.4, 12.9, 19.1, 19.6, 21.3, A.4
    viability A.4
    wall 3.6, 4.1, 4.5, 4.9
    wall Proteins 4.10
    embryo 4.12
    pistil interaction 9.10
Pollenkitt 4.9, 4.10
Pollinations 4.1, 7.1-7.3, 7.6, 7.9, 7.11, 8.1, 8.7, 8.8, 8.10, 8.12, 8.15, 9.1, 9.3, 9.4, 9.7, 9.11, 9.18, 11.1, 13.6, 13.8, 18.1, 18.10, 19.1, 19.2, 19.4, 19.9, 21.1, 21.6
Pollinators 7.7
Pollinia 4.3, 7.7, 8.10
Polyadelphous 2.4
Polyads 3.11
Polyamines 17.6
Polyembryony 8.16, 12.1, 12.6, 12.7, 15.2, 21.6
Polygamous 2.2
Polygonaceae 5.2, 5.5
Polygonum 6.3, 6.4
    type 6.1, 6.2, 6.7, 6.12, 15.7, 15.8, 15.9
Polyoxyethylene 22.2
Polypetalous 2.3, 2.4
Polyphyllous 2.4
Polyploids 13.8
Polyploidy 3.4, 6.1, 8.16, 10.16, 13.3
Polyribosomes 3.6
Polysaccharides 8.3, 8.6
Polysepalous 2.3
Polysiphonous 8.10
Polysomes 8.10
Polyspermy 8.16
Polyspory 3.10
Polytene 3.4
Pontederiaceae 5.3, 11.8
Populus tremuloides 21.9
Porogamy 8.13
Post-mitotic 8.16
Potamogetonaceae 11.8
Potassium 20.4
Potentilla 13.8
Potulaca 4.4
Pre-embryogenic Determined Cells' (PEDCs) 17.2
Precatorius 14.3
Primary archesporial cell 5.4, 5.9, 6.2
    parietal cell 6.2
    parietal layer 3.2
    sporogenous cell 5.5, 6.2
    sporogenous layer 3.2
Primexine 4.8, 4.8, 4.9

Primula 9.1, 9.3, 10.2
    vulgaris 7.9, 9.4
Primulales 5.3
Pro-bacula 4.6
Probaculum 4.8, 4.9
Proembryo 11.2, 11.4, 12.2, 12.6, 15.7, 20.1, 20.6
Proembryogenic masses (PEM) 17.5
Proline 4.13, 17.4
Pro-orbicular 3.7
Propagules 13.3
Prophase 3.4, 4.3
Protandrous 7.6
Protangiospermae 15.7
Proteaceae 10.5
Protein 4.1, 4.9-4.11, 6.9, 8.2, 8.3, 8.6, 8.7, 8.12, 9.8, 9.10, 10.12, 14.2, 14.4, 19.7, 21.7
Polysaccharides 8.3
Protoplast culture 19.5
Protosporopollenin 4.9
Pro-ubish 3.7
Prunus 8.10, 9.17
    avium 9.14, 9.16
Pseudo-crassunucellate 5.5
    embryo sac 11.14
Pseudoembryosac 15.5
Pyridoxine HCl 19.5
Pyrus 9.17

Q

Quercus 8.15, 11.9
Quinchamalium chilense 6.9
Quisqualis 7.6

R

Radicle 11.15, 14.3, 14.2, 14.11, 20.1
Raffinose 20.6
Rafflesiaceae 11.15
Ranunculaceae 15.5, 15.7, 18.1
Ranunculus 10.2, 13.5, 13.8, 14.9, 15.6
    auricomus 13.8
    cassbacfolius 13.8
    ficaria 11.15
    megacaropus 13.9
    sceleratus 12.10, 17.2, 17.3
Raphanus 9.4, 9.15
    cadatu 20.1
    landra 20.1
    sativa 20.1

Raphe 5.1, 14.4
Recognition and Rejection Reactions 9.10
Recognition pollens 9.14
Recombinant DNA technology 19.5
Recurrent Apomixis 13.1, 13.3
Regeneration A.6
RER 8.6, 8.10
Reseda 8.10
Resedaceae 5.3
Restitution nucleus 3.4, 13.3
Rhamnaceae 5.3
Rhamnose 8.3
Rheus 14.9
Rhizome 14.8
Rhizophora 14.11, 14.12
Rhonodendron 19.6
Ribes 9.8
Ribonuclease 4.10
Ribosome 9.6
Ribosomes 3.6, 8.2, 11.1, 11.11
Ricinus 10.1, 21.7, 21.8
    communis 14.12
RNA 8.12, 9.10, 19.7
RNAase 4.14
Robert's medium A.3
Rosa 20.7
Rosaceae 8.2, 13.1
Roxburgii 2.3
Rubiaceae 11.11, 15.7
Rubus 9.17, 13.8
Rudbeckia
    laneiniata 13.8
    speciosa 13.8
    sullivantie 13.8
Ruminate Endosperm 10.11
Runners 13.2
Rutaceae 12.5, 17.1

S

Saccharum officinarum 8.10
Sagina procumbens 11.6, 11.7
Sagittaria 8.16
Salicaceae 5.5
Salvia 7.2, 7.6
Santalaceae 6.9, 10.4, 12.2, 14.3, 15.7, 21.7
Santalum album 17.1, 17.5
Sapotaceae 14.8

Sarcocolla 6.5
Saxiflora granulate 11.6
Saxifragraceae 10.4, 15.5
Scabiosa succesa 11.15
Scarification 14.8
Schisandra chinensis 6.4
Scilla 6.4
Scurrula 5.9
Scutellum 14.2
Secale cereale 12.9
Sechium 20.6
Secondary 6.7
    haustoria 10.7-10.9
    nucleus 6.6, 6.7, 8.1, 8.15, 8.17, 10.1, 14.1, 21.4
Secretary or Glandular 3.4
Secretary zone 8.7
Secretory Tapetum 3.7
Sedum acre 11.11, 11.12
Seed 9.7, 12.1, 13.2, 14.1, 14.3, 14.4, 14.7, 15.7, 19.1, 19.3, 19.4
    coat 5.5, 14.1, 14.3, 14.8, 14.9, 21.1
    dispersal 14.4, 14.5
    dormancy 14.8, 21.3
    in Petunia 9.18
Self-pollination or autogamy 7.1, 7.2, 7.3, 7.9, 7.11, 9.19
Self-sterility or Self-incompatibility 7.8, 9.1, 9.5, 9.6, 9.9, 9.10, 9.12, 9.16-9.19, 19.4, 19.9, 21.1
Semigamy 13.8
Sepal 2.3
Sepaloid 2.3
Serine 4.6, 17.4, 18.5, 18.8
Sexual incompatibility 9.1
Sexual reproduction 13.1
SH medium 22.5
Shorea 14.6
    robusta 14.7
Simultaneous type 3.9, 3.10
Sodium alginate 22.3
    hypochloride 21.3, A.2
    pectate 22.3
Solanaceae 9.4, 18.13, A.2
Solanaceous A.2
Soland Type 11.3, 11.6, 15.8
Solanum melongena 21.5
Solanum tuberosum 12.9
Solid styles 8.3, 8.6

Somatic 22.2
    apospory 13.3-13.5
    embryo 17.7, 22.3
    embryogenesis 17.1, 17.2, 17.4, 17.6
    embryos 17.2, 17.6, 22.2, 22.6
Sonneratia 6.12
Sophora flavescens 11.13
Sorghum vulgare 8.10
Sparganiaceae 11.8
Sperm 4.1, 12.3, 13.8, 18.1, 19.6, 19.7, 19.9
Spherosomes 10.13
Spigelia 10.11
Spinacia 3.4
Spindles 4.2
Spiranthes cernua 12.5
Sporogenesis 6.13
Sporogenous 3.8
    cell 5.6, 5.9
    tissues 3.7, 5.5, 5.9, 15.2
Sporophytic 4.10
    budding 13.1
    generation 13.1
    Self-incompatibility (SSI) 8.3, 9.4
Sporopollelin 3.7, 4.6, 4.7, 4.9
SSI 9.7
    system 8.3, 9.10
    type 9.6
Stamen 2.1, 2.4, 3.1, 3.7, 15.6
Staminate 2.2
Staminode 2.4, 3.1
Starch 4.1, 4.8, 6.12, 8.10, 14.4, 21.7
Statica 8.16
Stellaria media 11.13
Sterculia 2.3
Stigma 4.4, 7.1, 7.5, 7.7, 7.9, 7.10, 8.2, 8.3, 8.4, 8.7, 8.8, 8.9, 8.10, 9.3, 9.4, 9.6, 9.16, 9.12, 10.11, 21.3
    papillae 8.2
    receptivity 8.9
Stigmatic exudates 8.2
    pollination 19.2, 9.10, 19.3
Stipa 14.9
Stomata 5.3
Stratification 14.9
Stress Tolerant Plants 19.10
Strychnos nuxvomica 14.12
Stub Pollination 9.16, 19.1

Stylar canal 8.7, 8.8
   tissue 5.8
Style 7.8, 7.10, 8.7, 8.12, 9.4, 9.6, 9.7, 9.10, 9.17, 19.1, 19.6
Successive type 3.9
Suckers 13.2
Sucrose 17.6, 18.8, 19.5, 20.1, 20.5, 21.8
   density gradient technique 19.7
Sugars 9.18, 18.5
Sunflower 2.5
Superior 2.5
Suspension culture 16.5, 17.2
Suspensor 11.7, 11.10, 11.11, 11.13, 12.5, 13.8, 20.7
   haustoria 11.11, 11.12
Sympetalae 5.2
Synandrium 3.7
Synandrous 2.4
Syncarpous 2.5
Syncytium 3.7, 4.13
Synergids 6.1, 6.6, 6.9, 6.12, 8.12, 8.16, 12.2-12.4, 12.7, 12.8, 18.1
Syngamy 8.15, 10.15, 13.1
Syngenesious 2.4
Synthetic seeds (synseed or artificial seed) 12.11, 17.7, 22.1

T
T-shaped 3.10
T. laevigatum 13.4
T. spelta 19.4
T. vestitus 21.8
Tacca 10.2
Tagetes 7.1
Tap root 17.2
Tapetal cells 4.8, 4.10, 4.13
Tapetum 3.3, 3.7, 3.8, 4.6-4.8, 4.10, 4.11, 9.8, 15.3
Taraxacum 13.8
   albidum 13.3
   koksaghys 8.15
   Type 13.3
Taxillus 21.8
   vestitus 21.7
Tectate exine 4.5
   grains 4.10
Tectum 4.5-4.7, 4.8, 4.10
Tegmen 14.3
Telophase-I 3.6
Tenuinucellate 5.4, 5.5, 5.7

   ovules 5.7, 5.9
Tepal 2.4
Tephrosia 14.9
Terminalia 14.6
Terniola 11.14
Test tube fertilization 19.2
Testa 14.2-14.5, 14.9
Tetrad 3.6, 3.8, 3.11, 4.7, 5.9
Tetradynamous 2.4
Tetrahedral 3.10
Tetraploid 6.6, 21.9
Tetrasporangiate 3.7
Tetrasporic Embryo Sac 6.1, 6.5
Tetrazolium Test A.4
Thalamus 2.2, 7.6
Themeda 14.9
Theobroma cacao 14.12
Thesium 6.4
   alpinum 10.14
Thiamine HCl 19.5
Threonine 17.4
Thrum 9.4
Thunbergia alata 10.6
Tilia 10.2
Torenia 6.12, 19.9
   fournieri 19.9
Torus 2.2
Transgenic plants 4.14
Translator mechanism 7.7
Transmitting tissue 8.6, 8.11, 8.10, 8.12
Trapa 15.5
Trapaceae 10.1, 15.5, 21.7
Trianthema 5.3, 14.4
Trianthema monogyna 14.4
Tribulus 14.8
Trichosanthes anguina 21.6
Trifolilium 2.7, 9.17, 20.3
Trigonella 14.8
Trillium 14.8
   ovatum 14.4
   undulatum 12.6
Trimorphic flowers 9.1
Trimorphic heterostyly 7.9
Triplets 12.8, 12.9
Triploid 6.7, 14.1, 21.7
   Culture 21.9
Tristyly 9.2

Triticale 3.2, 8.9
Triticum 10.1, 18.2, 21.7
    aestivum 18.10, 18.11, 19.4, 19.9
    compactum 18.1
    vulgare 12.9
Tropaeolum 11.13
    majus 11.11, 11.12
True Polyembryony 12.1
Tryphine 4.9
TSCP-Tissue specific complementaryproteins 9.11
Tubers 14.8
Tulipa 6.7
Tunica saxifragra 11.13
Twins 12.8 12.9
Type 15.9
Tyrosine 9.10, 16.4

U
U. glabra 12.4
Ubisch bodies 3.7, 4.8
Ulmus 8.16
Ulmus americanum 6.6, 12.4
Umbelliferae 5.7
Umbilicus intermedius 6.4
Unitegmic 5.3
Unreduced Parthenogenesis 13.8
Unreduced Pseudogamy 13.8
Urticaceae 5.2
UV rays 4.9, 4.10

V
Vallisneria 3.7, 4.4, 7.4
    spiralis 7.4
Vegetative cell 4.1
    nucleus 4.8, 8.14, 8.15
    reproduction 13.1
Verbascum 10.11
Vermiform appendage 10.5
Veronica 10.7
Vesicles 5.8
Viscin thread 3.11
Vigna unguiculata 8.7
Viola 7.2
Viscaceae 15.9

Viscin thread 3.11
Viscoideae 15.9
Vitamin B1 (thiamine) 16.4
Vitamins 14.4, 16.3, 16.5
Vitis 8.7
Vitis vinifera 8.10, 20.9, 21.6
Vivipary 13.3, 13.9
    Germination 14.11

W
Wet Stigma 8.2, 8.3, 8.9, 9.10
White's media 20.4
White's medium 20.6, 21.8, 22.3
Woodfordia 7.9

X
X-bodies 8.14, 8.16
X-body 8.16
X-rays 9.17
Xanthium 14.8
Xenia 10.11
Xenogamy 7.1, 7.3

Y
YE 21.3
Yeast extract (YE) 16.5, 21.8
Young root 17.2

Z
Z-layer 4.5
Zea 10.1, 21.7
    mays 4.4, 8.10, 8.17, 10.11, 12.8, 14.2, 14.8, 17.4-17.6, 18.10
Zeatin 16.5, 17.5
Zeuxine 7.9, 12.5
Zeylanica 4.4
Zeylanidium 11.14
Zingiberaceae 4.10
Zone of Inhibition 9.3, 9.6, 9.15
Zoochory 14.6, 14.7
Zoophily 8.1
Zostera 3.7, 7.4, 3.10
Zwischenkorper (z-layer) 4.5
Zygomorphic 2.4
Zygotic embryo 12.3, 12.5

Stylar canal 8.7, 8.8
    tissue 5.8
Style 7.8, 7.10, 8.7, 8.12, 9.4, 9.6, 9.7, 9.10, 9.17, 19.1, 19.6
Successive type 3.9
Suckers 13.2
Sucrose 17.6, 18.8, 19.5, 20.1, 20.5, 21.8
    density gradient technique 19.7
Sugars 9.18, 18.5
Sunflower 2.5
Superior 2.5
Suspension culture 16.5, 17.2
Suspensor 11.7, 11.10, 11.11, 11.13, 12.5, 13.8, 20.7
    haustoria 11.11, 11.12
Sympetalae 5.2
Synandrium 3.7
Synandrous 2.4
Syncarpous 2.5
Syncytium 3.7, 4.13
Synergids 6.1, 6.6, 6.9, 6.12, 8.12, 8.16, 12.2-12.4, 12.7, 12.8, 18.1
Syngamy 8.15, 10.15, 13.1
Syngenesious 2.4
Synthetic seeds (synseed or artificial seed) 12.11, 17.7, 22.1

T

T-shaped 3.10
T. laevigatum 13.4
T. spelta 19.4
T. vestitus 21.8
Tacca 10.2
Tagetes 7.1
Tap root 17.2
Tapetal cells 4.8, 4.10, 4.13
Tapetum 3.3, 3.7, 3.8, 4.6-4.8, 4.10, 4.11, 9.8, 15.3
Taraxacum 13.8
    albidum 13.3
    koksaghys 8.15
    Type 13.3
Taxillus 21.8
    vestitus 21.7
Tectate exine 4.5
    grains 4.10
Tectum 4.5-4.7, 4.8, 4.10
Tegmen 14.3
Telophase-I 3.6
Tenuinucellate 5.4, 5.5, 5.7

    ovules 5.7, 5.9
Tepal 2.4
Tephrosia 14.9
Terminalia 14.6
Terniola 11.14
Test tube fertilization 19.2
Testa 14.2-14.5, 14.9
Tetrad 3.6, 3.8, 3.11, 4.7, 5.9
Tetradynamous 2.4
Tetrahedral 3.10
Tetraploid 6.6, 21.9
Tetrasporangiate 3.7
Tetrasporic Embryo Sac 6.1, 6.5
Tetrazolium Test A.4
Thalamus 2.2, 7.6
Themeda 14.9
Theobroma cacao 14.12
Thesium 6.4
    alpinum 10.14
Thiamine HCl 19.5
Threonine 17.4
Thrum 9.4
Thunbergia alata 10.6
Tilia 10.2
Torenia 6.12, 19.9
    fournieri 19.9
Torus 2.2
Transgenic plants 4.14
Translator mechanism 7.7
Transmitting tissue 8.6, 8.11, 8.10, 8.12
Trapa 15.5
Trapaceae 10.1, 15.5, 21.7
Trianthema 5.3, 14.4
Trianthema monogyna 14.4
Tribulus 14.8
Trichosanthes anguina 21.6
Trifolilium 2.7, 9.17, 20.3
Trigonella 14.8
Trillium 14.8
    ovatum 14.4
    undulatum 12.6
Trimorphic flowers 9.1
Trimorphic heterostyly 7.9
Triplets 12.8, 12.9
Triploid 6.7, 14.1, 21.7
    Culture 21.9
Tristyly 9.2

Triticale 3.2, 8.9
Triticum 10.1, 18.2, 21.7
    aestivum 18.10, 18.11, 19.4, 19.9
    compactum 18.1
    vulgare 12.9
Tropaeolum 11.13
    majus 11.11, 11.12
True Polyembryony 12.1
Tryphine 4.9
TSCP-Tissue specific complementaryproteins 9.11
Tubers 14.8
Tulipa 6.7
Tunica saxifragra 11.13
Twins 12.8 12.9
Type 15.9
Tyrosine 9.10, 16.4

U
U. glabra 12.4
Ubisch bodies 3.7, 4.8
Ulmus 8.16
Ulmus americanum 6.6, 12.4
Umbelliferae 5.7
Umbilicus intermedius 6.4
Unitegmic 5.3
Unreduced Parthenogenesis 13.8
Unreduced Pseudogamy 13.8
Urticaceae 5.2
UV rays 4.9, 4.10

V
Vallisneria 3.7, 4.4, 7.4
    spiralis 7.4
Vegetative cell 4.1
    nucleus 4.8, 8.14, 8.15
    reproduction 13.1
Verbascum 10.11
Vermiform appendage 10.5
Veronica 10.7
Vesicles 5.8
Viscin thread 3.11
Vigna unguiculata 8.7
Viola 7.2
Viscaceae 15.9

Viscin thread 3.11
Viscoideae 15.9
Vitamin B1 (thiamine) 16.4
Vitamins 14.4, 16.3, 16.5
Vitis 8.7
Vitis vinifera 8.10, 20.9, 21.6
Vivipary 13.3, 13.9
    Germination 14.11

W
Wet Stigma 8.2, 8.3, 8.9, 9.10
White's media 20.4
White's medium 20.6, 21.8, 22.3
Woodfordia 7.9

X
X-bodies 8.14, 8.16
X-body 8.16
X-rays 9.17
Xanthium 14.8
Xenia 10.11
Xenogamy 7.1, 7.3

Y
YE 21.3
Yeast extract (YE) 16.5, 21.8
Young root 17.2

Z
Z-layer 4.5
Zea 10.1, 21.7
    mays 4.4, 8.10, 8.17, 10.11, 12.8, 14.2, 14.8, 17.4-17.6, 18.10
Zeatin 16.5, 17.5
Zeuxine 7.9, 12.5
Zeylanica 4.4
Zeylanidium 11.14
Zingiberaceae 4.10
Zone of Inhibition 9.3, 9.6, 9.15
Zoochory 14.6, 14.7
Zoophily 8.1
Zostera 3.7, 7.4, 3.10
Zwischenkorper (z-layer) 4.5
Zygomorphic 2.4
Zygotic embryo 12.3, 12.5

DATE DUE